D0333450

Ecotourism Policy and Planning

Edited by

David A. Fennell
Brock University
Canada

and

Ross K. Dowling
Edith Cowan University
Australia

CABI *Publishing*

338
.4791
E

CABI *Publishing* is a division of CAB *International*

CABI Publishing
CAB International
Wallingford
Oxon OX10 8DE
UK

CABI Publishing
44 Brattle Street
4th Floor
Cambridge, MA 02138
USA

Tel: +44 (0)1491 832111
Fax: +44 (0)1491 833508
E-mail: cabi@cabi.org
Web site: www.cabi-publishing.org

Tel: +1 617 395 4056
Fax: +1 617 354 6875
E-mail: cabi-nao@cabi.org

© CAB *International* 2003. All rights reserved. No part of this publication may be reproduced in any form or by any means, electronically, mechanically, by photocopying, recording or otherwise, without the prior permission of the copyright owners.

A catalogue record for this book is available from the British Library, London, UK.

Library of Congress Cataloging-in-Publication Data
Ecotourism policy and planning / edited by D.A. Fennell
    and R.K. Dowling.
      p. cm.
    Includes bibliographical references.
    ISBN 0-85199-609-4 (alk. paper)
      1. Ecotourism.   I. Fennell David A., 1963-
    II. Dowling, Ross Kingston.
    G156.5.E26 E355 2003
    338.4'791--dc21

2002009262

ISBN 0 85199 609 4

Typeset by Columns Design Ltd, Reading
Printed and bound in the UK by Cromwell Press, Trowbridge

ECOTOURISM POLICY AND PLANNING

DEDICATION

To our wives – Julie Fennell and Wendy Dowling

Contents

Contributors

Thomas Bauer, *School of Hotel and Tourism Management, The Hong Kong Polytechnic University, Hung Hom, Kowloon, Hong Kong SAR, China*

Kelly S. Bricker, *Division of Forestry, West Virginia University, PO Box 6125, Morgantown, WV 26506-6125, USA*

Anna Carr, *Department of Tourism, University of Otago, PO Box 56, Dunedin, New Zealand*

David Crouch, *School of Tourism & Hospitality Management, University of Derby, Kedleston Road, Derby DE22 1GB, UK*

Dimitrios Diamantis, *Les Roches Management School, Bluche, Crans-Montana, Valais, CH-3975, Switzerland*

Ross Dowling, *School of Marketing, Tourism and Leisure, Edith Cowan University, Joondalup, WA 6027, Australia*

Dianne Dredge, *School of Environmental Planning, Griffith University, Nathan, QLD 4111, Australia*

Stephen N. Edwards, *Ecotourism Department, Conservation International, 1919 M Street, NW, Suite 600, Washington, DC 20036, USA*

David Fennell, *Department of Recreation & Leisure Studies, Brock University, St Catherines, Ontario, Canada*

Nicola Foster, *Centre for Tourism, School of Sport & Leisure Management, Sheffield Hallam University, 1127 Owen Building, City Campus, Sheffield S1 1WB, UK*

C. Michael Hall, *Department of Tourism, University of Otago, PO Box 56, Dunedin, New Zealand*

Sam H. Ham, *Center for International Training & Outreach (CITO), College of Natural Sciences, Department of Resource Recreation & Tourism, University of Idaho, Moscow, ID 83844-1139, USA*

James Higham, *Department of Tourism, University of Otago, PO Box 56, Dunedin, New Zealand*

Christopher Holtz, *Conservation International, 1919 M Street, NW, Suite 600, Washington, DC 20036, USA*

Jeff Humphreys, *Humphreys Reynolds Perkins, Planning and Environment Consultants, Level 20, 344 Queen Street, Brisbane, QLD 4000, Australia*

John Jenkins, *Department of Leisure and Tourism Studies, University of Newcastle, Callaghan, NSW 2308, Australia.*

Colin Johnson, *Ecole Hoteliere De Lausanne, Le Chalet-à-Gobet, 100 Lausanne 25, Switzerland*

Fung Mei Sarah Li, *Tourism Programme, University of Tasmania, Locked Bag 1–340G, Launceston, TAS 7250, Tasmania*

Scott McCabe, *School of Tourism & Hospitality Management, University of Derby, Kedleston Road, Derby DE22 1GB, UK*

Tanja Mihalič, *Faculty of Economics, University of Ljubljana, Kardeljeva pl. 17. 1000 Ljubljana, Slovenia*

William J. McLaughlin, *Department of Resource Recreation & Tourism, University of Idaho, Moscow, ID 83844-1139, USA*

Ken Simpson, *School of Management and Entrepreneurship, UNITEC Institute of Technology, Private Bag 92-025, Auckland, New Zealand*

Trevor H.B. Sofield, *Tourism Programme, University of Tasmania, Locked Bag 1-340G, Launceston, TAS 7250, Tasmania*

Karen Thompson, *School of Leisure, Hospitality and Food Management, University of Salford, Frederick Road, Salford M6 6PU, UK*

Stephen Wearing, *School of Leisure, Sport and Tourism, University of Technology, Sydney, Kuiring-gai Campus, PO Box 222, Lindfield NSW 2070, Australia*

Heather Zeppel, *School of Business, James Cook University, PO Box 6811, Cairns, QLD 4870, Australia*

About the Editors

David Fennell is Associate Professor and Chair of the Department of Recreation and Leisure Studies at Brock University, Ontario, Canada. His main research focus is ecotourism, which he has explored since the 1980s. As a consequence of his involvement in this area, David has undertaken research, lectures, and conducted workshops in many countries around the world. He is the author of *Ecotourism: an Introduction*, which is a general text on ecotourism, and *Ecotourism Programme Planning*. He is also the editor-in-chief of the *Journal of Ecotourism*. Other research interests include carrying capacity, tourist group movement in space and time, and ecological impacts. He has a PhD in geography (Western Ontario), and MA in recreation (Waterloo), and an undergraduate degree in Environmental Science and Biology. Dr Fennell teaches ecotourism and outdoor recreation at Brock University, Canada.

Ross Dowling is Foundation Professor of Tourism and Head of the School of Marketing, Tourism and Leisure, Edith Cowan University, Perth, Western Australia. He is an international speaker, author, researcher and consultant on ecotourism. In 1993 he was awarded a PhD in Environmental Science from Murdoch University, Western Australia, for his ecotourism thesis 'An Environmentally Based Approach to Tourism Planning'. His model has since been applied in Australia, Canada, Indonesia, Malaysia and the USA.

He has over 25 years experience working in the field of the environment and he has been a tour guide in wilderness areas and national parks in Australia and New Zealand. Dr Dowling is Chairman of

EcoPlanet which is currently planning the building of two ecolodges in Western Australia. For his contributions to the environment he has been awarded a Mobil Environmental Prize and a New Zealand Conservation Foundation Citation.

Preface

There are many existing environmental books as well as tourism books that address policy and planning issues. However, there are relatively few texts that include environmental and tourism policy and planning aspects. This book focuses exclusively on this key area of ecotourism policy and planning.

As the global tourism industry continues a trend of sustained growth, moving more people and generating more domestic and foreign revenues, it often does so at the expense of the social and ecological integrity of destination regions. In consequence of this growth, tourism policy makers, particularly governments, have been forced to consider a variety of new approaches (for example, information, regulation, accreditation, and so on) to ensure that the environment, local people, tourists and business remain unaffected by the negative impacts of the industry. Despite mounting concern over such impacts, little has been accomplished, especially by government, to actively stimulate policy development or to enforce weak policies that are currently in use.

Policy is especially relevant to the ecotourism industry, because of what this 'type' of tourism is said to value (for example, ethical approaches to management, local communities and the protection of natural heritage). An absence of sound policy and planning, coupled with the fact that ecotourism is hailed as one of the fastest growing sectors of the world's largest industry, demonstrates an impending need for better industry organization. Unfortunately, however, ecotourism policy has only recently risen to prominence, as a consequence of insufficient consensus on what constitutes appropriate ecotourism development. The nature of the industry (strong advocates representing parks, the environment, non-government organizations, government, industry and local people) is one that demands an effective balance between development and conservation, supply and demand, benefits and costs, and people and the environment.

In a recent study by Fennell (1999) of over 60 regional tourism offices in North America, it was found that most had not instituted ecotourism policies, despite the fact that there was overwhelming consensus on the value of policy to the industry. A significant factor constraining policy development for the industry is the lack of agreement on how to define the concept and identify a process in which to classify ecotourism products. The central aim of this book, therefore, is to examine ecotourism policies and planning through a variety of case studies from around the world. This approach provides an objective overview of the extent of global ecotourism policy and demonstrates the need for further refinement of existing policy as well as for the creation of new dynamic policies geared towards the evolution of a successful ecotourism industry in the new millennium.

A key objective of the book is to highlight the importance of balancing social, ecological and economic factors in the development of policy for the ecotourism industry. Thus sustainability issues are addressed from a variety of approaches at a range of levels. Emerging aspects include regulation, accreditation and interpretation of the biophysical environment so that stakeholders such as local communities, tourists and businesses do not generate adverse impacts on it. Policies inform planning which in turn drives management. Thus an understanding of policy and planning for tourism development in natural areas is essential to the well being of the environment and those people which derive their living from these resources.

The book has been compiled for a broad audience including natural area tourism professionals, planners and managers; government and business decision-makers; and students from a wide range of disciplines seeking sound information on ecotourism development. We hope that you enjoy it.

Ross Dowling
Perth, Australia

David Fennell
St Catherines, Canada

November 2002

McDonald also helped to keep me going with their professional input into the running of a large and busy School.

I also wish to thank a number of close academic colleagues from around the world who have in some small way contributed to my own thoughts on ecotourism through discussion, debate and dialogue over time. They are Dallen Timothy (Arizona State University, USA), Allan Fyall (Bournemouth University, England), Michael Hall (University of Otago, New Zealand), Bruce Prideaux (University of Queensland, Australia), Stephen Page (University of Stirling, Scotland) and Shalini Singh (Brock University, Canada).

Thanks also to Tony Charters (Tourism Queensland, Australia) who is without doubt my ecotourism mentor. He has contributed more to the development of Australian ecotourism then any other person and he is always a great source of knowledge and professionalism. To him I owe a big debt of gratitude for his support over the past decade.

Finally I wish to thank my wife Wendy for her unfailing love and support through this my fourth book in the last 2 years. I could not have achieved this without her. I also wish to acknowledge my children Aurora, Francis, Jayne, Mark, Simon and Tobias as well as my two grand-daughters Shenee and Paige. This book is partly for you all.

The Context of Ecotourism Policy and Planning

Ross K. Dowling[1] and David A. Fennell[2]

[1]*School of Marketing, Tourism and Leisure, Edith Cowan University, Joondalup, WA 6027, Australia;* [2]*Department of Recreation and Leisure Studies, Brock University, St Catherines, Ontario, Canada*

Ecotourism is exploding around the world yet little is known about its possible and/or projected impacts and implications. A number of general books have been written recently on the subject of ecotourism (e.g. Fennell, 1999; Weaver, 2002; Page and Dowling, 2002). All have included some contributions to our collective understanding of the policies and strategies associated with ecotourism development. Conversely, there have been a number of general contributions made on tourism policy and planning issues (e.g. Gunn, 1994; Hall *et al.*, 1997; Hall, 2000). These books have included some principles and practices relevant to ecotourism. However, to the editors' knowledge no specific book exists which investigates and interrogates the specific topic of ecotourism policy and planning. Hence our desire to fill this niche with this edited volume of specially requested contributions from around the world.

Tourism

Tourism may be loosely defined as travel outside one's normal home and workplace, the activities undertaken during the stay and the facilities created to cater for tourist needs (Mathieson and Wall, 1982: 1). Mathieson and Wall argued that tourism can also be described as a system with an originating area (the market or demand element) and a destination area (the attraction or supply side) with a travel component linking the two. Overlying this approach are the characteristics of tourists and destinations as well as a consequential component (impacts).

Tourism can also be viewed as a global activity providing service sector employment, revenue and general economic impacts. On a large scale this is generally referred to as 'mass' tourism. However, over recent years a number of types of tourism have arisen as an alternative to mass tourism, which collectively are referred to as 'alternative' tourism. This has been broadly defined as forms of tourism that set out to be consistent with natural, social and community values and which allow both hosts and guests to enjoy positive and worthwhile interactions and shared experiences (Wearing and Neil, 1999: 3). Alternative tourism fosters sustainability through the process of selective marketing in order to attract environmentally conscious tourists who show respect for the natural and cultural components of tourism destinations and are conservation minded and culturally sensitive in their use of them. Such an alternative approach of appropriate tourism, which embraces strategies considered preferable to mass tourism, have been fostered for over two decades (Britton, 1980).

Alternative tourism, and indeed a whole range of nature tourism options such as farmhouse tourism and kibbutz guesthouses, offer ideal vehicles for sustainable development. Romeril (1989a) suggested that they can represent a new order of tourism development to parallel the hoped for new economic order. Alternative tourism has also been referred to in whole or in part as 'defensive' tourism (Krippendorf, 1982, 1987), 'green' tourism (Jones, 1987), 'nature-oriented' tourism (Durst and Ingram, 1988), 'conscious' or 'soft' tourism (Mäder, 1988) and ecotourism (Boeger, 1991).

Cox (1985: 6) suggests that the positive features of alternative tourism typically include:

1. Development within each locality of a special sense of place, reflected in architectural character and development style, sensitive to its unique heritage and environment.
2. Preservation, protection and enhancement of the quality of resources, which are the basis of tourism.
3. Fostering development of additional visitor attractions with roots in their own locale and developed in ways which complement local attributes.
4. Development of visitor services which enhance the local heritage and environment.
5. Endorsement of growth when and where it improves things, not where it is destructive, or exceeds the carrying capacity of the natural environment or the limits of the social environment, beyond which the quality of community life is adversely affected.

As a form of alternative tourism, ecotourism adopts many of the characteristics inherent in the above description, especially in regard to carrying capacities, preservation and local development. It differs from other forms of alternative tourism, however, on the basis of the primary focus of participants (natural history), the settings in which these activities take place (primarily natural areas, although other areas may play a part in ecotourism) and the focus on education (environmental). To Buckley (1994), ecotourism occurs along four main dimensions, including a nature base, support for conservation, sustainable management and environmental education. These may be perceived as the root characteristics of the concept. Other definitions have been much more elaborate, focusing on a number of related themes. For example, Weaver (2002: 15) writes that:

> Ecotourism is a form of tourism that fosters learning experiences and appreciation of the natural environment, or some component thereof, within its associated cultural context. It has the appearance (in concert with best practice) of being environmentally and socio-culturally sustainable, preferably in a way that enhances the natural and cultural resource base of the destination and promotes the viability of the operation.

In fact, over 80 different definitions of ecotourism have been examined in the tourism literature (see Fennell, 2001a). Consequently, it is important to keep in mind that, with such a wealth of definitions in circulation, no one stands out as a definitive example, with few prospects of ever achieving consensus as a result of a diversity of setting and situational dynamics.

Sustainability

The advent of mass tourism in the second half of the 20th century was paralleled by the rise of the environmental movement globally. With the increase in tourists visiting natural areas it was clear that at some stage the environmental movement would meet tourism development and object to the increased adverse impacts caused by mass tourists. This occurred in the 1980s and became a major focus for disenchanted environmentalists, who were rallying against the environmental destruction caused by rapid growth.

In an effort to solve the situation the World Commission on Environment and Development (WCED) published a report entitled *Our Common Future*, generally referred to as 'The Brundtland Report' after its chairperson, Gro Harlem Brundtland, the Prime Minister of Norway (WCED, 1987). The report examined the world's critical environmental and developmental problems and concluded that only through the sustainable use of environmental resources will long-term economic growth be achieved. Hence, the term 'sustainable development' was coined and in a relatively short space of time it became the new driving force of global development. Five basic principles of sustainability are proffered by the report (Bramwell and Lane, 1993). They are:

1. The idea of holistic planning and strategy making.
2. The importance of preserving essential ecological processes.
3. The need to protect both human heritage and biodiversity.
4. The need to develop in a manner that fosters long-term productivity sustainable for future generations.
5. The goal of achieving a better balance of equity among nations.

Overall, the concept of sustainable development is that it meets the needs of the present without compromising the ability of future generations to meet their own needs (Jordan, 1995: 166).

Tourism and Sustainability

The link between tourism and sustainability was fostered by a number of advocates in the 1980s (e.g. Mathieson and Wall, 1982; Farrell and McLellan, 1987). They suggested that the environment and tourism should be integrated in order to maintain environmental integrity and successful tourism development. They also advanced the notion that a symbiosis between tourism and the physical environment is the second strand of a dual braid of concern, the first being the contextual integration of both physical and social systems (Farrell and McLellan, 1987). They further argued that:

> the true physical environment is not the ecosystem, the central core of
> ecology. This is an environment (better still an analogue model) perceived
> by those occupying a subset of the scientific paradigm, and their
> viewpoint is not exactly the same as the abiotic vision of landscape
> perceived by the earth scientist or the more balanced landscape or region,
> the core of the geographer's study.
>
> (Farrell and McLellan, 1987: 12)

Their reasoned appeal for a more holistic view was advanced with the need for the integration of community concern and involvement in tourism development as contended by Murphy (1983, 1985). According to Gunn (1987: 245) this integrative approach is one in which the 'resource assets are so intimately intertwined with tourism that anything erosive to them is detrimental to tourism'. Conversely, support of environmental causes, by and large, is support of tourism. It is this view which has begun to shape tourism generally, and ecotourism specifically, in recent decades.

The underlying concept of sustainable tourism development is the equating of tourism development with ecological and social responsibility. Its aim is to meet the needs of present tourists and host regions while protecting and enhancing environmental, social and economic values for the future. Sustainable tourism development is envisaged as leading to the management of all resources in such a way that it can fulfil economic, social and aesthetic needs while maintaining cultural integrity, essential ecological processes, biological diversity and life support systems. According to GLOBE 90 (1990: 2), the goals of sustainable tourism are:

1. To develop greater awareness and understanding of the significant contributions that tourism can make to the environment and the economy.
2. To promote equity in development.
3. To improve the quality of life of the host community.
4. To provide a high quality of experience for the visitor.
5. To maintain the quality of the environment on which the foregoing objectives depend.

Achieving the fifth goal of environmental conservation includes providing for intergenerational equity in resource conservation (Witt and Gammon, 1991). It also includes avoiding all actions that are environmentally irreversible, undertaking mitigation or rehabilitation actions where the environment is degraded, promoting appropriate environmental uses and activities, and cooperating in establishing and attaining environmentally acceptable tourism.

However, it has been argued that, although 'the concept of ecotourism is still often used synonymously with that of sustainable tourism, in reality, ecotourism fits within the larger concept of sustainable tourism' (Ceballos-Lascurain, 1998: 8). And herein lies a conundrum. We would argue that ecotourism is a niche form of tourism which fosters sustainable development principles. That is, the former is a 'type' of tourism to which the latter is an approach, or it is a 'process' which drives tourism. Thus ecotourism encompasses sustainability principles and in fact should be regarded as the exemplar of the sustainability approach within tourism generally.

Policy and Planning Issues

Policies are the plan of action adopted or pursued by governments or businesses and so on whereas strategies represent the steps to achieve them. According to Hall *et al.* (1997: 25) the focus of government activity is public policy; thus it reflects 'a consequence of the political environment, its values and ideologies, the distribution of power, institutional frameworks and of decision-making processes'. Policies are made at a range of levels, from the micro (site scale), through to the medium (regional, state or provincial) and macro (national, supranational and global) scales. In addition there are numerous groups which influence policies. These groups can include pressure groups, such as tourism industry associations, conservation groups, community groups, community leaders and significant individuals (e.g. local government councillors and groups spokespeople), government employees and representatives (e.g. employees of tourism departments and regional boards as well as members of parliament), academics and consultants (Hall *et al.*, 1997).

The terms policy and planning are intimately related (Hall, 2000). Plans embrace the strategies with which policies are implemented. Planning is predicting and therefore requires some estimated perception of the future. Although it is reliant on observation and deduction from research conclusions it also relies heavily on values (Rose, 1984). Planning should provide a resource for informed decision making. Hall (2000) suggests that planning is a part of an overall planning–decision–action process.

Unplanned, uncontrolled tourism growth can destroy the very resource on which it is built (Pearce, 1989). Tourism planning is a process based on research and evaluation, which seeks to optimize the potential contribution to human welfare and environmental quality (Getz, 1987). Getz identifies four broad approaches to tourism planning. They are boosterism, economic, physical/spatial and community oriented. In this approach tourism planning is regarded as an integrated activity which incorporates economic, social and environmental components, spatial (accessibility) concerns and temporal (evolutionary stage) implications. In addition it recognizes the basic components of demand (markets) and supply (destinations) linked by transport and communications. To these four approaches a fifth has been added: 'sustainable tourism planning' (Hall, 1995).

Since the introduction of the *World Conservation Strategy* (IUCN, 1980) with its emphasis on 'ecodevelopment' there has been a strong move towards recognizing the interdependencies that exist among environmental and economic issues. This led to the Brundtland Commission's 'sustainable development' concept which equates development with environmental and social responsibility. This approach was advanced by Travis (1980) who suggested that taking actions which ensure the long-term maintenance of tourist resources (natural or human-made) is good economics, as it can mean long-term economic returns from their use. This was endorsed by Romeril in his study of tourism and the environment symbiosis when he concluded that 'tourism's strong dependence on quality natural resources makes such a goal (of sustainable development) not just a desired ideal but an economic necessity' (Romeril, 1985: 217). While it was being argued that it made good economic sense to look after the environment, it was also advocated that 'the environment should no longer be viewed pri-

marily in negative terms as a constraint, but as a resource and an exciting opportunity for compatible human use' (Pigram 1986: 2).

The call for the application of the sustainable development
approach to tourism has been reflected in its suggested incorporation
into planning procedures. Among the first advocates were Mathieson
and Wall (1982) who had compiled their treatise on tourism's economic, physical and social impacts. They stated 'planning for tourist
development is a complex process which should involve a consideration of diverse economic, environmental and social structures'
Mathieson and Wall (1982: 178). The same conclusion was drawn by
Murphy (1985) in his advocacy of a community approach to tourism
planning. He concluded that tourism planning needs to be restructured
so that environmental and social factors may be placed alongside economic considerations. Getz (1986) approached the situation from his
investigation of tourism planning models and indicated that reference
to theoretical models will remind tourism planners not to act in isolation from other social, economic and environmental planning.

During the late 1980s the sustainable development approach to
tourism planning was advanced by a number of authors (Inskeep,
1987, 1988; Gunn, 1987, 1988; Pearce, 1989; Romeril, 1989a,b).
Inskeep (1988) suggested that tourism planning cannot be carried out
in isolation but must be integrated into the total resource analysis and
development of the area with possible land and water conflicts
resolved at any early stage. He noted that recently prepared tourism
plans gave much emphasis to socio-economic and environmental factors and to the concept of controlled development.

Pearce (1989) indicated that the recognition of tourism's compos-
ite nature and multiplicity of players involved in its development are
critical in planning for tourism. This was endorsed by Romeril (1989a)
who stated that a strong emphasis of many strategies is their integrated nature where tourism is one of a number of sector and land-use
options. In deciding national and regional policies, a matrix of all sectors of activity are assessed and evaluated: positive and negative economic effects, positive and negative social effects, positive and
negative environmental effects, and so on. Thus tourism and environmental resource factors are not taken in isolation, nor at the remote
end of a decision-making process.

Goals

The goals of tourism plans will inevitably determine their role for
environmental protection or conservation. Murphy (1983) argued that
most tourism goals and planning were oriented towards business
interests and economic growth. This was echoed 3 years later by Getz

(1986) who asserted that a review of tourism models suggested that tourism planning is predominantly project and development oriented.

However, the goals of tourism planning are changing. The major goal of one approach, the Products' Analysis Sequence for Outdoor Leisure Planning (PASOLP), is to integrate tourism planning into a region or country's wider political, economic, social and environmental context (Baud Bovy, 1982). A comprehensive list of 13 aims of planning tourism development includes several oriented to the environment. These are: to minimize erosion of the very resources on which tourism is founded and to protect those which are unique; and to ensure that as far as practicable the image presented by the destination is matched by the extent of environmental protection and facilities provided (Lawson and Baud Bovy, 1977).

The planning goals of both McIntosh (1977) and Gunn (1979) have always included environmental aspects but have changed over time to incorporate social aspects. For example, the original goals of McIntosh (1977) encompassed tourism development within a community framework. They include the following:

1. To provide a framework for raising the living standard of local people through the economic benefits of tourism.
2. To develop an infrastructure and provide recreation facilities for both visitors and residents.
3. To ensure that the types of development within visitor centres and resorts are appropriate to the purposes of these areas.
4. To develop a programme that is consistent with the cultural, social and economic philosophy of the government and people of the host area.

In a later edition of McIntosh's (1977) book, McIntosh and Goeldner (1990) added a fifth goal, 'optimizing visitor satisfaction', which becomes the plan's zenith point and is Gunn's (1979) first tourism planning goal. The goals of Gunn (1979) originally included user satisfaction, increased rewards to ownership and development, and the protection of environmental resource assets. Murphy (1983: 182) later noted that 'while Gunn's first two goals were distinctly business oriented his final goal recognizes the symbiotic relationship between a successful tourism industry and a protected environment' – the early stirrings of a renewable resource philosophy. Gunn (1988) later added a fourth goal of 'local adaptation' in which tourism is integrated into the total social and economic life of a community. No doubt the inclusion of this goal is in part a reflection of the advocacy of tourism as a community industry by Murphy (1983, 1985).

The major goal of Mill and Morrison's (1985) model is to preserve and enhance unique destination attractions in order to maintain tourism as a long-term economic activity. To achieve this primary goal they list five subsidiary aims:

1. To identify alternative approaches to tourism marketing and development.
2. To adapt to the unexpected in economic and other external situations.
3. To maintain uniqueness of product.
4. To create the desirable in destination marketing and organization.
5. To avoid the undesirable such as negative economic, social or environmental impacts.

In summary, the goals of area development tourism planning models have shifted away from an emphasis on economic considerations to include community concerns (Gunn, 1988; McIntosh and Goeldner, 1990) and environmental aspects; for example, protection of resources (Gunn, 1988) and reduction of adverse impacts (Mill and Morrison, 1985).

Levels

Tourism planning can occur at a variety of levels including intranational, involving two or more countries from the same region (Pearce 1989), national, regional, local and site scale (WTO, 1980). National tourism planning incorporates economic, social and environmental aspects and details policies, strategies and phases commensurate with overall national planning goals. A physical structure plan includes identification of the major tourist attractions, designation of tourist regions, transport access to and within a country, as well as touring patterns. National plans also recommend development, design and facility standards and the institutional elements to effectively implement and operate tourism. Such plans are usually based on projections of demand and represent 5 or 10 year policies which are subject to periodic review.

The regional level of planning identifies appropriate regional policies and strategies, the major tourist access points, the internal transport network, primary and secondary tourist attraction features, specific resort and other tourism sites, types of urban tourism development needed, and regional tour patterns (Inskeep, 1988). It also usually incorporates economic, social and environmental factors. Social factors include the need for public participation in the preparation of a regional tourism plan. These factors form part of the community approach advocated by Murphy (1985). Environmental concerns at the regional level include the need for adequate zoning to encourage the concentration or dispersal of tourist activity. Areas of concentration should be those with highly resistant environments or should have been hardened to protect the environment. Dispersal allows for the distribution of small-scale developments throughout the region in order to reduce environmental pressures in any particular spot.

An environmentally oriented regional tourism policy includes the use of tourism to promote conservation; strategic market segmentation of conservation conscious tourists; and industry growth at a pace commensurate with the needs of adequate planning, implementation and monitoring of changes (Inskeep, 1987: 122). Of all the levels of tourism planning it is the regional level which appears to offer the best opportunity for achieving both tourism and environmental protection goals.

Ecotourism Policy and Planning

As the global tourism industry continues a trend of sustained growth, moving more people and generating more domestic and foreign revenues, it often does so at the expense of the social and ecological integrity of destination regions. As a consequence of this growth, tourism policy makers, particularly government, have been forced to consider a variety of new approaches (e.g. information, regulation, accreditation) to ensure that the environment, local people, tourists and business remain unaffected by the negative impacts of the industry. Despite the mounting concern over such impacts, little has been accomplished, especially by government, to actively stimulate policy development (Lickorish, 1991) or to enforce weak policies that are currently in use.

Policy is especially relevant to the ecotourism industry, because of what this 'type' of tourism is said to value (ethical approaches to management, local people, the protection of natural heritage, and so on). An absence of sound policy and planning, coupled with the fact that ecotourism is the fastest growing sector of the world's largest industry (upwards of 20% of the world travel market as ecotourism (WTO, 1998)), demonstrates an impending need for better industry organization. Unfortunately, however, ecotourism policy has only recently come about, as a consequence of insufficient consensus on what constitutes appropriate ecotourism development. The nature of the industry (strong advocates representing parks, the environment, NGOs, government, industry and local people) is one that demands an effective balance between development and conservation, supply vs. demand, benefits vs. costs and people vs. the environment.

A study by Fennell (2001b) of over 60 regional tourism offices in North America found that most had not instituted ecotourism policies, despite the fact that there was overwhelming consensus on the value of policy to the industry. A significant factor constraining policy development for the industry is the lack of agreement on how to define the concept and identify a process in which to classify ecotourism products. The central aim of this book is to critically examine ecotourism (definitions, products and policies) through a variety of case studies from around the world. This approach provides an objective overview of the extent of global ecotourism policy. It demonstrates the need for further refinement of existing policy, and for the creation of new, dynamic policies geared towards the evolution of a successful ecotourism industry at the start of the new millennium.

Although there are many environmental planning models as well as numerous tourism planning approaches, there are few ecotourism planning frameworks. Obviously the relationship between the two needs to be better understood. The few planning processes for ecotourism at the regional level that have already been proposed include the ecological approach of Van Riet and Cooks (1990) and the regional strategic tourism framework of Gunn (1988). Underlying these frameworks for the environment and tourism is the intrinsic belief that tourism developments must not only maintain the natural and cultural resources but also sustain them. To achieve this goal, strategic regional land-use planning should include such components as resource protection, agriculture, pastoral use, urban areas and mining to be established in a carefully planned and controlled manner which sets conditions on growth and maintains or enhances environmental quality. Outstanding natural features can continue to be significant tourist attractions only if they are conserved. If there is any doubt that the natural environment cannot be protected or enhanced then tourism development should not be allowed to proceed.

Ecotourism planning involves aspects of both environmental planning and tourism planning. Components of the former include environmental protection, resource conservation and environmental impact assessment while tourism planning provides aspects of area development and social assessment. The need for further research into ecotourism planning has been articulated for both the general planning framework arena (Farrell and McLellan, 1987) and the evaluation of impacts (Mathieson and Wall, 1982). Other aspects which require further research include case studies on areas such as the tropics, arid lands and small islands (Inskeep, 1987) as well as the link between environmental and social aspects of tourism development (Murphy, 1985).

An analysis of the more than 1600 tourism plans inventoried by the WTO (1980) study found that: (i) approximately one-third were not implemented; (ii) few plans integrated tourism with broader socio-economic development objectives and those whose 'social aspects' have priority over direct profitability are even more exceptional; and (iii) few examples were found of plans that made firm and specific provision for protecting the environment (quoted in Pearce, 1989: 276–277).

One aspect of tourism planning which is often fostered is the need for the integration of tourism in area development. However, as has been argued earlier in this introduction there are fewer approaches that advocate the need for the integration of tourism and environmental protection. Yet a real need exists for this to take place if the symbiotic link between the two is to be transferred from concept to reality. Inskeep (1987: 128) asserts that tourism planning of natural attractions 'should be closely coordinated and integrated with park and conservation planning at the national, regional and local levels with respect to both geographic distribution and intensity of the tourism development'. Achieving environmental–tourism compatibility in natural areas is best undertaken at the regional level where it is suggested that tourism planning can provide one of the best opportunities for attaining environmental goals (UNDP and WTO, 1986). This has also been supported from regional land-use planning (Sijmons, 1990).

Despite their different goals, tourism planning and environmental planning also share a common spatial framework. Within a tourism destination zone, Gunn (1988) identifies attraction clusters, the service community and linkage corridors, whereas in the ABC approach to environmental planning, Smith *et al.* (1986) identify cultural activity nodes, hinterlands and corridors. A final similarity concerns the integration of social values in each of the two planning approaches. The role of people as part of the ecosystem is central to emerging ecological approaches just as the incorporation of social values forms part of recent tourism planning processes.

Ecotourism planning can be carried out at a range of levels. At the site scale the planning of ecoresorts, lodges and associated facilities includes locational analysis, financial feasibility, environmental assessment and site planning. This last factor incorporates environmentally sensitive architectural and engineering designs as well as landscaping (Gunn, 1987). Ecoethics, which encourage the environmentally sensitive design of tourist developments in natural areas, should be included for every ecotourism development (WATC and EPA, 1989). Ecolodges should emphasize a high degree of creative, positive interaction with natural features. In the planning, development and operation of an ecotourism facility, rapport and empathy with the site should be fostered and its ambience maintained and enhanced harmoniously (Page and Dowling, 2002).

National ecotourism planning incorporates economic, social and environmental aspects and details policies, strategies and phases commensurate with overall national tourism planning goals. An ecotourism plan includes identification of the major ecotourism attractions, designation of the ecotourism regions, transportation access to and within a country, as well as ecotouring patterns. National plans also recommend development, design, and facility standards and the institutional elements to effectively implement and operate ecotourism. Such plans are usually based on projections of demand and represent 5 or 10 year policies that are subject to periodic review.

The regional level of ecotourism planning identifies appropriate regional policies and strategies, the major tourist access points, and the internal transportation network, primary and secondary eco-tourism attraction features, specific ecoresorts and other ecotourism sites, and regional ecotour pattern. It also usually incorporates eco-nomic, social and environmental factors. Social factors include the need for public participation in the preparation of a regional ecotourism plan. Environmental concerns at the regional level include the need for adequate zoning to encourage the concentration or dis-persal of ecotourist activity. Areas of concentration should be those with highly resistant environments or should have been hardened to protect the environment. Dispersal allows for the distribution of small-scale ecotourism developments throughout the region so as to reduce environmental pressures in any particular spot. Of all the lev-els of ecotourism planning it is the regional level that appears to offer the best opportunity for achieving both tourism and environmental protection goals.

Local Communities

Probably the most prominent benefits of ecotourism policies and plan-ning are to foster developments that provide benefits for local commu-nities and their natural environments. These include new jobs, businesses and additional income; new markets for local products; improved infrastructure, community services and facilities; new skills and technologies; increased cultural and environmental awareness, conservation and protection; and improved land-use patterns.

Ecotourism at the community level should be developed within the context of sustainable, regional, national and even international tourism development. Sustainable development principles that can be applied to regional ecotourism development include (from Page and Dowling, 2002):

1. Ecological sustainability which ensures that development is com-patible with the maintenance of essential ecological processes, biolog-ical diversity and biological resources.
2. Social and cultural sustainability which ensures that development increases people's control over their own lives, is compatible with the

culture and values of people affected by it, and maintains and strengthens community identity.

3. Economic sustainability which fosters development that is economically efficient and so that resources are managed so that they can support future generations.

Ecotourism can provide the opportunity to present a region's natural areas, promoting an identity that is unique. It can create new and exciting tourism experiences, promote excellence in tourism, present and protect natural areas, benefit local communities and encourage commercially successful and environmentally sound tourism operations (Page and Dowling, 2002). The vision for regional ecotourism development is for a vibrant and ecologically, commercially and socially sustainable ecotourism industry that leads the way in tourism development (Dowling and James, 1996).

The key to capitalizing on the potential benefits offered through ecotourism development is to maximize the opportunities and minimize the adverse impacts through environmentally appropriate policies and planning. If this is carried out then a sound base will have been established for ecotourism to develop and flourish in harmony with the natural environments and cultural settings on which it depends.

Management Strategies

There are a wealth of strategies and actions for managing tourism in natural areas (Newsome *et al.*, 2002). Strategies are often viewed as representing the mechanisms and processes by which objectives are achieved; for example, through the reservation of an area as a national park. Once reservation has occurred, zoning generally follows. The management of ecotourism is generally carried out through either site management or visitor management. Site management actions rely on manipulating infrastructure, where visitors go and what they do. Some examples include campsite and trail design and management.

Visitor management concentrates on managing visitors by regulating numbers, group size and length of stay. It also provides information and education as well as the enforcement of regulations. Ecotourism management includes the incorporation of a number of voluntary strategies such as codes of conduct, accreditation and best practice. Strategies employed by government organizations are generally more regulatory in manner and include licensing and leases.

Outline of the Book

The book is divided into five main sections focusing on the theoretical considerations of ecotourism policy and planning (Section 1) followed by case studies at a range of levels including regions (2), countries (3) and continents (4), ending with some brief conclusions (5).

Section 1 on 'Understanding Ecotourism Policies' includes five chapters. Chapter 2 examines the institutional arrangements for ecotourism policy (Michael Hall, New Zealand). Hall argues that ecotourism policy does not occur in a vacuum but instead is the outcome of a shared process reflecting a combined set of stakeholders' interests and values. He outlines a range of ecotourism policy scales (e.g. local, state, national and international) and instruments (e.g. regulatory, voluntary, expenditure and financial).

Christopher Holtz and Stephen Edwards (USA) describe the synergy between biodiversity conservation and sustainable tourism in Chapter 3. The authors note that while this can be achieved through appropriate planning and management of tourism, to date this has not occurred to any large extent. They argue that tourism planning should include biodiversity conservation at all levels in order to sustain the resource base upon which it is built.

Heather Zeppel (Australia) explores the relationship between ecotourism and indigenous people (Chapter 4). She asserts that indigenous ecotourism ventures focus on the cultural significance of the natural environment. Indigenous tours educate visitors on their environmental values combined with the sustainable use of natural resources. Zeppel concludes that, in Australia, indigenous ecotourism ventures remain peripheral to the ecotourism industry of the country.

The wider aspect of culture and ecotourism is investigated by David Crouch and Scott McCabe (England). The chapter (5) seeks to explore the issues surrounding the labelling of ecotourism as a set of ideas and practices. It examines ecotourism from the standpoint of tourist consumption and production practices and investigates the effects that these may have on the creation of ecotourism policy development. The authors conclude that ecotourists should be made fully aware of their contribution towards development and their role in affecting the cultures of their destination hosts.

Tanja Mihalič (Slovenia) outlines the economic instruments of ecotourism policy derived from environmental theories (Chapter 4). Mihalič concludes that market, fiscal and administrative instruments can be used in ecotourism to minimize or prevent environmental damage.

Section 2 presents two case studies of ecotourism development policies in regions in Australia and the People's Republic of China. The former examines policy and strategy issues for ecotourism in Australia's tropical rainforests (Diane Dredge and Jeff Humphreys). The chapter (7) examines the approach of the Douglas Shire, North Queensland, in regard to the development of ecotourism in the Daintree River area. The discussion includes an examination of a number of environmental, social and economic issues arising from ecotourism and demonstrates the complexity of local government's role in ecotourism management 'especially where overlapping jurisdictions and responsibilities give rise to complex policy making environments'.

Chapter 8 by Trevor Sofield and Sarah Li, explores the complexities of ecotourism policy formulation in Yunnan Provinces, southwest China. Some of the issues canvassed by the authors include a lack of fit with existing government priorities, the politics of the situation and the power exercised by vested (often competing) interests in the recipient society, and a lack of cultural values which conflict with those imported by the plan. Another issue is that planning had to take account of the Chinese desire to visit reserves in extremely large numbers, instead of the idealized 'Western' concept comprising small-scale, low impact ecotourism development.

Section 3 comprises five country case studies. In Chapter 9 Karen Thompson and Nicola Foster (England) examine ecotourism development and government policy in Kyrgyzstan. The authors note that the country remains at a relatively primitive stage of tourism development and is hindered by unresolved political and economic difficulties. However, they argue that the ecological and cultural resources present a strong base for its emerging ecotourism market.

Ecotourism development in Fiji is investigated by Kelly Bricker (USA). Like Kyrgyzstan, Fiji is politically unstable, thus presenting another level of challenge to the establishment of ecotourism policies and plans. Nevertheless the country has a national professional association ecotourism plan and advisory council. Overall the seeds of a sound future have been planted but, until confidence is restored in Fiji's political situation, this seed will lie dormant.

Ecotourism in Australia is described by John Jenkins and Stephen Wearing (Chapter 11). Despite the fact that the country is upheld as one of the world's leaders in ecotourism development, the authors present a number of issues, which are critical to its future growth. These include public sector reforms influencing the management of tourism in protected areas, the imposition of user fees and other charges, and the role of the private sector in protected areas. They conclude that there is a continued need for scientific research in ecotourism planning and management as well as the increased protection of protected areas and greater conservation measures.

Ecotourism policies in New Zealand are outlined in two chapters. The first, by James Higham and Anna Carr (Chapter 12) describes the scope and scale of the industry and provides a critical review of current policy initiatives to develop high quality and sustainable ecotourism in New Zealand. Policy directions advanced include redefining the country's definition of ecotourism, the establishment of an accreditation scheme and enhancing the place of interpretation within ecotourism operations.

Ken Simpson examines broader policy issues related to the protection of the environment and the development of tourism in New Zealand's national parks and other protected areas (Chapter 13). Ecotourism policy development is investigated both in theory as well as in practice in relation to the government departments of conservation and tourism. Simpson concludes that the foundations for effective ecotourism policy are in place but that major problems are inherent in these departments' roles. The Department of Conservation advocates both environmental protection and tourism development, while Tourism New Zealand calls for maximizing international tourist visitation to the country with little responsibility for them once they arrive.

The fourth section comprises three continental case studies of Europe, the Americas and Antarctica.

Dimitrios Diamantis and Colin Johnson (Switzerland) investigate the economic role of ecotourism development within central and eastern Europe (Chapter 14). They particularly highlight the importance of biosphere reserves in ecotourism. Tourism markets are likely to focus on the high value added, environmentally aware niche of Western tourists combined with the traditional mass demands from more general tourists. To cope with both it is proposed that a key element in the successful management of ecotourism attractions such as biosphere reserves is the introduction of carrying capacity levels.

Ecotourism policies in North and South America are investigated by Stephen Edwards, William McLaughlin and Sam Ham (Chapter 15). The study describes how governmental tourism agencies define ecotourism and foster its development through legislation, plans, reports and discussion documents, speeches and the range of tourism policy roles. The authors find that the countries on the two continents still lack clearly defined ecotourism policies. They argue that defining ecotourism is a necessary first step in the ecotourism policy development process. Once this has occurred then ecotourism acts as a positive force in conservation, benefits host communities and promulgates environmental awareness.

Chapter 16 examines ecotourism policies in Antarctica. The authors, Thomas Bauer (China) and Ross Dowling (Australia) outline the growth and impacts of tourism on the continent.

The conclusions of Section 5 (Chapter 17) are formulated as a synthesis of what is viewed to be the most salient issues emerging from the book. In particular, the discussion touches upon the importance of stakeholder groups in policy; management actions; and policy development, complexity and governance. An attempt is made to examine both the macro perspective of policy (i.e. the role that sustainable development and governance play in policy) and the micro perspective, which involves how ecotourism operators might better structure their services to fit into a broader policy environment.

References

Baud Bovy, M. (1982) New concepts in planning for tourism and recreation. *Tourism Management* 3 (4), 308–313.
Boeger, E. (1991) Ecotourism/the environment: or the immense potential and importance of ecotourism. *Travel and Tourism Research Association Newsletter* 12 (3), 2, 5–6.
Bramwell, B. and Lane, B. (1993) Sustainable tourism: an evolving global approach. *Journal of Sustainable Tourism* 1 (1), 6–16.
Britton, R. (1980) Alternatives to conventional mass tourism in the Third World. Paper presented to the 76th Annual Meeting of the Association of American Geographers, Louisville, USA.
Buckley, R. (1994) A framework for ecotourism. *Annals of Tourism Research* 21 (3), 61–665.
Ceballos-Lascurain, H. (1998) Introduction. In: Lindberg, K., Epler Wood, M. and Engeldrum, D. (eds) *Ecotourism: Guide for Planners & Managers*, Vol. 2. The Ecotourism Society, Vermont, pp. 7–10.
Cox, J. (1985) The resort concept: the good, the bad and the ugly. Keynote paper presented to the National Conference on Tourist Resort Development: 4–11 November Kuring-gai College of Advanced Education, Sydney, Australia, November.
Dowling, R.K. and James, K. (1996) The South West Ecotourism Strategy. In: Ritchins, H., Richardson, J. and Crabtree, A. (eds) *Proceedings of the Ecotourism Association of Australia National Conference Taking the Next Steps*, Alice Springs, 18–23 November 1995. Ecotourism Association of Australia, Brisbane, pp. 25–32.
Durst, P.B. and Ingram, C.D. (1988) Nature-oriented tourism promotion by developing countries. *Tourism Management* 9 (1), 39–43.
Farrell, B.H. and McLellan, R.W. (1987) Tourism and physical environment research. *Annals of Tourism Research* 14 (1), 1–16.
Fennell, D.A. (1999) *Ecotourism: an Introduction*. Routledge, London.
Fennell, D.A. (2001a) A content analysis of ecotourism definitions. *Current Issues in Tourism* 4(5), 403–421.
Fennell, D.A. (2001b) Anglo-America. In: Weaver, D.B. (ed.) *The Encyclopedia of Ecotourism*. CAB International, Wallingford, UK, pp. 107–122.
Getz, D. (1986) Models in tourism planning: towards integration of theory and practice. *Tourism Management* 7 (1), 21–32.

Getz, D. (1987) Tourism planning and research: traditions, models and futures. Paper presented to the Australian Travel Research Workshop, Bunbury, Western Australia, 3–6 November.

GLOBE '90 (1990) *An Action Strategy for Sustainable Development.* Tourism Stream, Action Strategy Committee, GLOBE '90. Vancouver, British Columbia.

Gunn, C.A. (1979) *Tourism Planning.* Crane-Russak, New York.

Gunn, C.A. (1987) Environmental designs and land use. In: Ritchie, J.R.B. and Goeldner, C.R. (eds) *Travel, Tourism and Hospitality Research: a Handbook for Managers and Researchers.* John Wiley & Sons, New York, pp. 229–247.

Gunn, C.A. (1988) *Tourism Planning,* 2nd edn. Taylor and Francis, New York.

Gunn, C.A. (1994) *Tourism Planning: Basics, Concepts, Cases,* 3rd edn. Taylor & Francis, New York.

Hall, C.M. (1995) *An Introduction to Tourism in Australia: Impacts, Planning and Development,* 2nd edn. Longman, Melbourne.

Hall, C.M. (2000) *Tourism Planning: Policies, Processes and Relationships.* Pearson Education, Harlow, UK.

Hall, C.M., Jenkins, J. and Kearsley, G. (eds) (1997) *Tourism Planning and Policy in Australia and New Zealand: Cases, Issues and Practice.* Irwin Publishers, McGraw-Hill Book Company, Sydney.

Inskeep, E. (1987) Environmental planning for tourism. *Annals of Tourism Research* 14 (1), 118–135.

Inskeep, E. (1988) Tourism planning: an emerging specialisation. *Journal of the American Planning Association* 54 (3), 360–372.

IUCN (1980) *World Conservation Strategy.* International Union for the Conservation of Nature and Natural Resources, Gland, Switzerland.

Jones, A. (1987) Green tourism. *Tourism Management* 26, 354–356.

Jordan, C.F. (1995) *Conservation: Replacing Quantity with Quality as a Goal for Global Management.* John Wiley & Sons, New York.

Krippendorf, J. (1982) Towards new tourism policies – the importance of environmental and sociocultural factors. *Tourism Management* 3 (3), 135–148.

Krippendorf, J. (1987) *The Holiday Makers: Understanding the Impact of Leisure and Travel.* Heinemann, Oxford.

Lawson, F. and Baud Bovy, M. (1977) *Tourism and Recreation Development.* Architectural Press, London.

Lickorish, L.J. (ed.) (1991) *Developing Tourism Destinations.* Longman, Harlow, UK.

Mäder, U. (1988) Tourism and environment. *Annals of Tourism Research* 15 (1), 274–277.

Mathieson, A. and Wall, G. (1982) *Tourism: Economic, Physical and Social Impacts.* Longman Scientific and Technical, Harlow, UK.

McIntosh, R.W. (1977) *Tourism: Principles, Practices, Philosophies,* 2nd edn. Corid Inc., Columbus, Ohio.

McIntosh, R.W. and Goeldner, C.R. (1990) *Tourism: Principles, Practices and Philosophies,* 6th edn. John Wiley & Sons, New York.

Mill, R.C. and Morrison, A.M. (1985) *The Tourism System: an Introductory Text.* Prentice-Hall, Englewood Cliffs, New Jersey.

Murphy, P.E. (1983) Tourism as a community industry: an ecological model of tourism development. *Tourism Management* 4 (3), 180–193.

Murphy, P.E. (1985) *Tourism: a Community Approach.* Methuen, New York.
Newsome, D., Moore, S.A. and Dowling, R.K. (2002) *Natural Area Tourism: Ecology, Impacts and Management.* Channel View Publications, Clevedon, UK.
Page, S. and Dowling, R.K. (2002) *Ecotourism.* Pearson Education, Harlow, UK.
Pearce, D.G. (1989) *Tourist Development,* 2nd edn. Longman Scientific and Technical, Harlow, UK.
Pigram, J.J. (1980) Environmental implications of tourism development. *Annals of Tourism Research* 7 (4), 554–583.
Pigram, J.J. (1986) Regional resource management: a case for development. Paper presented to the Conference on Planning and Development on the North Coast, Valla Park, NSW Australia, August.
Romeril, M. (1985) Tourism and the environment – towards a symbiotic relationship (introductory paper). *International Journal of Environmental Studies* 25 (4), 215–218.
Romeril, M. (1989a) Tourism and the environment – accord or discord? *Tourism Management* 10 (3), 204–208.
Romeril, M. (1989b) Tourism – the environmental dimension. In: Cooper, C.P. (ed.) *Progress in Tourism, Recreation and Hospitality Management.* Belhaven Press, London. pp. 103–113.
Rose, E.A. (1984) Philosophy and purpose in planning. In: Bruton, M.J. (ed.) *The Spirit and Purpose of Planning,* 2nd edn. Hutchinson, London, pp. 31–65.
Sijmons, D. (1990) Regional planning as a strategy. *Landscape and Urban Planning* 18, 265–273.
Smith, P.G.R., Nelson, J.G. and Theberge, J.B. (1986) *Environmentally Significant Areas, Conservation, and Land Use Management in Northwest Territories: a Method and Case Study in the Western Arctic.* Technical Paper No. 1, Heritage Resources Centre, University of Waterloo, Waterloo, Canada.
Travis, A.S. (1980) The need for policy action. In: *The Impact of Tourism on the Environment.* OECD, Paris, pp. 79–97.
UNDP and WTO (1986) *Bhutan Tourism Development Master Plan.* United Nations Development Programme and the World Tourism Organisation, Madrid.
Van Riet, W.F. and Cooks, J. (1990) An ecological planning model. *Environmental Management* 14 (3), 339–348.
WATC and EPA (1989) *The Eco Ethics of Tourism Development.* Prepared by B.J. O'Brien for the WA Tourism Commission and the Environmental Protection Authority, Perth.
WCED (1987) *Our Common Future.* Report of the World Commission on Environment and Development (The Brundtland Commission). Oxford University Press, Oxford.
Wearing, S. and Neil, J. (1999) *Ecotourism.* Butterworth – Heinemann, Oxford.
Weaver, D. (2002) *Ecotourism.* John Wiley & Sons, Milton, Australia.
Witt, S.F. and Gammon, S. (1991) Sustainable tourism development in Wales. *The Tourist Review* 4, 32–36.
WTO (1980) *Physical Planning and Area Development for Tourism in the Six WTO Regions, 1980.* World Tourism Organisation, Madrid.
WTO (1998) Ecotourism now one-fifth of market. *World Tourism Organisation News* 1, 6.

Institutional Arrangements for Ecotourism Policy

2

C. Michael Hall

Department of Tourism, University of Otago, PO Box 56, Dunedin, New Zealand

Ecotourism policy does not occur in a vacuum. Ecotourism policies are the outcome of a policy-making process which reflects the interaction of actors' interests and values in the influence and determination of the tourism planning and policy processes. The focus of this chapter is on ecotourism public policy which, after Hall and Jenkins (1995), may be defined as whatever governments choose to do or not to do with respect to ecotourism. However, such a deceptively simple definition hides within it a multitude of issues which this chapter will seek to address. Most significantly the chapter focuses on public policy which 'stem from governments or public authorities ... A policy is deemed a public policy not by virtue of its impact on the public, but by virtue of its source' (Pal, 1992: 3).

The definition of ecotourism public policy described above covers government action, inaction, decisions and non-decisions, as it implies a deliberate choice between alternatives. For an ecotourism policy to be regarded as public policy, at the very least it must have been processed, even if only authorized or ratified, by public agencies. This is an important distinction because it means that the 'policy may not have been significantly developed within the framework of government' (Hogwood and Gunn, 1984: 23). Pressure and interest groups, community leaders, lobbyists, bureaucrats and others working inside and outside the 'rules of the game' established by government, influence and perceive public policies in significant and often markedly different ways. This latter observation is especially important for ecotourism policy because of the role that conservation interest groups, such as the International Ecotourism Society and the

Wilderness Society, have played in determining government policies. Moreover, ecotourism policy is also considerably complicated by virtue of the fact that it is multi-scalar. Ecotourism policies are developed at multiple scales of public governance, from the international through to the local, with policy at each level affecting the determination and implementation of policies at other levels. Therefore, the components of the ecotourism policy making process (Fig. 2.1) occur at a range of different levels of public governance, often simultaneously. For example, if an interest group fails to achieve its goals with respect to ecotourism at the regional level it may often seek to act and influence at the national level in order to achieve its aims.

Institutional Arrangements for Ecotourism

'Policy making is filtered through a complex institutional framework' (Brooks, 1993: 79). However, institutional arrangements have received little attention in the tourism literature (Hall and Jenkins, 1995). Institutions are 'an established law, custom, usage, practice, organization, or other element in the political or social life of a people; a regulative principle or convention subservient to the needs of an

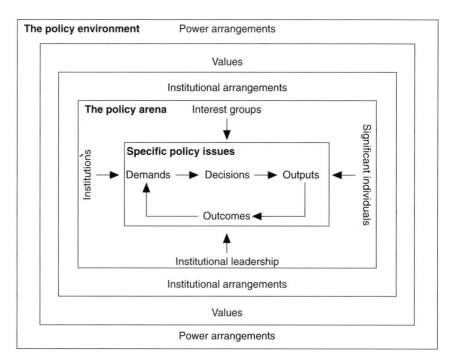

Fig. 2.1 Elements in the ecotourism policy-making process (Hall, 1994).

organized community or the general needs of civilization' (Scrutton, 1982: 225). Institutions may be thought of as a set of rules which may be explicit and formalized (e.g. constitutions, statutes and regulations) or implicit and informal (e.g. organizational culture, rules governing personal networks and family relationships). Thus institutions are an entity devised to order interrelationships between individuals or groups of individuals by influencing their behaviour. This chapter primarily focuses on the institutional arrangements for ecotourism in the context of developed countries. In the less developed countries the conditions that affect institutional arrangements are deeply influenced by several factors including financial resources and the availability of intellectual capital and expertise, as well as local administrative cultures, particularly as it may relate to issues of corruption within the administrative process and the role of political appointees. However, with respect to corruption it should be noted that more developed countries are not immune from such pressures.

As a concept and as an aspect of tourism policy-making, institutions cast a wide net and are extensive and pervasive forces in the policy system. In a broad context, O'Riordan (1971: 135) observed that:

> One of the least touched upon, but possibly one of the most fundamental, research needs in resource management [and indeed, tourism management] is the analysis of how institutional arrangements are formed, and how they evolve in response to changing needs and the existence of internal and external stress. There is growing evidence to suggest that the form, structure and operational guidelines by which resource management institutions are formed and evolve clearly affect the implementation of resource policy, both as to the range of choice adopted and the decision attitudes of the personnel involved.

Institutions therefore 'place constraints on decision makers and help shape outcomes ... by making some solutions harder, rather than by suggesting positive alternatives' (Simeon, 1976: 574). As the number of check points for policy increase, so too does the potential for bargaining and negotiation. In the longer term, 'institutional arrangements may themselves be seen as policies, which, by building in to the decision process the need to consult particular groups and follow particular procedures, increase the likelihood of some kinds of decisions and reduces that of others' (Simeon, 1976: 575). For example, new government departments may be established as part of the growth in the activity and influence of government, particularly as new demands, such as environmental concerns, reach a higher priority on the political agenda.

> The setting up of entirely new government departments, advisory bodies or sections within the existing administration is a well established strategy on the part of governments for demonstrating loudly and clearly

that 'something positive is being done' with respect to a given problem. Moreover, because public service bureaucracies are inherently conservative in terms of their approach to problem delineation and favoured mode of functioning... administrative restructuring, together with the associated legislation, is almost always a significant indicator of public pressure for action and change.

(Mercer, 1979: 107)

In this context conservation and ecotourism concerns provide a very good example of the manner in which institutional arrangements may change in the light of shifts in public values and the activities of interest groups. For example, the growth in environmental concerns in the Western world in the 1960s and early 1970s led not only to the development of new environmental legislation and policies but also to the establishment of Environmental Protection Agencies whose task it was to implement such laws and provide an institutional framework within which environmental protection matters could be managed. Similarly, in the late 1980s and 1990s the integration of environmental and tourism interests under the umbrella of sustainability led to the establishment of the first government units with a mandate to promote and develop ecotourism (e.g. Department of Conservation and the Environment, 1992; Department of Tourism (Commonwealth), 1994). However, while a number of tourism departments and agencies now often have ecotourism sections or units within them (e.g. Tourism Queensland's Environmental Unit or the West Australian Tourism Commission's Environmental Unit) there are no Departments of Ecotourism *per se*. Therefore, the institutional arrangements for ecotourism need to be seen within the wider institutional context for tourism and the environment and the nature of ecotourism itself.

Developing Ecotourism Institutions and Policies

The tourist industry is diverse, fragmented and dynamic. It has also proved to be hard to define. Such issues are not merely academic as it is difficult for government to develop policies and design institutions for a policy area that is difficult to determine. Tourism public policies are enmeshed in a dynamic, ongoing process, and it has become increasingly evident that governments struggle to comprehend the tourism industry, its impacts and future, and how they should intervene (Pearce, 1992; Jenkins, 1993; Jenkins and Pigram, 1994). Until recently, basic information concerning visitor flows and expenditures has often been lacking, and in some countries and regions such data are still far from comprehensive or even accurate. In other words, quality information concerning the tourist industry is limited. Hall and Jenkins (1995) even hypothesize that there is an element of inexperience in tourism policy formulation and implementation. Much government activity in the tourist industry is relatively recent compared with other traditional concerns of government, such as economics, manufacturing and social welfare. Hall and Jenkins suggest that tourism public policies are therefore likely to be ad hoc and incremental. For example, Hall (2000), in a review of the role of government in New Zealand tourism, identified three government agencies with primary responsibilities with respect to tourism policy and over 30 agencies with secondary responsibilities; there was typically very little formal tourism policy coordination between the various agencies. The situation becomes even more complicated when one notes the range of tourism policy instruments that are available to government (Table 2.1).

Such a situation should not be surprising, as the nature of tourism means that it cuts across a range of government responsibilities. This makes policy coordination inherently difficult unless a lead agency is clearly identified. The position of ecotourism within the institutional arrangements of government is perhaps even more difficult to determine than that of tourism because, as recognized elsewhere in this book, ecotourism is a contested concept leading to a range of definitions. This may, therefore, lead to a range of agencies taking the lead on ecotourism matters depending on the national institutional context. For example, in some jurisdictions, tourism agencies have taken the lead on developing ecotourism policies while in others it has come from national parks and conservation agencies. Furthermore, the integrative nature of ecotourism, in that it aims to integrate elements of conservation with that of tourism, also means that in policy terms it is related as much to conservation and environmental concerns as it is to those seeking tourism development. This situation has therefore led to the development of an extremely complex array of institutional arrangements for ecotourism which have had a dramatic effect on ecotourism policy development and implementation.

Regulatory Influences on Ecotourism

One of the best means to illustrate the extent to which institutional arrangements impact ecotourism is with respect to international law (Hall, 2000). International law helps to proscribe the extent to which agreements undertaken between nations at the international level affect domestic arrangements. According to Hall (2000) international law may be described as either 'hard' or 'soft'. Hard international law refers to firm and binding rules of law such as the content of treaties and the provisions of customary international law to which relevant nations are bound as a matter of obligation. Soft law refers to regulatory conduct which, because it is not provided for in a treaty, is not as

Table 2.1. Tourism policy instruments (after Hall, 2000).

Categories	Instruments	Examples
Regulatory instruments	1. Laws	Planning laws can give considerable power to government to encourage particular types of tourism development through, for example, land use zoning.
	2. Licences, permits and standards	Regulatory instruments can be used for a wide variety of purposes especially at local government level; e.g. they may set materials standards for tourism developments, or they can be used to limit the number of visitors at any given time.
	3. Tradable permits	Often used in the United States to limit resource use or pollution. However, the instrument requires effective monitoring for it to work.
	4. Quid pro quos	Government may require businesses to do something in exchange for certain rights; e.g. land may be given to a developer below market rates if the development is of a particular type or design.
Voluntary instruments	1. Information	Expenditure on educating the local public, businesses or tourists to achieve specific goals, e.g. appropriate visitor behaviour.
	2. Volunteer associations and non-governmental organizations	Government support of community tourism organizations is very common in tourism. Support may come from direct grants and/or by provision of office facilities. Examples of this type of development include local or regional tourist organizations, industry associations, conservation groups or ecotourism associations.
	3. Technical assistance	Government can provide technical assistance and information to businesses and professional bodies with regard to planning and development requirements.
Expenditure	1. Expenditure and contracting	This is a common method for government to achieve policy objectives as government can spend money directly on specific activities. This may include the development of infrastructure, such as roads, or it may include conservation programmes. Contracting can be used as a means of supporting existing local businesses or encouraging new ones.
	2. Investment or procurement	Investment may be directed into specific businesses or projects, while procurement can be used to help provide businesses with a secure customer for their products or for training and education.
	3. Public enterprise	When the market fails to provide desired outcomes, governments may create their own businesses, e.g. rural or regional development corporations or enterprise boards. If successful, such businesses may then be sold off to the private sector.

	4. Public–private partnerships	Government may enter into partnership with the private sector in order to develop certain products or regions. These may take the form of a corporation which has a specific mandate to attract business to a certain area, for example.
	5. Monitoring and evaluation	Government may allocate financial resources to monitor economic, environmental and socio-economic indicators. Such measures may not only be valuable to government to evaluate the effectiveness and efficiency of tourism development objectives but can also be a valuable source of information to the private sector.
	6. Promotion	Government may spend money on promoting a region to visitors either with or without financial input from the private sector. Such promotional activities may allow individual businesses to reallocate their own budgets by reducing expenditures on promotion.
Financial incentives	1. Pricing	Pricing measures may be used to encourage appropriate behaviour or to stimulate demand; e.g. use of particular walking trails, lower camping or permit costs.
	2. Taxes and charges	Governments may use these to encourage appropriate behaviours by both individuals and businesses, i.e. pollution charges. Taxes and charges may also be used to help fund infrastructure development, e.g. regional airports.
	3. Grants and loans	Seeding money may be provided to businesses to encourage product development or to encourage the retention of landscape features.
	4. Subsidies and tax incentives	Although subsidies are often regarded as creating inefficiencies in markets they may also be used to encourage certain types of behaviour with respect to social and environmental externalities, e.g. landscape conservation, that are not taken into account by conventional economics.
	5. Rebates, rewards and surety bonds	Rebates and rewards are a form of financial incentive to encourage individuals and businesses to act in certain ways. Similarly, surety bonds can be used to ensure that businesses act in agreed ways; if they don't then the government will spend the money for the same purpose.
	6. Vouchers	Vouchers are a mechanism to affect consumer behaviour by providing a discount on a specific product or activity, e.g. to visit a specific attraction.
Non-intervention	1. Non-intervention (deliberate)	Government deciding not to directly intervene in sectoral or regional development is also a policy instrument, in that public policy is what government decides to do and not do. In some cases the situation may be such that government decides that policy objectives are being met so that their intervention may not add any net value to the tourism development process and resources could be better spent elsewhere.

binding as hard law, and which therefore does not require actions by
signatories. Examples of soft law include recommendations or declarations made by international conferences or organizations (Lyster,
1985). For example, the Convention on Biological Diversity adopted at
the United Nations Conference on Environment and Development
(UNCED) in June 1992 in Rio de Janeiro may be regarded as hard
international law. The recommendations of the same conference are
examples of soft international law.

According to Hall (2000) soft law is particularly important in the
area of international conservation and environmental law because
treaties and conventions often require parties to attend regular meetings which make recommendations for implementation. For example,
the World Heritage Convention has annual meetings of its members to
discuss the progress of the implementation of the treaty. Agreed procedures under the Antarctic Treaty, the Man and Biosphere
Programme and the World Conservation Strategy are all examples of
soft environmental law that arose out of United Nations conferences
and which have substantially affected ecotourism development, planning and policy in various countries throughout the world. However,
ecotourism *per se* is actually little mentioned in such procedures and
the debates that surround such procedures. Indeed, here is one of the
key lessons to be learnt in examining the institutional arrangements
for ecotourism: many of the most significant arrangements do not
define or mention ecotourism in their wording although their wording
may act as a significant constraint or boost for ecotourism activities in
terms of their regulatory influence.

One of the central issues in the enactment of treaties and conventions is the obligation that the international agreement places on the
signatory. International law cannot be enforced in the same manner as
domestic law, because nations can only rarely be compelled to perform their legal obligations (i.e. through the use of force). However,
the moral obligations that accrue to members of the international
diplomatic community and the norms of international relations are
usually sufficient to gain compliance from nations. Soft law fixes
norms of behaviour which nations should observe, but which cannot
usually be enforced. As Lyster (1985: 14) observed, 'states [i.e.
nations] make every effort to enforce a treaty once they have become
party to it: it is in the interests of almost every state that order, and not
chaos, should be the governing principle of human life, and if treaties
were made and freely ignored chaos would soon result.' Matters of
international concern, for example those covered by soft international
law, do not necessarily have to be the subject of international treaties.
However, the existence of a treaty, a convention or an agreed declara-
tion may serve to provide evidence for such concern in domestic
political life. For example, the World Heritage Convention does

appear to oblige signatories to protect World Heritage property on their territory and has substantially influenced the development and promotion of ecotourism in World Heritage areas (Shackley, 1998; Thorsell and Sigaty, 1998; Hall and Piggin, 2001).

The notion of hard and soft law, however, can be applied over a much wider range of scales than those suggested by Hall (2000). Indeed, the concept can usefully describe the difficulties of getting actors to meet the obligations stated in policy statements and strategies. Table 2.2 indicates a number of different examples of hard and soft law at different scales. As noted above, at the global international scale the World Heritage Convention can be regarded as a very significant piece of international law that carries obligations to signatory states to implement the convention effectively. This is contrasted with the recommendations of the conferences, meetings and symposia of the Organization of World Heritage Cities (http://www.ovpm.org/index.asp) which, while indicating areas of international concern, does not oblige member states to act in certain ways to meet an international legal requirement. At the supranational or regional international level the example of the Association of South-East Asian Nations (ASEAN) is useful. ASEAN member states have signed a legal agreement on the conservation of nature and natural resources which binds its members to act in certain ways and which encourages the development of domestic legislation to meet the aims of the agreement. In contrast, the resolutions and recommendations of the ASEAN environment forum held in Hanoi in 1999, while serving to establish programmes and cooperation between member countries and thereby influence the regulation and management of environmental activities, such as ecotourism, did not have status in international law.

At the national level an example is provided by the Australian World Heritage Properties Conservation Act which can be used to prevent road construction or other activities that threaten the 'universal significance' of an area; including, for example, inappropriate tourism developments. In contrast, the resolutions of an Australian ICOMOS (International Council on Monuments and Sites) conference, which may, for example, highlight the cultural significance of Aboriginal sites in a major ecotourism destination such as Kakadu National Park, carry no legal obligation, even though the organization is usually involved in the research process which leads to World Heritage nominations from Australia. Similarly, at the state or provincial level, an example is provided by Idaho, which has a legislative basis for conservation of its wild rivers while the activities of Idaho Rivers United, an umbrella group which encourages the listing of wild rivers under the act, only carry moral and political and not legal influence on permissible activities on Idaho's wild rivers. Finally, at the local scale, an example is provided by Napier in New Zealand which has art deco as

Table 2.2. Hard and soft ecotourism-related law at various scales of application.

Scale	Hard ←	→ Soft
International	World Heritage Convention	Resolutions of the World Heritage Cities annual conference
Supranational	Agreement on the Conservation of Nature and Natural Resources (ASEAN)	Resolutions and recommendations of the ASEAN environment forum
National	Australian World Heritage Properties Conservation Act	Resolutions of the Australian ICOMOS conference
State/ Provincial	Idaho Comprehensive Water Planning and Protected Rivers Act 1998	Strategic plan of Idaho Rivers United
Local	Napier City Council planning regulations	Napier City Council recommendations regarding design and colours of heritage properties

a core tourist attraction. Although Napier City Council has a number of local planning regulations which apply to permissible activities, much of the retention of the art deco character of the city has been retained through recommendations rather than use of legal sanctions.

The differentiation between the hard and soft aspects of legal institutional arrangements is extremely important for ecotourism, as the majority of ecotourism specific policies and codes of conduct regarding tourist and organizational activities are soft environmental law with no specific legal sanction beyond that which might accrue to members of a specific business association. Nevertheless, such sanctions may be significant if membership is significant in the attraction of visitors; for example if an operator is a member of organizations such as the Ecotourism Association of Australia (EEA), the International Ecotourism Society (TIES) or the International Association of Antarctica Tour Operators (IAATO). The characteristics of the latter NGO are discussed below in order to illustrate the role of such organizations within a specific tourism setting.

The International Association of Antarctica Tour Operators (IAATO) (www.iaato.org) was founded in August 1991 by seven charter members (Enzenbacher, 1992) and now includes most of the main cruise lines which operate in the Antarctic. In 2001 IAATO had 14 full members, six provisional members, one probational member and 14 associate members. IAATO members meet annually in conjunction with the National Science Foundation/Antarctic Tour Operators Meeting; attendance is compulsory as memberships, by-laws and other important issues are discussed. It is estimated that IAATO members carry approximately 70% of all Antarctic tourists (Enzenbacher,

1995). As Claus (1990 in Enzenbacher, 1995: 188) noted, 'Over the past few years we have been involved in Antarctic policy meetings, US Congressional hearings and scientific conferences, not only in the US but in Australia and New Zealand as well, where we have taken a leading role in the environmental protection of Antarctica'. IAATO has two sets of guidelines, the first is addressed to Antarctica tour operators (IAATO, 1993a), the second to visitors to Antarctica (IAATO, 1993b). IAATO tour operator guidelines are intended for crew and staff members of Antarctic tour companies. The agreed principles contained within them aim at increasing awareness and establishing a code of behaviour that minimizes tourism impacts on the environment. The willingness of industry members to cooperate with Antarctic Treaty Parties in regulating tourism is crucial to the protection of the Antarctic environment given that the Antarctic is transnational space within which domestic laws are complicated in their application (Keage and Dingwall, 1993; Hall and Johnston, 1995). Tour operators maintain that current IAATO guidelines are adequate, noting that tourists often serve as effective guardians of the wildlife and environment. Yet, as Enzenbacher (1995: 188) noted, 'it is not clear that self-regulation sufficiently addresses all issues arising from tourist activity as no neutral regulatory authority currently exists to oversee all Antarctic operators'. Infractions of IAATO guidelines by members have been documented, but it is not known to what extent the environment was seriously affected by them (Enzenbacher, 1992).

Awareness of Institutional Arrangements

The discussion of hard and soft environmental law at different scales also indicates how much ecotourism operators and ecotourists themselves are often unaware of the extent to which a set of institutional arrangements exist around their activities. For example, the Shark Bay region in Western Australia is a major international ecotourism destination. The region is the most westerly point of the Australian continent. Shark Bay is the largest enclosed marine embayment in Australia and contains an unusual and varied blend of geological, biological, cultural and climatic factors that combine to form an environment with many features of scientific, environmental, historical and tourist interest. Although perhaps most well known for its dolphins, Shark Bay is a World Heritage site that hosts a substantial number of ecotourism operations. However, visitors and operations are subject to a range of institutional influences (Table 2.3) which interact with each other. For example, the activities of the Shark Bay Shire Council are governed by the constitution of the State of Western Australia. In relation to issues that affect the World Heritage significance of the region

Table 2.3. Institutional arrangements for ecotourism in Shark Bay, Western Australia.

Scale	Laws	Organizations
International National	World Heritage Act World Heritage Properties Conservation Act	World Heritage Secretariat Australian Heritage Commission Department of the Environment and Heritage
State	Conservation and Land Management Act	Department of Conservation and Land Management
	WA Planning Commission Act	State Planning Commission
	WA Tourism Commission Act	WA Tourism Commission
Local	Local planning regulations	Shark Bay Shire

the state is subject to federal law, while Australia itself is restricted by its international treaty obligations. This layer cake of institutional arrangements therefore covers all ecotourism activity although operators at the local level will often be unaware of the influence of institutional arrangements at the international level as it seemingly does not impinge on their operational activities.

The Formation of Institutional Arrangements and Policies

Figure 1.1 outlined the key components of the ecotourism policy making process. Central to this process is the policy arena within which institutions and institutional leadership along with interest groups and significant individuals determine policy. By including institutions and their leadership in the process we are recognizing that they are both an outcome of, as well as an influence on, the public policy process. Indeed, one of the key lessons which has been learned from studying public institutions is that once they are established they are very difficult to abolish completely. They may transform themselves in order to survive but there is still continuity in the institution. Such a situation led Kaufman (1976) to ask, *Are Government Organizations Immortal?*

Several factors combine to influence the longevity of government institutions:

- Administrative agencies are commonly established by statute or accorded statutory recognition. So long as the statute is in place the institution has a legal foundation to its existence.
- Legislative committees that oversee agencies may develop protective attitudes towards them.

- Government spending is now so substantial that it is very difficult to analyse its budget allocation. Once an agency has received an appropriation then 'it is apt to be borne along by the sheer momentum of the budgetary process' (Kaufman, 1976: 7).
- Some agencies have been designed to have little political control so that they can act on a more independent basis. In some jurisdictions this has been an important role of environmental protection agencies, although the ecotourism units which have been established around the world do not have such independence.
- The motivations of institutional leaders to preserve their agencies can be quite strong.
- Their clientele are important allies who may be their most ardent supporters. This situation may apply both at an institutional level (e.g. tourism commissions, promotion boards, regional tourist organizations) and at the sub-institutional level (e.g. ecotourism or indigenous tourism units).
- Professional or trade associations will often be important supporters of institutional arrangements which meet their interests.

In the case of government tourism organizations several of the above factors have been extremely important. With respect to ecotourism, for example, existing agencies and organizations have taken over this policy area rather than encouraging the development of new ecotourism specific agencies. Such incremental change in organizational direction is especially significant for ecotourism as it may determine the set of functions, linkages and relationships that ecotourism units have within the government tourism bureaucracy. Nevertheless, the list of factors noted above does not necessarily reflect on the effectiveness of any institution although the application of such evaluative criteria is recognized as being significant in tourism programming and policies (Hall and McArthur, 1998).

It has also been argued by several commentators that members of the tourism industry often have a vested interest in supporting the existence and funding of national and provincial/state tourism organizations because the agency serves to undertake marketing, promotion and development activities for which the industry does not have to pay (e.g. Craik, 1990, 1991a,b; Hall and Jenkins, 1995). This situation can therefore potentially lead to creation of close relationships between tourism industry associations and government tourism agencies which may serve to exclude other stakeholders from the tourism policy making process or even the definition of policy problems. Unfortunately, the validity of partnership arrangements between the public and private sectors in tourism has only received a limited amount of analysis (e.g. Craik, 1990; Dombrink and Thompson, 1990; *Journal of Sustainable Tourism*, 1999; Lovelock, 1999). Nevertheless, these studies, along with the review by Hall and Jenkins (1995) of the role of interest groups in the tourism policy making process, strongly suggest that business groups tend to dominate the policy process to the exclusion or detriment of other interests.

Concern over undue industry influence on the tourism policy making process may be as much to do with lack of understanding of tourism as it is over the real strength of industry influence. For example, in a study of the attitudes of tour operators, management staff and public interest group leaders in Victoria's Alpine National Park, McKercher (1997) noted that an undercurrent of suspicion and fear about the power of the tourism industry existed, even among proponents of tourism. 'While few people involved in the political debate over tourism in the Alpine National Park expressed a feeling of overt conflict with tourism, many felt that the potential exists for indirect conflict'. Similarly, Feick and Draper (2001), in a study of tourism and World Heritage in Banff National Park in Canada, noted the perceived influence of the tourism industry on policy and planning. However, it is inappropriate to conceive of the tourism industry as a monolithic entity with a series of completely shared values and attitudes. Although the profit motive and the desire to stay in business will clearly be a common factor, attitudes towards the environment and the development of appropriate environmental operation standards may diverge substantially, particularly with respect to ecotourism. For example, McKercher (1997) identified two groups of tour operators based on their product range offered and shared attitudes to the Alpine National Park. Horse and four-wheel-drive tour operators offered the most intense park experiences, taking their clients into back country areas of the park, often for extended periods. These tour operators tended to share an anthropocentric attitude to the role and purpose of the Alpine National Park. The 'Other' tour operator category included cross-country skiing and bushwalking tour operators, bus tour, educational tour, water-based tour operators and ecotour operators. They, on the other hand, expressed a more biocentric approach to the Alpine National Park.

Ecotourism operators will react to regulatory policies and structures depending on their individual and collective interests. Indeed, operators may simultaneously oppose government policies in one area, that is, taxation and fees, while supporting policies in another, that is, tourism promotion and conservation measures (Fennell, 1999). Given the nature of ecotourism it is possible to suggest that ecotourism operators may be more likely to suggest regulation, limits to tourist numbers or other policy settings as a way to manage tourism in destinations that have high environmental values and/or areas which attract ecotourists. However, the goals of ecotourism operators may prove at odds with other sectors of the tourism industry that may be pursuing different business strategies. Some evidence for this position is supplied from a survey of tourism operators in New Zealand which examined industry attitudes towards sustainability (Kearsley, 1998). In terms of overall industry concerns business issues were paramount. These included staffing, product positioning, product development, seasonality and simple economic viability and survival. The second set of concerns had to do with market conditions, such as exchange rates, changing leisure patterns and market decline in some sectors. Government policy was the next concern, usually in the context of compliance costs and border control issues. There were also concerns about the tourism industry itself, with regard to promotional activity, coordination and diversification. The fifth concern had to do with the environment and, while it was only mentioned by 5%, raised concerns about the quality of the natural environment, environmental impacts, who should fund environmental management and protection, and the funding levels of the Department of Conservation. Sustainability as an overtly specific issue was only raised by 2%.

Nevertheless, it was those tourism operators that could be classified as ecotourism operators which were most concerned about the environment and sustainability questions. Yet, from a policy perspective, their influence on the tourism policy process overall may be limited because their particular agendas may be lost in the overall industry voice presented by a general tourism industry association. Indeed, in some instances it is possible to argue that conservation groups have exerted more influence on ecotourism policies in areas of high natural values than the ecotourism operators although they may at times form a policy coalition to advance mutual interests (e.g. McKercher, 1993). In the case of Australia, for example, the desire to advance ecotourism operator interests led to the establishment of the Ecotourism Association of Australia which has developed a range of policies and practices to advance ecotourism and which has clearly influenced successive federal and state governments in terms of their ecotourism policies and funding for operators. Nevertheless, in the wider context of Australian tourism policy the Ecotourism Association of Australia has arguably had much less influence.

Conclusions

The chapter has provided an overview of the role of institutional arrangements in ecotourism. It has emphasized the importance of recognizing that the institutional arrangements for ecotourism are multi-scalar in character and that they have significant regulatory impact on ecotourism operations. However, operators themselves may not necessarily appreciate the influence of institutional arrangements at the global scale. In addition, the chapter has emphasized that the institutional and policy framework for ecotourism needs to be seen within the wider tourism context, particularly as governments throughout the world are grappling with the formation of appropriate institutional arrangements for tourism given that as a policy issue it cuts across traditional government administrative boundaries. Different jurisdictions see different agencies acting as the lead institution for ecotourism policy development.

Notwithstanding several of the chapters in this volume, it is true to say that our understanding of the political processes by which ecotourism policy is formed is extremely poor even though there are now a very significant number of public policies existing with respect to ecotourism (e.g. Fennell, 1999; Honey, 1999; WWF, 2000; Font and Buckley, 2001). This is not to say that the policies and institutional arrangements which have been established with respect to ecotourism may not be useful. However, given that politics, of which institutional arrangements and policies are an outcome, is about who gets what, where, when and why, we often do not fully appreciate the winners and losers of the ecotourism policy process. If ecotourism is to make a genuine contribution to sustainable forms of development it is vital that consideration of the social capital component of sustainability includes a more thorough examination of the policy making process than has hitherto been the case.

References

Brooks, S. (1993) *Public Policy in Canada.* McClelland and Stewart, Toronto.

Craik, J. (1990) A classic case of clientelism: the Industries Assistance Commission Inquiry into Travel and Tourism. *Culture and Policy* 2 (1), 29–45.

Craik, J. (1991a) *Resorting to Tourism: Cultural Policies for Tourist Development in Australia.* Allen & Unwin, St Leonards.

Craik, J. (1991b) *Government Promotion of Tourism: the Role of the Queensland Tourist and Travel Corporation.* The Centre for Australian Public Sector Management, Griffith University, Brisbane.

Department of Conservation and the Environment (1992) *Ecotourism Victoria Australia.* Department of Conservation and the Environment, Melbourne.

Department of Tourism (Commonwealth) (1994) *National Ecotourism Strategy.* Commonwealth Department of Tourism, Canberra.

Dombrink, J. and Thompson, W. (1990) *The Last Resort: Success and Failure in Campaigns for Casinos.* University of Nevada Press, Reno, Nevada.

Enzenbacher, D.J. (1992) Antarctic tourism and environmental concerns. *Marine Pollution Bulletin* 25 (9–12), 258–265.

Enzenbacher, D.J. (1995) The regulation of Antarctic tourism. In: Hall, C.M. and Johnston, M. (eds) *Polar Tourism: Tourism in the Arctic and Antarctic Regions.* John Wiley & Sons, Chichester, UK, pp. 179–216.

Feick, J. and Draper, D. (2001) Valid threat or tempest in a teapot? An historical account of tourism development and the Canadian Rocky Mountain Parks World Heritage site designation. *Tourism Recreation Research* 26 (1), 35–46.

Fennell, D. (1999) *Ecotourism: an Introduction.* Routledge, London.

Font, X. and Buckley, R.C. (eds) (2001) *Ecolabels in Tourism.* CAB International, Wallingford, UK.

Hall, C.M. (1994) *Tourism and Politics: Power, Policy and Place.* John Wiley & Sons, Chichester, UK.

Hall, C.M. (2000) *Tourism Planning.* Prentice Hall, Harlow, UK.

Hall, C.M. and Jenkins, J.M. (1995) *Tourism and Public Policy.* Routledge, London.

Hall, C.M. and Johnston, M. (eds) (1995) *Polar Tourism: Tourism in the Arctic and Antarctic Regions.* John Wiley & Sons, Chichester, UK.

Hall, C.M. and McArthur, S. (1998) *Integrated Heritage Management.* Stationery Office, London.

Hall, C.M. and Piggin, R. (2001) Tourism and world heritage in OECD countries. *Tourism Recreation Research* 26(1), 103–105.

Hogwood, B. and Gunn, L. (1984) *Policy Analysis for the Real World.* Oxford University Press, Oxford.

Honey, M. (1999) *Ecotourism and Sustainable Development: Who Owns Paradise?* Island Press, Washington, DC.

IAATO (1993a) *Guidelines of Conduct for Antarctica Tour Operators as of November 1993.* International Association of Antarctica Tour Operators, Kent.

IAATO (1993b) *Guidelines of Conduct for Antarctica Visitors as of November 1993.* International Association of Antarctica Tour Operators, Kent.

Jenkins, J.M. (1993) Tourism policy in rural New South Wales: policy and research priorities. *Geojournal* 29 (3), 281–290.

Jenkins, J.M. and Pigram, J. (1994) Rural recreation and tourism: policy and planning. In: Mercer, D. (ed.) *New Viewpoints in Australian Outdoor Recreation Research and Planning.* Hepper Marriott and Associates, Williamstown, pp. 119–128.

Journal of Sustainable Tourism (1999) Special Issue on Collaboration and Partnerships. *Journal of Sustainable Tourism* 7 (3/4), 179–397.

Kaufman, W. (1976) *Are Government Organizations Immortal?* The Brookings Institution, Washington, DC.

Keage, P.L. and Dingwall, P.R. (1993) A conservation strategy for the Australian Antarctic Territory. *Polar Record* 29 (170), 242–244.

Kearsley, G.W. (1998) Perceptions of sustainability in the New Zealand tourism industry. In: *Tourism and Hospitality Research Conference, 1988, Akaroa.* Lincoln University, Lincoln.

Lovelock, B. (1999) Stakeholder relations in the 'national park-tourism' domain: the influence of macro-economic policies. In: Mitchell, R. and Hall, C.M. (eds) *Tourism Policy and Planning, Proceedings of the IGU Study Group on the Geography of Sustainable Tourism Regional Conference.* Centre for Tourism, University of Otago, Dunedin, New Zealand.

Lyster, S. (1985) *International Wildlife Law: an Analysis of International*

Treaties Concerned with the Conservation of Wildlife. Grotius
   Publications, Cambridge.
Mercer, D. (1979) Victoria's Land Conservation Council and the alpine region.
   Australian Geographical Studies 17 (1), 107–130.
McKercher, B. (1993) Tourism in parks: the perspective of Australia's national
   parks and conservation organizations. *GeoJournal* 29 (3), 307–313.
McKercher, B. (1997) Benefits and costs of tourism in Victoria's Alpine
   National Park: comparing attitudes of tour operators, management staff
   and public interest group leaders in Victoria's Alpine National Park. In:
   Hall, C.M., Jenkins, J. and Kearsley, G.W. (eds) *Tourism Planning and
   Policy in Australia and New Zealand.* Irwin, Sydney, pp. 99–109.
O'Riordan, T. (1971) *Perspectives on Resource Management.* Pion, London.
Pal, L.A. (1992) *Public Policy Analysis: an Introduction.* Nelson Canada,
   Scarborough, Ontario.
Pearce, D.G. (1992) *Tourist Organisations.* Longman Scientific and Technical,
   Harlow, UK.
Scrutton, R. (1982) *A Dictionary of Political Thought.* Pan Books, London.
Shackley, M. (ed.) (1998) *Visitor Management: Case Studies from World
   Heritage Sites.* Butterworth-Heinemann, Oxford.
Simeon, R. (1976) Studying public policy. *Canadian Journal of Political
   Science* 9 (4), 558–580.
Thorsell, J. and Sigaty, T. (1998) *Human Use of World Heritage Natural Sites –
   a Global Overview.* IUCN, Gland, Switzerland.
WWF-UK (2000) *Tourism Certification: an Analysis of Green Globe and Other
   Certification Programs.* Synergy Ltd for World Wide Fund for Nature UK,
   London.

Linking Biodiversity and Sustainable Tourism Policy

3

Christopher Holtz[1] and Stephen Edwards[2]

[1]*Regional Strategic Planning, Conservation International, 1919 M Street, NW, Suite 600, Washington, DC 20036, USA;* [2]*Americas Region, Ecotourism Department, Conservation International, 1919 M Street, NW, Suite 600, Washington, DC 20036, USA*

Introduction

Increasing the contribution made by tourism to biodiversity conservation is proving to be more challenging than many of us in NGOs, government or the private sector ever anticipated. For example, ecotourism was thought to be an obvious ally of the conservation movement because there appeared to be a convergence of interests among major conservation players that could be nurtured and strengthened to support conservation, including:

- conservation organizations seeking to support new and better managed protected areas and reserves, often by using tourism activities to create economic incentives for conservation among communities in and around these reserves;
- government at various levels wishing to conserve biodiversity and ensure continued ecosystem services like watershed protection through networks of parks and protected areas, as well as nurturing alternative development activities, such as tourism related employment, for those who might lose economic opportunities through the protection of these areas;
- ecotourism's private sector looking to protect the natural resources that form its core product base.

It has been argued by many, including the authors of this chapter, that these interests can be accommodated and supported through well-

planned and managed tourism, yet this seems to have been the exception rather than the rule in most cases. And so a fundamental question
must be addressed: Why, if our assumptions regarding the interests of
these sectors are correct, has the promise of tourism not lived up to
expectations? What piece of the puzzle might be missing? How can
tourism be a more effective tool for conservation of biodiversity?

The authors of this chapter believe that lack of supporting public
policies at all scales of governmental involvement – local, regional,
national and even global – is one possible answer to this question. We
suggest that if tourism policy is reviewed at any or all of these scales
and takes biodiversity conservation into consideration, the contribution
of tourism to conservation will increase. The same holds for natural
resource management and policy – the real and potential impact (both
positive and negative) of tourism must be incorporated into environmental policy. This assertion will be expanded upon in this chapter by:

- examining the relationship between tourism policy and biodiversity conservation;
- identifying the stakeholders involved in this relationship;
- discussing tourism policy and the role it plays in biodiversity conservation at different spatial scales.

This chapter draws upon peer-reviewed literature, 'grey' or self-pub-
lished literature, web pages, presentations and draft documents (used
with permission), as well as our own experience working on ecotourism development as part of an international conservation organization. We hope that the conclusions and recommendations contained
herein will provoke further discussion about the challenges of linking
sustainable tourism policy and conservation and, ultimately, contribute to sustainable tourism fulfilling its promise as an effective tool
in the international effort to conserve global biodiversity.

Linking Sustainable Tourism Policy and Biodiversity Conservation

Before advancing with any discussion, it is important to have a clear
understanding of what we mean by the many terms that are commonly used in the tourism and conservation arena. For this chapter,
we focus briefly on defining ecotourism, sustainable tourism, policy
and biodiversity.

Ecotourism

There are extensive sources of information about the ongoing debate on
definitions of ecotourism, and clearly understanding and defining eco-

tourism is critical from a policy perspective. For the purposes of this chapter we follow the definition of The International Ecotourism Society that ecotourism is 'responsible travel to natural areas that conserves the environment and sustains the well being of local people' (Epler Wood, 2002). Ecotourism should provide some level of contribution to conservation of biodiversity, benefits to local communities, economic and social benefits, and have an education and awareness component.

Sustainable tourism

The World Tourism Organization (WTO) uses the following definition of sustainable tourism:

> Sustainable tourism development meets the needs of present tourists and host regions while protecting and enhancing opportunity for the future. It is envisaged as leading to management of all resources in such a way that economic, social, and aesthetic needs can be fulfilled while maintaining cultural integrity, essential ecological processes, biological diversity, and life support systems.
>
> (WTO, in WTO, WTTC and EC, 1996)

We consider ecotourism as a 'subset' or market segment of sustainable tourism, and that ecotourism policy is necessarily a part of the sustainable tourism policy discussion. For this chapter we often mention ecotourism, but are primarily examining the role of sustainable tourism.

Sustainable tourism policy

We consider policy to be the range of actions and statements by government indicating their areas of activity, priority and focus (Steinberger, 1995). Examples of sustainable tourism policy can include: (i) actions governments are engaged in as they carry out policy (e.g. hiring consultants or using staff to develop reports, bringing together organizations and enterprises in their country to try to organize tourism, doing studies of ecotourism markets, drafting regulations, developing promotional materials, proposing legislation); (ii) policy outputs that governments or their partnerships have developed (e.g. strategic plans, marketing plans, guidelines, regulations); (iii) identifiable organizational mechanisms that address sustainable tourism (e.g. commissions, new divisions within organizations, new positions, partnerships); and (iv) participation in and/or endorsement of international initiatives such as the Convention on Biological Diversity, Agenda 21 and free trade agreements (Edwards *et al.*, 1999).

Biodiversity and tourism

Biodiversity is described by Pulitzer Prize winning author and famed naturalist Edward O. Wilson as 'the totality of the inherited variation of all forms of life across all levels of variation, from ecosystem to species to gene' (Wilson, 1993). Biodiversity, and the natural landscapes where it is most abundant, provide a wide range of services required by the tourism sector generally, and the ecotourism sector specifically.

The concept of biodiversity is particularly relevant to the tourism sector as it deals with the interface between nature, commerce and social processes. In describing the relationship between tourism and biodiversity, Preece and van Oosterzee (1995) wrote:

> In a positive light, the relationship between biodiversity and [tourism and ecotourism] can and should be mutually reinforcing. On the one hand, the declared and publicly promoted protection of natural features, ecosystems, and biodiversity acts as a strong attractor for the tourism trade and provides a vehicle for the development of national and regional economies. On the other hand, there are opportunities–and indeed a strong obligation – for the tourism trade to promote and contribute to biodiversity conservation.

Most illustrative of this tourism and biodiversity nexus is the sheer number of international tourists who travel to enjoy and experience nature. Fillion *et al.* (1992) analysed inbound tourist motivations to different international tourism destinations and found that 40–60% of all international tourists are nature tourists and 20–40% are wildlife-related tourists. They defined 'nature tourists' as those interested in enjoying and experiencing nature and 'wildlife-related tourists' as those with a specific interest in observing wildlife, and there is a noticeable increase in both subsectors (Fillion *et al.*, 1992).

Table 3.1. Tourism growth in megadiversity countries.

Selected megadiversity countries	International tourist arrivals in 1997	Percentage increase in 1997	Revenue generated in 1997 (US$)
Australia	4,318,000	44	9,026,000
Brazil	2,850,000	81	2,595,000
Ecuador	529,000	12	290,000
Indonesia	5,034,000	16	5,437,000
Mexico	19,351,000	18	7,594,000
Peru	649,000	139	805,000
Philippines	2,222,000	62	2,831,000
Venezuela	796,000	101	1,086,000

Tourism is also a particularly fast growing economic sector in developing countries with globally significant biological diversity; that is megadiversity countries – those countries, 17 in total, that account for some 60–70% of total global biodiversity (Mittermeier *et al.*, 1998). The magnitude of tourist interest in nature and wildlife suggests that there are roles for both public institutions and private institutions in developing policies that affect the management of biodiversity and the development of the tourism sector. It is only through working together that the private sector (needing to protect its core products), the public sector (mandated with managing and conserving natural resources) and other important stakeholders can make the policy decisions required to protect these resources.

Stakeholders in Sustainable Tourism and Biodiversity Conservation Policy

It is essential to have a solid understanding of the important stakeholders that need to be engaged for the development of sustainable tourism policy. The following discussion describes what we consider to be the critical players in lobbying for, writing, implementing and establishing tourism policy that takes biodiversity conservation into consideration, and vice versa for natural resource management policy. While tourism and conservation affect many stakeholders, we are primarily focused on those that play a significant role in policy creation.

In developing the discussion paper that served as the foundation for this chapter we conducted a stakeholder analysis of the tourism sector and its relationship with biodiversity conservation, and concluded that six primary stakeholder groups interact to influence tourism policy and development patterns in areas with globally significant biodiversity (Edwards *et al.*, 2000):

- the public sector (local, national, regional and global governance bodies);
- the private sector;
- multilateral and bilateral donors;
- non-governmental organizations (NGOs);
- local communities and indigenous people;
- consumers.

The public sector

The public sector – in the form of action by local and national governments – is at the centre of the divergent interests of tourism development and biodiversity conservation, and serves as the primary link between all the sectors involved. The public sector holds this challenging position by virtue of its dual responsibilities for protecting and regulating the use of natural resources and promoting the economic development of its citizens.

Most public sector agencies at the national, regional and local level probably have an interest in the tourism sector, as tourism potentially interacts with a range of public policy issues, including those relating to: transport, infrastructure, investment, employment, natural resource use and many more. In terms of implementing policies, particularly those that impact environmental (including biodiversity) issues, the important players on the public sector side of the tourism sector are municipal authorities and local natural resource managers who must deal with the day-to-day dilemmas of inadequate infrastructure, waste management, limited resources and expertise, and land use conflicts between the private sector and communities. National government agencies (e.g. ministries of tourism, natural resource management agencies, public works agencies, etc.) are tasked with initiating, developing and implementing tourism and natural resource policy at a local, national and often international level. Due to their role which impacts tourism and conservation at multiple scales of influence, we consider national government agencies to be critical to tourism and conservation policy development and implementation.

Specific public sector duties and policy spheres that impact tourism development and biodiversity conservation include, but are not limited to:

- serving as the official signatory to treaties and conventions;
- land use planning and regulatory enforcement;
- protected area management;
- business licensing;
- infrastructure development;
- transport capacity;
- investment promotion and tax structures;
- tourism promotion.

Private sector

Tourism's private sector includes a broad range of businesses such as accommodations of all types and sizes from small bed & breakfasts to 1000 room resorts, airlines, real estate and time-share developers, inbound and outbound tour operators, rental car agencies, cruise ship companies, restaurants, bars and others in the food and beverage businesses, retail establishments and many other enterprises. On a secondary level, the scope of tourism's private sector is even more varied and ranges from the construction companies that build hotels and resorts to the farmer who might sell his vegetables to a small inn.

The private sector is the dominant stakeholder in terms of financial resources, but tends to stand outside the centre of the tourism development versus biodiversity conservation conflict, in terms of a formal policy making role. It does have enormous influence over the tourism sector's impact on biodiversity. The activities of tourism's private sector, particularly in the form of hotel/resort development and tour operations within areas of high biodiversity, can be significant threats to biodiversity. These diverse public and private sector groups often may be in direct conflict during policy debates that influence the location and intensity of tourism development. Despite these differences, there are broad common interests shared by both groups which represent a foundation for cooperation. As Honey (1999) points out, the tourism sector has an interest in protecting the world's cultural and natural resources that provide the base for a diverse range of tourism products and destinations.

The importance of tourism to the economies of many developing countries and the enormous financial resources of many tourism investors leads to a familiar but unfortunate dynamic of the private sector circumventing or ignoring government policies and regulations to the detriment of biodiversity. There are, however, some signs that this dynamic may be changing and that the private sector is more willing to recognize the importance of biodiversity to the sustainability of their businesses.

These hopeful signs are in the form of nascent voluntary programmes that promote the use of environmentally sound technologies, adoption of environmental management systems and adherence to appropriate codes of conduct (UNEP, 1998). Organizations such as the International Hotels Environment Initiative and programmes like the Green Globe 21 Hotel Certification scheme offer evidence that the private sector is making an effort to move in the direction of more sound practices. Despite promising initiatives, the tourism industry often resists enhanced international or national regulatory regimes of its activities or mechanisms that would support an integral costing approach to environmental protection. For example, in June 1997 at a Special Session of the United Nations General Assembly, the World Travel and Tourism Council successfully lobbied against a proposed international airline transport tax that would have generated revenue for environmental protection (Greenglobe, 1999; www.greenglobe.org/oldsite/members/certification.htm).

Other important stakeholders

Our stakeholder analysis concluded that three other key stakeholders – donors, NGOs and local communities and indigenous people – play increasingly important but still subordinate roles in tourism policy development. The impact of donors, NGOs and local communities on integrating biodiversity considerations into tourism policy is varied, nuanced and thus difficult to characterize definitively. All three can act as powerful 'drivers' in favour of biodiversity interests within tourism policy, but all three can, individually or in partnership, take actions that counter biodiversity interests.

Examples of these stakeholders acting in support of biodiversity interests within the tourism sector include:

- donors funding tourism planning and policy-making projects, protected area management and ecotourism development; or donors' influence over the public sector through grants and loans and its support of NGOs;
- the NGO community's involvement as an advocate for policy change and its important role in supporting economic development, social change and environmental initiatives within local communities and indigenous people;
- conservation NGOs monitoring environmental impacts of tourism, initiating community-based ecotourism projects and acting as advocates that lobby government to support biodiversity interests within tourism policies;
- local communities, often indigenous peoples within protected areas, playing primary roles as stewards of the land through community-based ecotourism and resource management.

Conservationists have an interest in encouraging tourism development that contributes to biodiversity conservation by: (i) providing economic alternatives for communities that engage in destructive livelihood activities (e.g. dynamite 'blast' fishing over coral reefs, poaching or indiscriminate logging); (ii) creating new revenue streams to support conservation through user fee systems and other mechanisms; and (iii) building constituencies that support conservation priorities by exposing tourists to the value of protecting unique pristine ecosystems.

Despite these general areas of parallel interests, initiatives from the tourism sector and the international conservation community that collectively initiate policies or programmes that provide significant benefits to the environment in general and biodiversity conservation in particular are still in their infancy.

Taken individually, the impact of a donor, NGO or local community may not represent a significant influence over the public sector's tourism policy, but these stakeholders, as advocates for biodiversity, have ample opportunities to create partnerships that are prominent in promoting the integration of biodiversity considerations into the tourism sector.

Consumers

Our stakeholder assessment concluded that consumers primarily influence public policy via the private sector, whose impact was limited to affecting the behaviour of the larger tourism industry, and not often as a stakeholder group with direct sway over or impact on tourism policy that affects biodiversity. This preliminary assessment is perhaps somewhat contentious as conventional wisdom holds that tourist demand for environmentally and culturally sensitive tourism is a major factor in the move towards the 'greening' of the industry. This view does have significant, if not overwhelming, support. A recent study of consumer attitudes towards environmental and social responsibilities, as perceived by the tourism industry, found that a majority of industry participants believe that their clients prefer businesses to operate in an environmentally and socially responsible manner; 29% rated it as very important and 47% considered it somewhat important. These perceived consumer attitudes are likely to become more influential within the industry in the near future as the findings for perceived importance of these issues in 5 years revealed that travel and tourism businesses believed that 46% of their clients would see environmental operations as very important and 38% would see them as somewhat important. The study data suggested that the industry was aware of a trend towards heightened consumer awareness of environmental issues in the future (IITS, 1999).

Despite this data and the evidence of increased consumer demand for environmentally and socially responsible tourism, we believe that the impact of this increased awareness among consumers has been felt in the tourism industry's voluntary initiatives – some of which are legitimate and some of which are merely public relations efforts – and not yet in the public sector policy-making realm. The impact of consumers on government sustainable tourism policy requires more exploration.

Tourism and Biodiversity Conservation Policy Spatial Scales

While the stakeholder assessment offers a simplified presentation of a complex array of interests and interactions, it does represent our conclusion that public policy and the inter-organizational dynamics of the policy-making process are an appropriate focal point of efforts to reduce the adverse impact of the tourism sector on globally significant biological diversity. But what is the best way to explore reducing the negative impact and improving the positive impact of tourism development on globally significant biodiversity, given the stakeholder relationships that have been identified and the wide range of tourism development modes (ecotourism vs. mass tourism) and related policies that have an effect on biodiversity. We believe this should be approached by examining the impact of tourism policies on biodiversity at the different spatial scales at which tourism policy making and biodiversity conservation efforts function.

- at the local/sub-national level;
- at the national level;
- at the international regional level;
- at the global level.

Clearly the policy issues faced when creating community-based ecotourism development policies in the buffer zone of a protected area are very different from those faced when creating regional or national policies to manage resort development in a region with hundreds of miles of coastline. There is no question that, while these common scenarios represent different policy challenges, they both require tourism

Table 3.2. Tourism and biodiversity policy: four levels of management.

Level	Tourism management framework	Biodiversity conservation framework
At the local/sub-national level	Municipal/regional destination management	Parks and protected areas/ individual ecosystems
At the national level	National tourism policies and master plans	Park and protected area networks/aggregated and integrated ecosystems
At the international regional level	Regional tourism organizations, e.g. Caribbean Tourism Organization, Tourism Council of the South Pacific	Biosphere/transboundary ecosystems/corridors
At the global level	International tourism organizations, e.g. World Tourism Organization, World Travel and Tourism Council	Global environment/conventions and treaties, e.g. Convention on Biological Diversity

policies that integrate biodiversity considerations in order to conserve globally significant biodiversity and ensure the sustainability of the important economic benefits generated by tourism.

Exploring sustainable tourism and conservation policy through these four perspectives is appropriate because, in many ways, both tourism and biodiversity exist and are already managed on these levels. These parallels are illustrated below.

This framework can yield useful results because by recognizing the policy challenges and their similarities and differences in dealing with tourism and biodiversity issues, relevant policy solutions are likely to be identified and implemented relative to their area of influence and applicability.

These four levels will be examined in more depth by reviewing the policy challenges faced in integrating biodiversity considerations into the tourism sector.

At the local/sub-national level

We consider the local and sub-national level to include local, municipal, state, county and provincial government, local natural resource management agencies, individual parks and protected areas, and other agencies that have authority at the local and sub-national scale.

Policy-making initiatives that integrate biodiversity consideration into the tourism sector at the local or site level tend to take two forms: (i) municipal and regional government programmes of action to make tourism more sustainable; and (ii) protected area management plans that incorporate the tourism sector.

Critical policy challenges where tourism and biodiversity interests converge and are felt at the local level include:

- managing individual protected areas and networks of reserves in biodiversity corridors;
- planning and zoning for appropriate land use;
- resolving resource use conflicts among stakeholders;
- creating alternative employment opportunities for local residents engaged in environmentally destructive activities;
- educating tourists about acceptable behaviour while visiting parks and protected areas.

It is at the local level that one sees the success or failure of policy implementation and that the negative impacts from biodiversity loss are felt most sharply. Making tourism policy at the local level that integrates biodiversity considerations is an enormous challenge because resources for implementation are often limited and conflicts between stakeholder groups over the use of resources can tend to

reach boiling point. Local level tourism policy that integrates the conservation of biodiversity is perhaps most relevant within the venue of
a park or protected area as part of a comprehensive protected area
management plan. Successful tourism policy within areas with high
concentrations of unique and globally significant biodiversity will
need to be established and managed with close cooperation of the
public sector, NGOs, local communities and the private sector. An
additional challenge faced at the local level is often the lack of control
over national and international level policy that affects tourism.

At the national level

The national level is represented by country level agencies that are the
primary institutions mandated with establishing tourism and conservation policy. Discussion of tourism policy making often focuses on
the national level as it is generally considered the responsibility of
national governments to make the decisions that determine the mode,
location and intensity of tourism development (Edwards *et al.*, 1999).
It also generally falls upon national governments – through international treaty obligations such as the Convention on Biological
Diversity – to protect biodiversity resources through the creation of
protected area networks. Thus, there has been an increase in the creation of national tourism policies and master plans with ecotourism
built around national parks and other reserves as a core priority.
These policies and plans typically address the following issues:

- creating national strategies for ecotourism, nature tourism or sustainable tourism development;
- developing effective environmental impact assessment and regulatory frameworks;
- creating and managing protected areas;
- promoting investment in tourism facilities that are consistent with
  sustainable development objectives;
- promoting investment in transport infrastructure and systems that
  support sustainable development objectives.

As discussed previously, there are also issues that require local and
national stakeholders to work together in order to be successful. These
include:

- licensing only those tourism businesses that comply with environmental laws and regulations;
- providing a reliable supply of energy that has minimal environmental impact;
- treating wastewater and sewage;

- building and maintaining a transport infrastructure;
- developing human capacity for ecotourism, nature tourism or sustainable tourism development;
- marketing individual destinations – including flagship parks and reserves.

However, national tourism policies that refer specifically to biodiversity considerations are still uncommon, but based upon our experience, there does seem to be a recognition by national tourism policy makers that conserving nature is important for maintaining tourism. On the flip side, natural resource use and conservation policies are often designed without considering the positive role that the tourism sector might play, and we believe that this is an area where enhanced communication between government agencies and other stakeholders could be beneficial.

At the international regional level

We loosely define the international regional level as the collaboration between multiple countries in a given region. Examples include regional free trade agreements, like the North American Free Trade Agreement (NAFTA), which currently incorporates Canada, the USA and Mexico, MERCOSUR, the European Union and CARICOM; other attempts at linking tourism at the regional scale, such as the Caribbean Tourism Organization (CTO), the Pacific Asia Travel Association (PATA) and the Mundo Maya Organization; and also include conservation collaboration attempts like the Mesoamerican Biological Corridor.

The international regional level is the area of tourism policy making that has seen the least progress in implementing initiatives that integrate biodiversity consideration, but perhaps has the most potential for yielding benefits. Regional, and even global, trade agreements are increasingly common, and these agreements have served to increase the flow of tourism within and between regions. However, there has been little progress in developing corresponding policies that address the impact of this increase in tourism on biodiversity. Existing regional tourism institutions are often focused on marketing and promotion, but generally are not effective policy makers and have made little progress in developing corresponding policies that address the impact of tourism development on biodiversity.

Working at the regional level is appealing to those interested in integrating tourism policy with biodiversity considerations because: (i) regional tourism organizations that could potentially exert a degree of influence over policy already exist; and (ii) biodiversity conservation efforts are often initiated at the regional level due to the complex nature of large ecosystems. In order to make progress at this scale, the following issues must be addressed through collaboration between national and regional authorities. They include:

- creating protected area networks that incorporate transboundary ecosystems;
- developing training systems that build capacity for sustainable tourism development;
- advocating the support of regional multilateral donors in making sustainable tourism a funding priority.

At the global level

The integration of biodiversity considerations into the tourism sector at the global level is an emerging area. International organizations have produced agreements in principle on issues of tourism and its impact on the environment, but little has been done to actually implement enforceable global tourism policies that integrate biodiversity considerations.

We see at least five global level policy challenges for integrating biodiversity considerations into the tourism sector. Two of these challenges appear most relevant at the global level and three others seem to require involvement of both regional and global entities. The global challenges are:

- raising international awareness of the threats to global biodiversity, the consequences of this loss, and the role the tourism sector plays as a threat and, importantly, as a partner in conservation;
- creating databases of best practice in sustainable tourism from around the world that focus on biodiversity specific issues, such as managing tourism in and around national parks.

The challenges that appear to have both regional and global implications are:

- creating effective environmental certification programmes for the tourism industry and linking these programmes to park and protected area planning and management;
- advocating global multilateral and bilateral donor support of sustainable tourism;
- promoting the use of new technologies within the tourism industries that minimize environmental impacts.

We have found this four-tiered framework to be very useful because it explicitly identifies and links the similarities and differences among tourism policy and biodiversity conservation at various spatial scales.

Organizing the policy challenges in this way will lead, we believe, to a more effective pursuit of further research and even potential interventions, as the degree of influence and applicability of specific tourism policy issues relative to biodiversity is made more clear.

Conclusions and Recommendations

This chapter has outlined the major issues involved in the integration of biodiversity considerations into sustainable tourism policy at the local, national, regional and global levels. We hope that this will be the basis for generating new thinking about the relationship between tourism and biodiversity. The following summarizes the conclusions we reached about the issues that are at the heart of the sustainable tourism policy and biodiversity conservation debate.

1. We determined that public policy and the inter-organizational dynamics of the policy-making process at the local, national, regional and global levels are central to strengthening linkages between the interests of tourism development and biodiversity conservation.
2. The public sector, being at the centre of the policy-making process, can be a catalyst for partnership in developing sustainable tourism policies with important stakeholders. Thus, it is the appropriate focus for policy initiatives in the effort to reduce the adverse impact of the tourism sector on globally significant biological diversity.
3. Further research on the role of tourism in all its forms in creating and strengthening protected areas is needed. This should include not only ways in which tourism that directly affects parks and reserves is managed, but also how tourists who may never visit a reserve derive benefits from it indirectly, e.g. by having clean water available from the watershed protection service the reserve offers, and can contribute to its effective management. This and other policy challenges that deal with tourism development where the link to parks and reserves is less direct but no less important is a potentially exciting avenue for further development of the tourism and biodiversity link.

Acknowledgements

The Global Environment Facility (GEF), the United Nations Environment Programme (UNEP) and Conservation International (CI) generously provided funding and support to prepare the discussion paper that served as the foundation for this chapter. The opinions expressed herein are those of the authors and not of the GEF, UNEP or CI.

References

Edwards, S., McLaughlin, W. and Ham, S. (1999) *Comparative Study of Ecotourism Policy in the Americas: 1998,* Vols I – III. Contribution #872 of the Idaho Forest, Wildlife and Range Experiment Station, CNR, University of Idaho. Organization of American States and the University of Idaho, USA.

Edwards, S., Holtz, C. and Hillel, O. (2000) *Integrating Biodiversity Considerations into the Tourism Sector through Public Policy,* Discussion Document. Conservation International and United Nations Environment Programme, Washington, DC.

Epler Wood, M. (2002) *Ecotourism: Principles, Practices and Policies for Sustainability.* United Nations Environment Programme, Division of Technology, Industry and Economics, Paris.

Fillion, F.L., Foley, J.P. and Jacquemot, A.J. (1992) The economics of global ecotourism. Paper presented at the *Fourth World Congress on National Parks and Protected Areas, Caracas, Venezuela.*

Honey, M. (1999) *Ecotourism and Sustainable Development.* Island Press, Washington, DC.

IITS (1999) *Going Green? The Emerging Importance of Environmental Issues to the Global Travel and Tourism Industry.* International Institute of Tourism Studies, George Washington University, Washington, DC.

Mittermeier, R.A., Gil, P. and Mittermeier, C. (1998) *Megadiversity – Earth's Biologically Wealthiest Nations.* Conservation International and CEMEX, Washington, DC.

Preece, N. and van Oosterzee, P. (1995) *Two Way Track – Biodiversity Conservation and Ecotourism: An Investigation of Linkages, Mutual Benefits and Future Opportunities.* Biodiversity Series, Paper No. 5. Department of Environment, Sport and Territories, Canberra, Australia.

Steinberger, P. (1995) Typologies of public policy: meaning construction and the policy process. In: McCool, D. (ed.) *Public Policy Theories, Models, and Concepts, an Anthology.* Prentice-Hall, Englewood Cliffs, New Jersey, pp. 220–233.

UNEP (1998) *Ecolabels in the Tourism Industry.* United Nations Environment Programme, Division of Technology, Industry and Economics, Paris.

Wilson, E.O. (1993) *Diversity of Life.* Harvard University Press, Cambridge, Massachusetts.

WTO, WTTC and EC (1996) *Agenda 21 for the Travel and Tourism Industry: Towards Sustainable Development.* World Tourism Organization, World Travel and Tourism Council, and the Earth Council, USA.

Sharing the Country: Ecotourism Policy and Indigenous Peoples in Australia

4

Heather Zeppel

School of Business, James Cook University, PO Box 6811, Cairns, QLD 4870, Australia

'The land is our being, The rivers our blood' (Harry Nanya Tours)

Introduction

This chapter describes and reviews indigenous participation in Australian ecotourism policy. It focuses on the changing roles of Australian indigenous people in ecotourism policy-making and indigenous control of ecotourism ventures mainly in protected areas. The extent and effectiveness of indigenous participation in Australian ecotourism policy is also evaluated. The roles of indigenous people in Australian ecotourism now include Native Title holders, traditional owners, land managers, park rangers, tourism operators and guides. This chapter first discusses tourism policy-making and Aboriginal or indigenous tourism in Australia, including ecotourism ventures. Next, various Aboriginal tourism strategies are reviewed with reference to indigenous input and roles in ecotourism. The policy and management roles for indigenous people are also identified in nature-based tourism strategies and ecotourism plans in several Australian states. A case study highlights the role of rainforest Aboriginal people in policy making and in a nature tourism strategy for the Wet Tropics World Heritage Area of North Queensland. This chapter suggests that ecotourism policies address indigenous cultural heritage and environmental relationships, but have limited means for indigenous

© CAB *International* 2003. *Ecotourism Policy and Planning*
(eds D.A. Fennell and R.K. Dowling)

participation in the control and management of ecotourism as Native Title holders and traditional land owners.

Defining Ecotourism and Aboriginal Tourism

The Ecotourism Association of Australia (2000: 4) defines ecotourism as: 'Ecologically sustainable tourism with a primary focus on experiencing natural areas that fosters environmental and cultural understanding, appreciation and conservation.' Nature-based tourism, however, involves 'any sustainable tourism activity or experience that relates to the environment' (South Australian Tourism Commission, 2000: 4). In both definitions, there is a primary focus on the natural environment with a secondary emphasis on cultural heritage, including indigenous cultures. Aboriginal or indigenous tourism has been defined as 'a tourism product which is either: Aboriginal owned or part owned, employs Aboriginal people, or provides consenting contact with Aboriginal people, culture or land' (SATC, 1995: 5). It includes cultural heritage, rural and nature-based tourism products or accommodation owned by indigenous operators. Most tourism agencies though consider Aboriginal tourism and ecotourism as separate niche or special interest areas of nature-based tourism. The Wilderness Society further defines indigenous cultural tourism as 'responsible, dignified and sensitive contact between indigenous people and tourists which educates the tourist about the distinct and evolving relationship between indigenous peoples and their country, while providing returns to the local indigenous community' (The Wilderness Society, 1999). Indigenous ecotourism, therefore, involves nature-based attractions or tours owned by indigenous people, and also indigenous interpretation of the natural and cultural environment. The next section reviews Aboriginal tourism and a range of indigenous ecotourism ventures in Australia.

Aboriginal Tourism in Australia

Aboriginal-owned tourism ventures are a growing segment of the Australian tourism industry, mainly since the 1990s (Office of Northern Development, 1993; Commonwealth Department of Tourism, 1994a; Sykes, 1995; Aboriginal and Torres Strait Islander Commission, 1997a; Pitcher *et al.*, 1999; Zeppel, 1998a,b,c,d, 1999, 2001; DISR, 2000). The range of Aboriginal-owned tourism products includes cultural tours, art and craft galleries, cultural centres, accommodation, boat cruises and other visitor facilities. Well-known Aboriginal tourist ventures include the Tjapukai Aboriginal Cultural Park in Cairns

(QLD), 50% owned by the local *Djabugay* people; and Tiwi Tours on Bathurst and Melville Islands (NT), fully owned by the Tiwi Tourism Authority since 1995. Aboriginal culture has mainly been promoted in the Northern Territory, North Queensland and the Kimberley (WA), but other states (e.g. NSW, SA, Victoria) are also developing Aboriginal tourism products and cultural attractions.

There are around 200 indigenous tourism businesses in Australia, with Aboriginal cultural tourism earning Aus$5 million a year while mainstream Aboriginal tourism enterprises generate Aus$20–30 million (ATSIC, 1997a). In the Northern Territory, Aboriginal land owners also derive income from licensing, leasing, renting and tourism concessions operating on Aboriginal lands (Sykes, 1995; Pitcher *et al.*, 1999). In the remote Kimberley region of Western Australia, Aboriginal communities on the Dampier Peninsula, north of Broome, provide accommodation and charge access fees for tour groups and private vehicles (Western Australian Tourism Commission, 1999). National Parks such as Uluru, Kakadu and Nitmiluk (NT), and Mutawintji (NSW), jointly managed with Aboriginal landowners, also provide a variety of Aboriginal-owned tours (Mercer, 1994; Sutton, 1999). Most nature-based Aboriginal tourism ventures are located on Aboriginal lands or jointly managed national parks in northern and central Australia.

Indigenous ecotourism ventures

Indigenous ecotourism ventures include boat cruises, nature-based accommodation, cultural ecotours and wildlife attractions operating on Aboriginal lands, National Parks and in traditional tribal areas (see Table 4.1). These indigenous-owned ecotourism enterprises present unique indigenous perspectives on the natural and cultural environment, promote nature conservation and provide real benefits (i.e. employment) for local indigenous people (ANTA, 2001). Hence, these indigenous products meet the key criteria of ecotourism as nature-based, environmentally educative and sustainably managed or conservation supporting tourism (Blamey, 1995). Indigenous nature conservation or 'caring for country' involves traditional land owners/custodians 'looking after the environmental, cultural and spiritual well being of the land' (Aboriginal Tourism Australia, 2000). Looking after Aboriginal sites, landscapes or natural resources and educating visitors about 'country' often motivates indigenous conservation ethics in ecotourism or land management. Nganyintja, a *Pitjantjatjara* Elder working with Desert Tracks stated that: 'carefully controlled ecotourism has been good for my family and my place Angatja' (cited in James, 1994: 12). Most Aboriginal tours are marketed as cultural rather than ecotours, emphasizing the links between indigenous operators and their lands.

Table 4.1 Indigenous ecotourism ventures in Australia.

Boat tours
Yellow Water Cruises, Kakadu, NT[a]
Guluyambi Aboriginal Cultural Cruise, Kakadu, NT[a]
Nitmiluk Cruises, Katherine, NT[a]
Darngku Heritage Cruises, Geikie Gorge, Kimberley, WA[a]

Accommodation
Pajinka Wilderness Lodge, Cape York, QLD[b]
Seisia Resort and Campground, Cape York, QLD[b]
Kooljaman at Cape Leveque, Kimberley, WA[b]

Cultural ecotours
Tiwi Tours, NT[b]
Umorrduk Safaris, Arnhem Land, NT[b]
Manyallaluk, NT[b]
Anangu Tours, Uluru, NT[a]
Lilla Aboriginal Tours, NT[a]
Wallace Rockhole, NT[b]
Desert Tracks, Pitjantjatjara Lands, SA[b]
Camp Coorong, SA
Iga Warta, SA
Karijini Walkabouts, WA[a]
Mimbi Caves Experience, WA
Brambuk Living Cultural Centre, VIC
East Gippsland Wilderness Tours, VIC
Harry Nanya Tours, NSW
Mutawintji Heritage Tours, NSW[a]
Tobwabba Tours, NSW
Umbarra Cultural Tours, NSW
Yarrawarra Aboriginal Cultural Centre, NSW
Native Guide Safari Tours, QLD
Munbah Aboriginal Culture Tours, QLD[b]
Kuku-Yalanji Dreamtime Walks, QLD

Wildlife attractions
Djungan Nocturnal Zoo, Kuranda, QLD
Whale Watching, Yalata Aboriginal Land, SA[b]

[a]Indigenous-owned cruises or tours operating in Aboriginal-owned and/or jointly managed National Parks.
[b]Nature-based accommodation, cultural tour or wildlife attraction located on indigenous land.

Indigenous ecotourism ventures focus on indigenous relationships with the land and the cultural significance of the natural environment. This includes indigenous use of bush foods and traditional medicine, rock art, landscape features with Dreamtime significance, creation stories, totemic animals, traditional artefacts and ceremonies, and contemporary land use. Such tours educate visitors on indigenous environmental values, sustainable use of natural resources and 'caring for country.' As Tom Trevorrow, a *Ngarrindjeri* operator of Camp Coorong noted: 'We have to look after the environment and we teach visitors the importance of this' (cited in ATSIC, 1996: 29). Indigenous interpretations of nature are important for the maturing ecotourism market (Office of National Tourism, 1999). Aboriginal operators, however, resent 'outsiders' setting up tours in their traditional area, permits to visit sites in their own country, and ecotourism certification when 'Aboriginal "accreditation" involves approval from elders' (Bissett, Perry and Zeppel, 1998: 7).

Research on indigenous ecotourism

Research on indigenous issues in ecotourism includes sustainable development and Aboriginal tourism (Burchett, 1992; Altman and Finlayson, 1993); environmental impacts of tourism (Ross, 1991; Miller, 1996); Aboriginal heritage sites and cultural interpretation (Bissett *et al.*, 1998); and Aboriginal tourism in national parks (Mercer, 1994; Pitcher *et al.*, 1999; Sutton, 1999). Research on industry issues includes ecotourism education and training for Aboriginal people (Weiler, 1997; ANTA, 2001); Aboriginal tourism strategies (Zeppel, 1998a,b, 2001); Aboriginal control of tourism (Pitcher *et al.*, 1999); indigenous involvement in Australian ecotourism (Dowling, 1998); and Aboriginal nature-based tourism products (Zeppel, 1998d). Other international research has addressed the benefits of ecotourism for indigenous communities (Zeppel, 1997, 1998c), especially women (Scheyvens, 1999, 2000), and potential conflicts between ecotourism and indigenous hunting or use of natural resources (Grekin and Milne, 1996; Hinch, 1998). Recent research has reviewed indigenous wildlife tourism in Australia (Muloin *et al.*, 2000, 2001); and cultural interpretation by Aboriginal tour guides at Mutawintji National Park (Smith, 1999), and on nature tours in Queensland (Chalmers, 2000). There is a need to assess tourism policies and the linkages between indigenous tourism and nature/ecotourism strategies (Pitcher *et al.*, 1999; Whitford *et al.*, 2001). This chapter addresses the policy environment influencing ecotourism and indigenous people in Australia.

Policy Environment for Australian Ecotourism and Indigenous People

The Australian policy environment for ecotourism and indigenous issues reflects social values and attitudes toward indigenous people and public policy-making for tourism. In Australia, tourism policy is largely a public policy-making or government activity. Such tourism policies guide government actions, decision-making, funding and planning for tourism. Public tourism policy is a consequence of the political system; social values and principles; institutional structures; and the government's power to make policy decisions. To be recognized as official public policy, tourism policies are devised, processed, authorized and implemented by government agencies. However, tourism policies are also influenced by the social, economic and cultural characteristics of society (Hall, 1994). Community, industry and government groups influence and direct public tourism policy. Hence, tourism policies reflect the interests of government agencies, pressure groups (e.g. conservation groups), tourism industry associations, community leaders, significant individuals, bureaucrats, politicians and, more recently, indigenous groups.

Key elements in tourism policy-making include (Hall *et al.*, 1997):

- The policy environment: power arrangements, values, institutional arrangements;
- The policy arena: interest groups, institutions and their leadership, important individuals;
- Specific policy issues: demands, decisions, outputs (products), outcomes (or impacts).

Table 4.2 outlines key features of the policy environment and the policy arena shaping indigenous involvement in ecotourism and Aboriginal tourism. In Australia, indigenous people have only been included in tourism policies since the 1990s (Zeppel, 1997, 2001; Whitford *et al.*, 2001). Greater government recognition of the real need for Aboriginal economic and social advancement mainly derived from the recommendations contained in the 1991 Royal Commission into Aboriginal Deaths in Custody report. This led, in 1997, to the development of tourism, rural and cultural industry strategies for Aboriginal and Torres Strait Islander (ATSI) people as three key areas for indigenous economic progress. The other crucial factor was federal government legislation recognizing indigenous rights to land, through the *Aboriginal Land Rights (Northern Territory) Act 1976*, and the Commonwealth *Native Title Act 1993*. These laws recognize Aboriginal or Native Title rights and interests over traditional land areas in Crown lands and national parks.

In response to public policy-making, relevant tourism policies are also developed by the tourism industry and by non-government environmental agencies. The Tourism Council of Australia's statement on indigenous tourism focuses on industry issues and training needs (Tourism Council Australia, 1999). It supports Aboriginal Tourism Australia as the key industry body for indigenous tourism, providing

Table 4.2. Australian policy environment for ecotourism and indigenous people.

Policy environment

Power arrangements (government)

● Office of National Tourism, State/Territory Tourism Commissions
● Environment Australia, State/Territory National Park and World Heritage Agencies
● ATSIC, State/Territory Aboriginal Affairs Departments
● Legislation
  ATSI Heritage Protection Act 1984
  Native Title Act 1993
  World Heritage Properties Conservation Act 1983
  Environment Protection and Biodiversity Act 1999
  State/Territory Aboriginal Heritage and Aboriginal Land Acts
  State/Territory National Parks and Nature Conservation Acts

Institutional arrangements (industry)

● Aboriginal Tourism Australia
● Tourism Council Australia
● Ecotourism Association Australia
● Other tourism industry associations
● ATSI Commercial Development Corporation, Indigenous Land Corporation

Social values (attitudes toward indigenous people)

● ATSI cultural heritage/cultural custodians (consultation, negotiation)
● ATSI affinity with natural environment (ecological & spiritual relationships)
● ATSI native title landholders/traditional landowners (partnerships, control)

Policy arena

Interest groups
● Tourism industry associations, Ecotourism/Aboriginal tourism operators
● Aboriginal Land Councils, Native Title landholders, indigenous communities
● Non-government environmental agencies (e.g. ACF, The Wilderness Society)

Institutions and institutional leadership
● As above – Government and Industry
● Tourism education and training providers

Significant individuals
● Politicians (Federal/State Ministers for Tourism/Environment/Aboriginal Affairs)
● Aboriginal Leaders (Noel Pearson, Cape York)
● Environmental Leaders (Peter Garrett, ACF)
● Aboriginal tourism operators/officers (Joe Ross, WA; Glen Miller, Tourism Queensland)

ATSIC, Aboriginal and Torres Strait Islander Commission; ACF, Australian
Conservation Foundation.

indigenous input into marketing and tourism policy development
(ATA, 2000). The two key environmental organizations in Australia,
The Wilderness Society and Australian Conservation Foundation,
have specific policy documents on tourism and on indigenous rights
and interests in wilderness areas and in land and water management

(TWS, 1999; Australian Conservation Foundation, 2000). These policies support consultation and partnerships with indigenous people for tourism on indigenous lands and in national park management.

The Ecotourism Association of Australia (EAA) has no policy document on ecotourism and indigenous people. Instead, the EAA's main strategy document is the Nature and Ecotourism Accreditation Program (NEAP) which certifies genuine ecotour operators. This NEAP document (EAA, 2000) includes a cultural component recognizing indigenous consultation, interpretation and employment in ecotourism. Of 297 products accredited by December 2000, Tobwabba Tours (NSW) was the sole Aboriginal business certified by the NEAP. Only Desert Tracks and Tobwabba Tours are listed in the EAA's *Australian Ecotourism Guide 2001*. There are no indigenous members on the EAA executive or committee. At the 2000 national EAA conference, an Aboriginal keynote speaker, Gatjil Djerrkura, wanted 'Aboriginal enterprises to be given the opportunity to play contemporary roles in Australia's burgeoning ecotourism industry' (*Ecotourism News*, 2000: 6). These indigenous roles in (eco)tourism are identified below by reviewing strategies for Aboriginal tourism and nature-based tourism/ecotourism in Australia.

Aboriginal Tourism Strategies and Ecotourism

From 1995 to 1998, Aboriginal tourism strategies were produced for Australia as a whole (ATSIC, 1997a), for three states (SA, NT, Victoria) and for one region (Kimberley, WA). Ecotourism was not specifically addressed in any of these Aboriginal tourism strategies. However, Aboriginal cultural links to the environment and indigenous issues relevant to ecotourism were outlined in most of these strategies (see Table 4.3). The indigenous issues included cultural integrity; interpretation; access to Aboriginal land; developing Aboriginal tourism; and consultation or partnerships with Aboriginal people.

The South Australian strategy noted that 'Aboriginal communities live in or have cultural ties to (diverse natural) environments' (SATC, 1995: 6), with semi-traditional Aboriginal lifestyles a strong attraction in the Pitjantjatjara Lands, an area with restricted visitor entry. Aboriginal culture was also associated with the Flinders Ranges, Murray River and Coorong regions. Key aspects for tourism were indigenous cultural integrity; tour guides 'who can legitimately speak about Aboriginal history and culture or show sites' (SATC, 2000: 9), and gaining consenting contact with Aboriginal people or land. This strategy recommended tourism training with eco-awareness seminars and that Aboriginal tourism products be developed with strong tourism areas such as ecotourism and cultural tourism.

Table 4.3. Indigenous ecotourism issues in Australian Aboriginal tourism strategies.

Aboriginal tourism strategy, Area, Year	Indigenous ecotourism issues
Aboriginal Tourism Strategy, SA, 1995	Cultural integrity, Pitjantjatjara lands
Aboriginal Tourism Strategy, NT, 1996	Land access, partnerships, interpretation
Kimberley Aboriginal Cultural Tourism Strategy, WA, 1996[a]	Control and manage tourism, copyright, Joint management (National Parks), Cultural integrity, protect sites and landscapes
Aboriginal Economic Development in WA, WA, 1997	Developing Aboriginal tourism enterprises Kimberley, National Parks (e.g. Karijini)
Indigenous Tourism Product Development Principles, NSW, 1997	Consultation, interpretation, protocols, Cultural integrity, intellectual property
National Aboriginal and Torres Strait Islander Tourism Industry Strategy, Australia, 1997	Cultural interpretation, permits, employment
	National parks, ecotourism operators
Aboriginal Tourism Industry Plan, Victoria, 1998	None

[a] Non-government strategy prepared by Global Tourism and Leisure for the Kimberley Aboriginal Tourism Association, WA.

The Northern Territory (NT) strategy focused on developing partnerships between Aboriginal groups and the NT tourism industry. A task force including the tourism industry, Aboriginal Land Councils, tourism operators and government agencies (e.g. ATSIC, Office of Aboriginal Development) devised the strategy. In particular, it noted that Aboriginal people 'control large tracts of land (50% of NT), largely in a natural state and that they have control over access to those tracts of land' (Northern Territory Tourist Commission, 1996: 10). The opportunities identified for nature-based tourism were wilderness camps, tourism accommodation and development of natural attractions on Aboriginal land. It further noted that 'Aboriginal owned National Parks in the Northern Territory (e.g. Uluru, Kakadu) already play a crucial role in the tourist industry' (NTTC, 1996: 12). Aboriginal interpretive material was to be included within cultural centres in such parks. Aboriginal tourism opportunities within national parks included nature interpretation, cultural tours and joint management roles. One key aim of the NT strategy was to develop models of access to Aboriginal land or sea by tour operators through a system of permits and licences. Tourism ventures on NT Aboriginal lands are negotiated with the Aboriginal Land Councils.

A non-government strategy for the Kimberley region (WA) was developed 'primarily to enable Kimberley Aboriginal people to control and manage tourism on their land' (Global Tourism and Leisure,

1996). The strategy was prepared for the Kimberley Aboriginal Tourism Association and included actions for product development, cultural integrity and environmental issues. To this end, the strategy focused on protecting Aboriginal knowledge and cultural integrity; gaining access to traditional lands; and seeking greater involvement in management, operation and tourism joint ventures in national parks. This strategy emphasized the cultural relationship between Aboriginal people and their land and Kimberley Aboriginal communities deriving economic benefits from tourism. Specific programmes for managing tourism were not outlined in the strategy.

A strategy for *Aboriginal Economic Development in Western Australia* (Office of Aboriginal Economic Development, 1997) outlined government support for developing Aboriginal tourism businesses, focused on the Kimberley region and national parks. A key principle of this strategy was Aboriginal participation in national parks and tourism development. There was a strong focus on Aboriginal involvement in national parks and 'enabling visitors to experience Aboriginal heritage in the natural environment' (OAED, 1997: 8). The strategy noted the Aboriginal Tourism Unit in the Department of Conservation and Land Management was training Aboriginal people and developing Aboriginal heritage and tourism enterprises on WA national park lands. It also promoted indigenous partnerships with the tourism sector and tourism projects on Aboriginal land.

In New South Wales, the *Indigenous Tourism Product Development Principles* (Tourism NSW, 1997) were based on consultations with indigenous communities. These principles aimed to protect Aboriginal heritage and cultural integrity by setting out key protocols for tourism industry operators. No specific mention was made of Aboriginal involvement in ecotourism. Yet the strategy particularly stated that 'Government policy strongly supports the interpretation of Aboriginal culture by Aboriginal people' (Tourism NSW, 1997: 4). It also recommended consultation and negotiation with NSW Aboriginal Land Councils and Tribal Elders about access to and interpretation of Aboriginal heritage sites.

The Australian *National Aboriginal and Torres Strait Islander Tourism Industry Strategy* (ATSIC, 1997a) was guided by a Tourism Industry Advisory Committee including four indigenous members involved in tourism. The Strategy included a section on 'links with nature-based tourism' which focused on indigenous interpretation of the natural environment. It noted obstacles to Aboriginal involvement in nature-based tourism were a lack of commercial permits for Aboriginal tour operators and 'uncertainty on the part of ecotourist operators about how to involve indigenous people' (ATSIC, 1997a: 25). This Strategy highlighted the training and employment of indigenous people as interpreters in national parks, nature reserves and Aboriginal lands with conservation significance. It suggested that it was more feasible to develop indigenous tourism where an enterprise could 'be linked to an area of spectacular environmental quality' (ATSIC, 1997a: 22).

This national Strategy included three specific actions aimed to assist indigenous people in presenting their culture to visitors. These were allocating permits for indigenous environmental tours in protected areas; indigenous employment in national parks; and nature-based tour operators employing indigenous people as environmental and cultural guides or interpreters (ATSIC, 1997). The Strategy, however, did not mention government environmental or national park agencies as key stakeholders in Aboriginal nature-based tourism. Of 172 submissions received for the Draft Strategy, 30 were from Aboriginal cultural or community organizations and another 30 were from indigenous individuals. The Strategy had a strong cultural emphasis and had few specific suggestions for developing indigenous ecotourism ventures. Appendix D of the Strategy, however, reviewed indigenous tourism in Karijini National Park (Karijini Walkabouts).

The *National Aboriginal and Torres Strait Islander Rural Industry Strategy* (ATSIC, 1997b) reviewed rural tourism opportunities for indigenous communities. These options included recreational fishing and trophy hunting on Aboriginal lands, but not indigenous ecotourism. Environmental agencies were not mentioned as potential Aboriginal tourism partners, but controlling tourist numbers and impacts on Aboriginal lands or communities was addressed. This Rural Strategy also noted the need to 'develop policies and programmes which will manage travel through Indigenous owned lands' (ATSIC, 1997b: 36).

In Victoria, the *Aboriginal Tourism Industry Plan* simply noted that Aboriginal tourism appealed to visitors seeking 'quality experiences of nature and culture' (Tourism Victoria, 1998). There was no mention of Aboriginal ecotourism since the Plan focused on developing Aboriginal business skills and tourism industry networks. Aboriginal Tourism Australia and Aboriginal government agencies were included as partners in the Plan. Specific actions were to upgrade visitor access and interpretation at Brambuk Cultural Centre in the Grampians, and a cultural trail for the Murray Outback region. Given this limited emphasis on ecotourism in Aboriginal tourism strategies, the next sections review indigenous input and roles in nature tourism and ecotourism strategies.

Australian Ecotourism Strategies and Indigenous Involvement

From 1994 to 2000, various ecotourism and nature-based tourism strategies were devised in Australia. Indigenous tourism issues and

roles were addressed in ecotourism plans for Australia and
Queensland and in nature-based tourism strategies for WA and the
Wet Tropics of North Queensland (see Table 4.4). These key ecotourism issues included indigenous nature interpretation; intellectual
copyright; consultation with indigenous people; developing indigenous ecotourism; and ecotourism assets on Aboriginal lands.

The *National Ecotourism Strategy* (Commonwealth Department of
Tourism, 1994b) included a section on the 'Involvement of Indigenous
Australians' (Sect. 5.10, 42–45) in ecotourism. It recognized opportunities for the involvement of Aboriginals and Torres Strait Islanders
(ASTI) in ecotourism as 'land owners, resource managers and tourism
operators' (1994b: 3) and as 'site and intellectual property custodians'
(1994b: 8). Two (of seven) key actions to enhance opportunities and
encourage indigenous involvement in Australian ecotourism were as
follows (1994b: 44):

- Action 1: include ATSI communities and organizations in development and implementation of ecotourism programmes, and

Table 4.4. Indigenous issues in Australian nature-based/ecotourism strategies.

Nature/ecotourism strategy, Area, Year	Indigenous issues in ecotourism
National Ecotourism Strategy, Australia, 1994	Consultation and negotiation with ATSI communities Recognize ATSI intellectual property rights *Minimize social and cultural impacts on ATSI sites*
Ecotourism: Adding Value to Tourism in Natural Areas, Tasmania, 1994	Aboriginal products, heritage sites, consultation
Ecotourism: a Natural Strategy for South Australia, 1994	Increase in Aboriginal operators, Aboriginal lands
Queensland Ecotourism Plan, QLD, 1997	ATSI cultural perspectives of natural environment, Foster ATSI involvement in QLD ecotourism
Nature Based Tourism Strategy for WA, 1997	Aboriginal involvement in tourism, Aboriginal lands – natural and cultural assets, interpretation
Wet Tropics Nature Based Tourism Strategy, QLD, 2000	Partnerships, management of nature-based tourism Cultural values, tourism employment and training

[a]There is no Indigenous input in ecotourism mentioned in *Nature-Based Tourism
in Tasmania 1998–99 Update* (May 2000); *Nature Based Tourism Strategy* (SA,
2000); *Nature-based Tourism: Directions and Opportunities* (VIC, 2000); and
Ecotourism: a Natural Strength for Victoria – Australia (1992).

- Action 4: Encourage ATSI to participate across the full range of ecotourism development, planning, management, decision-making, regulation and implementation.

Specific measures to include ATSI people in ecotourism programmes were not addressed, though the Strategy recognized that 'many potential indigenous tourism products will be ecotourism based' (Commonwealth Department of Tourism, 1994b: 44). However, the entire indigenous input into the *National Ecotourism Strategy* consisted of just ten comments or submissions (out of 252) from four Aboriginal agencies and one Aboriginal tour operator, Desert Tracks. The Strategy also received input from an ATSI Tourism Resource Steering Committee. This Strategy, however, is no longer government policy.

The ecotourism strategy for South Australia sought to increase the number of Aboriginal tourism operators, noting that 20% of the state was held as Aboriginal freehold land. It featured Aboriginal culture and heritage as a key ecotourist attraction and included comments by ecotour operators. For example, Tom and George Trevorrow, the Aboriginal managers of Camp Coorong, stated: 'true ecotourism needs the meaningful involvement of Australia's indigenous people' (SATC, 1994). However, specific programmes to involve Aboriginal people in ecotourism were not outlined. In Tasmania, a discussion paper on ecotourism addressed the need for more Aboriginal heritage products and consultation with Aboriginal communities about tourism. It included a section on the 'Involvement of Tasmanian Aboriginals', focused on the needs of Aboriginal people in tourism and issues in presenting Aboriginal heritage sites to visitors (Foley, 1994).

In Queensland, Aboriginal interests and links with natural areas were recognized in the *Queensland Ecotourism Plan* (Department of Tourism, Small Business and Industry, 1997). This included the indigenous cultural significance of natural areas and Aboriginal-guided tours interpreting indigenous heritage in the natural environment. The Plan included sections on indigenous involvement in ecotourism and land management, and ATSI as stakeholders in ecotourism. It recognized that indigenous people could be involved in ecotourism as 'operators ... guides and trainers, or as participants in ecotourism planning, management and operation' (DSTBI, 1997: 33). ATSI people were listed ninth of ten key stakeholders in Queensland ecotourism. The Plan emphasized the development of indigenous ecotourism ventures and producing materials to support indigenous involvement in ecotourism. It highlighted the opportunities for ecotours with an indigenous cultural focus on indigenous lands and national park areas. In this Plan, indigenous people were considered 'integral to all stakeholder groups' for ecotourism (DSTBI, 1997: 52).

However, indigenous contributions to the ecotourism industry, government agencies and as natural resource managers were not specified.

Aboriginal cultural links with the natural environment and the benefits of Aboriginal involvement in ecotourism were also recognized in *The Nature Based Tourism Strategy for Western Australia* (WATC, 1997). The Strategy included sections on Aboriginal tourism and Aboriginal community involvement in nature-based tourism. It particularly noted that Aboriginal knowledge of the environment would 'provide an enormous resource for the development of nature based tourism products' (WATC 1997: 14). The Strategy acknowledged the unique relationship between Aboriginal people and the land. It further recognized that Aboriginal lands contained cultural and nature-based assets of great interest to ecotour operators. However, practical methods for involving Aboriginal communities in nature-based tourism planning and activities or in WA national parks were not outlined.

In contrast, Aboriginal involvement in rainforest-based tourism was central to the recent *Wet Tropics Nature Based Tourism Strategy* (Wet Tropics Management Authority, 2000a). This tourism strategy for the Wet Tropics World Heritage Area (WHA) of North Queensland aimed to 'facilitate Aboriginal involvement in (nature) tourism and tourism management' (WTMA, 2000a: 3). It also acknowledged the native title rights of rainforest Aboriginal people and their role as participants and partners in managing nature-based tourism in the Wet Tropics WHA. The Strategy included policy statements on 'Rainforest Aboriginal people's rights and interests,' including the cultural responsibilities of Native Title holders; visitor site management involving traditional Aboriginal owners; and Aboriginal involvement in nature-based tourism. The strategy outlined Aboriginal participation, employment and training in tourism; interpretation of natural and cultural values; and partnerships in tourism such as Aboriginal cultural tours at Mossman Gorge. The Bama Wabu Rainforest Aboriginal Association was listed as a key partner in Wet Tropics marketing guidelines; monitoring visitor sites; and setting accreditation levels for tour operators. They were not involved, however, in the permit systems for commercial tour operators in the Wet Tropics.

The Wet Tropics Strategy endorsed nature-based tourism in the WHA which promoted Aboriginal cultural heritage values and empowered Aboriginal people as participants in the tourism industry (WTMA, 2000a). ATSIC, Bama Wabu and local indigenous groups contributed to policies and principles for the Wet Tropics nature tourism strategy. Consultation with traditional Aboriginal owners about site planning and management was required for most of the popular visitor areas in the Wet Tropics. Rainforest Aboriginal interests in tourism or cultural interpretation were also mentioned for 18 sites, including Mossman Gorge, Cathedral Fig Tree, Lake Barrine, Lake Eacham, Tully Gorge and five scenic waterfalls – Murray, Josephine, Millstream, Blencoe and Wallaman Falls. Other sites were under review with traditional Aboriginal owners about tourism concerns (e.g. Roaring Meg Falls, Bare Hill, North Johnstone rafting sites and Lamins Hill Lookout). The Strategy also noted the need for assessment of the impacts of tourism on Aboriginal cultural landscapes in the Wet Tropics. It did not address the programmes or training required for developing rainforest Aboriginal tourism ventures at key visitor sites. The next section presents a case study of tourism and Aboriginal people in the Wet Tropics.

Ecotourism and Rainforest Aboriginal People in the Wet Tropics

Rainforest-based nature tourism and ecotourism is a major activity in the Wet Tropics WHA. In 1998, over 210 commercial tour operators had permits to operate in the Wet Tropics region. Half of all Queensland nature-based tour operators were located in Far North Queensland, with the majority visiting sites in the Wet Tropics (WTMA, 2000a,b). Aboriginal participation in Wet Tropics ecotourism includes Kuku-Yalanji Dreamtime Walks in Mossman Gorge; Native Guide Safari Tours conducted by Hazel Douglas in the Daintree; Munbah Aboriginal Cultural Tours near Hopevale; and guided rainforest walks at Malanda Falls with *Ngadjonji* elder, Ernie Raymont. During 2000, the Kuku Djungan Aboriginal Corporation bought the Kuranda Wildlife Noctarium and renamed it the Djungan Nocturnal Zoo. Local Aboriginal women were trained as tour guides at the zoo (closed in December 2001). Other rainforest Aboriginal participation in Wet Tropics tourism includes the construction of boardwalks and cultural interpretation in national parks.

Tourism in the Wet Tropics directly generates Aus$179 million while flow-on tourism expenditure in the local region is around Aus$753 million (Driml, 1997). At present, rainforest Aboriginal groups do not receive any licensing income from tourism operations in the Wet Tropics nor do visitors pay park entry fees at popular rainforest sites such as waterfalls, lakes and rainforest boardwalks. Apart from consulting with traditional owners about site management, there is limited Aboriginal participation in the planning and management of ecotourism in the Wet Tropics WHA. Indigenous issues are addressed by the Bama Wabu Association, representing Wet Tropics native title holders; and by three Aboriginal community liaison officers employed by the Wet Tropics Management Authority which is mainly a policy agency. Like other national park or tourism agencies, indigenous staff were not employed in planning or policy areas.

The involvement of rainforest Aboriginal communities is set out in Wet Tropics WHA legislation, the *Wet Tropics Management Plan 1998* and in the key policy document *Protection through Partnerships*. In the future, new protocols will guide consultations with Aboriginal people about permit applications in the Wet Tropics. For the past 10 years, however, rainforest Aboriginal people have sought to have the Wet Tropics officially relisted for its indigenous cultural values in order to jointly manage the WHA. With this official cultural recognition, 'they would become equal partners rather than seen as "stakeholders"' in Wet Tropics management, including tourism (WTMA, 2000b).

Indigenous Participation in Australian Ecotourism Policy: a Summary

Social values and attitudes toward indigenous people and indigenous relationships with the land influence public policy-making for tourism. The current need for partnerships or consultation with Aboriginal people about national park management and tourism ventures have mainly been driven by federal legislation for Aboriginal land rights and native title. Both community values and legislation are reflected in the policy arena for ecotourism and in strategies prepared for Aboriginal tourism and nature-based tourism in Australia. Public tourism policies address indigenous cultural heritage and environmental relationships, but have limited means for indigenous participation in the control and management of ecotourism (see Table 4.5). Tourism policies for industry bodies (e.g. EAA, TCA) and Aboriginal tourism strategies for southern states (i.e. NSW, VIC, SA) still focus on Aboriginal people as cultural heritage custodians instead of as land owners. They mainly discuss consultation and negotiation processes with indigenous people about cultural heritage, site management and appropriate use of Aboriginal culture in tourism. Other tourism policy focuses on indigenous knowledge of the natural environment (including ecological and spiritual relationships) recognized as a prime asset in nature interpretation for ecotourists. Ecotourism and nature tourism strategies for Australia, Queensland and WA mainly present Aboriginal people in terms of their affinity with the natural environment. They also include Aboriginal groups as stakeholders in ecotourism.

Policies for non-government environment agencies (i.e. ACF, TWS) and Aboriginal Land Councils, however, recognize indigenous people as landowners and native title holders with special rights and

Table 4.5. Community values and policies for Aboriginal tourism and ecotourism.

- *Cultural heritage (consultation)* Aboriginal heritage sites, cultural copyright
  Aboriginal Tourism Strategies (NSW, SA, VIC, Kimberley – WA)
  National ATSI Tourism Industry Strategy
  Ecotourism Association of Australia
  Tourism Council Australia

- *Natural environment (ecological relationship)* Environmental knowledge, interpretation
  National Ecotourism Strategy
  Ecotourism (QLD), nature tourism (WA)
  Aboriginal Tourism Australia
  Office of National Tourism
- *Native title holders/traditional owners (partnerships)* Aboriginal lands, national parks
  Aboriginal tourism (NT), Aboriginal economic development (WA)
  Wet Tropics WHA Nature Based Tourism (QLD) 2000
  The Wilderness Society, Australian Conservation Foundation
  Aboriginal Land Councils

ATSI, Aboriginal and Torres Strait Islander; WHA, World Heritage Area.

interests in land. Such policies promote indigenous partnerships in national park management and indigenous people controlling tourism on their lands. They recognize indigenous people as key managers of land areas rather than just heritage sites. Strategies for Aboriginal tourism (NT) and Aboriginal economic development (WA) address the need for beneficial tourism joint ventures with indigenous landowners. In North Queensland, the nature tourism strategy for the Wet Tropics WHA has specific policies and legislation regarding the input of traditional landowners in site management and tourism. Some 80% of the Wet Tropics WHA are claimable under Native Title.

Conclusions

Most indigenous ecotourism ventures are located on Aboriginal lands or jointly managed national parks in northern and central Australia. These enterprises include boat cruises; nature-based accommodation, cultural ecotours and wildlife attractions. Indigenous ecotourism includes unique indigenous perspectives of the natural and cultural environment, 'caring for country' and offering real benefits for local indigenous people. These indigenous tourism ventures, however, remain peripheral to the ecotourism industry in Australia. The Ecotourism Association of Australia has no policy on indigenous issues in ecotourism while southern Australian states (i.e. SA, VIC, NSW, TAS) have little indigenous input into ecotourism. In the NT, WA and Queensland, Aboriginal land rights and the *Native Title Act 1993* have been the main policy influences on tourism strategies, including indigenous groups as landowners and tourism partners. Indigenous input in ecotourism policies and organizations is still very limited. Public land policies recognizing native title rights and tourism industry positions are required for indigenous people to have contemporary roles in Australian ecotourism. Indigenous members of ecotourism committees and on permit-granting bodies for protected areas are also needed. These measures will assist indigenous involvement in managing ecotourism.

Acknowledgements

Staff at the state tourism organizations kindly forwarded copies of Aboriginal tourism and nature tourism/ecotourism strategies. Sue Muloin provided helpful comments on a draft of this chapter.

References

Altman, J. and Finlayson, J. (1993) Aborigines, tourism and sustainable development. *Journal of Tourism Studies* 4 (1), 38–50.
ACF (Australian Conservation Foundation) (2000) *ACF Policies.* ACF Website. (http://www.acfonline.org.au/asp/pages/policy/.asp)
ANTA (Australian National Training Authority) (2001) Indigenous Ecotourism Toolbox. ANTA Website. (http://staging.qantm.com.au/eco/)
ATA (Aboriginal Tourism Australia) (2000) Protocols. Aboriginal Tourism Australia Website. (http://www.ataust.org.au/Protocols.htm)
ATSIC (Aboriginal and Torres Strait Islander Commission) (1996) *On Our Own Terms: Promoting Aboriginal and Torres Strait Islander Involvement in the Australian Tourism Industry.* ATSIC, Canberra.
ATSIC (Aboriginal and Torres Strait Islander Commission) (1997a) *National Aboriginal and Torres Strait Islander Tourism Industry Strategy.* ATSIC and The Office of National Tourism, Canberra. (http://www.atsic.gov.au/programs/Economic/Industry_Strategies/Tourism_Industry_Strategy/default.asp)
ATSIC (Aboriginal and Torres Strait Islander Commission) (1997b) Rural tourism. In: *National Aboriginal and Torres Strait Islander Rural Industry Strategy.* ATSIC and The Department of Primary Industries and Energy, Canberra. (http://www.atsic.gov.au/programs/Economic/Industry_Strategies/Rural_Industry_Strategy/default.asp)
Bissett, C., Perry, L. and Zeppel, H. (1998) Land and spirit: Aboriginal tourism in New South Wales. In: McArthur, S. and Weir, B. (eds) *Australia's Ecotourism Industry: a Snapshot in 1998.* Ecotourism Association of Australia, Brisbane, pp. 6–8.
Blamey, R.K. (1995) *The Nature of Ecotourism.* Occasional Paper No. 21. Bureau of Tourism Research, Canberra.

Burchett, C. (1992) Ecologically sustainable development and its relationship to Aboriginal tourism in the Northern Territory. In: Weiler, B. (ed.) *Ecotourism Incorporating the Global Classroom.* Bureau of Tourism Research, Canberra, pp. 70–74.

Chalmers, D. (2000) Aboriginal involvement and the interpretation of Aboriginal culture within terrestrial nature-based tours in Queensland. BSc Honours thesis, School of Environmental and Applied Sciences, Griffith University Gold Coast, Australia.

Commonwealth Department of Tourism (1994a) *A Talent for Tourism: Stories about Indigenous People in Tourism.* Commonwealth Department of Tourism, Canberra.

Commonwealth Department of Tourism (1994b) *National Ecotourism Strategy.* AGPS, Canberra.

DISR (Department of Industry, Science and Resources) (2000) *National Indigenous Tourism Forum Proceedings Report: Tourism–the Indigenous Opportunity.* DISR, Canberra. (http//www.industry.gov.au/library/content _ library/forum.PDF)

Dowling, R. (1998) The growth of Australian ecotourism. In: Kandampully, J. (ed.) *Advances in Research: Proceedings of New Zealand Tourism and Hospitality Research Conference 1998. Part 1.* Lincoln University, Canterbury.

Driml, S. (1997) *Towards Sustainable Tourism in the Wet Tropics World Heritage Area.* Report to the Wet Tropics Management Authority, WTMA, Cairns.

DTSBI (Department of Tourism, Small Business and Industry) (1997) *Queensland Ecotourism Plan.* DTSBI, Brisbane. (http://www.tq.com.au/ ecotourism/qep/qep.htm)

EAA (Ecotourism Association of Australia) (2000) *NEAP: Nature and Ecotourism Accreditation Program,* 2nd edn. EAA, Brisbane. (http:// www.ecotourism.org.au/)

Ecotourism News (2000) Aborigines offer ecotourism more than just the didgeridoo. *Ecotourism News (EAA),* Spring 2000, 6.

Foley, J. (1994) *Ecotourism: Adding Value to Tourism in Natural Areas.* A Discussion Paper on Nature Based Tourism, July 1994. Department of Tourism, Sport and Recreation, Hobart.

Global Tourism and Leisure (1996) *Kimberley Aboriginal Cultural Tourism Strategy.* Kimberley Aboriginal Tourism Association.

Grekin, J. and Milne, S. (1996) Towards sustainable tourism development: the case of Pond Inlet, NWT. In: Butler, R. and Hinch, T. (eds) *Tourism and Indigenous Peoples.* Thomson Business Press, UK, pp. 76–106.

Hall, C.M. (1994) *Tourism and Politics: Policy, Power and Place.* John Wiley & Sons, Chichester, UK.

Hall, C.M., Jenkins, J. and Kearsley, G. (1997) *Tourism Planning and Policy in Australia and New Zealand: Cases, Issues and Practice.* Irwin Publishers–McGraw-Hill Australia, Sydney.

Hinch, T. (1998) Ecotourists and indigenous hosts: diverging views on their relationship with nature. *Current Issues in Tourism* 1 (1), 120–124.

James, D. (1994) Desert tracks. In: *A Talent for Tourism: Stories about Indigenous People in Tourism.* Commonwealth Department of Tourism, Canberra, pp. 10–12.

Mercer, D. (1994) Native peoples and tourism: conflict and compromise. In: Theobald, W.F. (ed.) *Global Tourism: the Next Decade.* Butterworth Heinemann, Boston, pp. 124–145.

Miller, G. (1996) Indigenous tourism – a Queensland perspective. In: Richins, H., Richardson, J. and Crabtree, A. (eds) *Ecotourism and Nature-based Tourism: Taking the Next Steps.* Ecotourism Association of Australia, Brisbane, pp. 45–57.

Muloin, S., Zeppel, H. and Higginbottom, K. (2000) Indigenous wildlife tourism in Australia: issues and opportunities. 4th New Zealand Tourism and Hospitality Conference Website. Auckland University of Technology, Auckland. (http://online.aut.ac.nz/conferences/NZTHRC/conf.nsf/programme? OpenView)

Muloin, S., Zeppel, H. and Higginbottom, K. (2001) *Indigenous Wildlife Tourism in Australia: Wildlife Attractions, Cultural Interpretation and Indigenous Involvement.* CRC for Sustainable Tourism, Gold Coast.

NTTC (Northern Territory Tourist Commission) (1996) *Aboriginal Tourism Strategy.* NTTC, Darwin.

OAED (Office of Aboriginal Economic Development) (1997) *Aboriginal Economic Development in Western Australia: A Strategy for Responsive Government Services and Programs.* OAED, Department of Commerce and Trade, Perth.

ONT (Office of National Tourism) (1999) *Ecotourism.* Tourism Facts No. 15. Department of Industry, Science and Tourism, Canberra. (http://www.industry.gov.au/library/content_library/Ecotourism_2001.pdf)

Office of National Tourism (2000) *Aboriginal and Torres Strait Islander Tourism.* Tourism Facts No. 11. Department of Industry, Science and Tourism, Canberra.

Office of Northern Development (ed.) (1993) *Indigenous Australians and Tourism: a Focus on Northern Australia.* ATSIC and Office of Northern Development, Darwin.

Pitcher, M., van Oosterzee, P. and Palmer, L. (1999) *'Choice and Control': The Development of Indigenous Tourism in Australia.* Centre for Indigenous Natural and Cultural Resource Management and CRC Tourism, Darwin.

Ross, H. (1991) Controlling access to environment and self: Aboriginal perspectives on tourism. *Australian Psychologist* 26 (3), 176–182.

SATC (South Australian Tourism Commission) (1994) *Ecotourism: a Natural Strategy for South Australia.* SATC, Adelaide.

SATC (South Australian Tourism Commission) (1995) *Aboriginal Tourism Strategy.* SATC, Adelaide.

SATC (South Australian Tourism Commission) (2000) *Nature Based Tourism Strategy.* SATC, Adelaide.

Scheyvens, R. (1999) Ecotourism and the empowerment of local communities. *Tourism Management* 20 (2), 245–249.

Scheyvens, R. (2000) Promoting women's empowerment through involvement in ecotourism: experiences from the Third World. *Journal of Sustainable Tourism,* 8 (3), 232–249.

Smith, B. (1999) Interpretation of Aboriginal culture: guide roles and responsibilities. BAppSc (Ecotourism) Honours thesis, School of Environmental and Information Sciences, Charles Sturt University, Albury, Australia.

Sutton, M. (1999) Aboriginal ownership of National Parks and tourism. *Cultural Survival Quarterly* 23 (2), 55–56.

Sykes, L. (1995) Welcome to our land. *The Geographical Magazine* 67 (10), 22–25.

TCA (Tourism Council Australia) (1999) Policy and research – indigenous tourism. TCA.

Tourism New South Wales (NSW) (1997) *Indigenous Tourism Product Development Principles.* Tourism NSW, Sydney.

Tourism Victoria (1998) *Aboriginal Tourism Industry Plan.* Tourism Victoria, Melbourne.

TWS (The Wilderness Society) (1999) *Tourism in Natural Areas Policy.* 16 May 1999. TWS Website. (http://www.wilderness.org.au/member/tws/projects/Policies/tourism.html)

WATC (Western Australian Tourism Commission) (1997) *Nature Based Tourism Strategy for Western Australia.* WATC, Perth.

WATC (Western Australian Tourism Commission) (1999) *Guide to Kimberley Aboriginal Product Experiences.* WATC, Broome.

Weiler, B. (1997) Meeting the ecotourism education and training needs of Australia's indigenous community. In: Nuryanti, W. (ed.) *Tourism and Heritage Management.* Gadjah Mada University Press, Yogyakarta, Indonesia, pp. 304–316.

Whitford, M., Bell, B. and Watkins, M. (2001) Indigenous tourism policy in Australia: 25 years of rhetoric and economic rationalism. *Current Issues in Tourism* 4 (2,3,4), 151–181.

WTMA (Wet Tropics Management Authority) (2000a) *Wet Tropics Nature Based Tourism Strategy.* WTMA, Cairns. (http://www.wettropics.gov.au/mlr/managing_tourism.htm)

WTMA (Wet Tropics Management Authority) (2000b) *Rainforest Aboriginal Heritage.* WTMA Website. (http://www.wettropics.gov.au/rah/heritage_home.htm)

Zeppel, H. (1997) Ecotourism and indigenous peoples. Issues Paper for Ecotourism Information Centre. Ecotourism Association of Australia & Charles Sturt University. (http://ecotour.csu.edu.au/ ecotour/ ecowwwhz. html)

Zeppel, H. (1998a) Selling the Dreamtime: Aboriginal culture in Australian tourism. In: Rowe, D. and Lawrence, G. (eds) *Tourism, Leisure, Sport: Critical Perspectives.* Hodder Headline, Sydney, pp. 23–38.

Zeppel, H. (1998b) Beyond the Dreaming: developing Aboriginal tourism in Australia. In: Kandampully, J. (ed.) *Advances in Research: Proceedings of New Zealand Tourism and Hospitality Research Conference 1998. Part 1.* Lincoln University, Canterbury.

Zeppel, H. (1998c) Land and culture: sustainable tourism and indigenous peoples. In: Hall, C.M. and Lew, A. (eds) *Sustainable Tourism: a Geographical Perspective.* Addison Wesley Longman, London, pp. 60–74.

Zeppel, H. (1998d) Tourism and Aboriginal Australia. *Tourism Management* 19 (5), 485–488.

Zeppel, H. (1999) *Aboriginal Tourism in Australia: a Research Bibliography.* CRC Tourism Research Report Series: Report 2. CRC for Sustainable Tourism, Gold Coast.

Zeppel, H. (2001) Aborriginal cultures and indigenous tourism. In: Douglas, N., Douglas, N. and Derrett, R. (eds) *Special Interest Tourism: Context and Cases*, John Wiley & Sons Australia, Brisbane, pp. 232–259.

Culture, Consumption and Ecotourism Policies

5

David Crouch and Scott McCabe

School of Tourism and Hospitality Management, University of Derby, Kedleston Road, Derby DE22 1GB, UK

Introduction

The consumption of tourism is changing rapidly. As people engage with tourism activity they seek to label their experiences (McCabe, 2001). In many respects such 'labelling' mirrors the ways in which academics and industrialists seek to label or typify behaviour for the purpose of defining and understanding tourists' activities. In such a process of labelling, places (and cultures) also become categorized and thus commodified by tourists, policy makers, academia and through representations constructed at least partly by the media. This chapter seeks to explore the issues surrounding the 'labelling' of ecotourism as a set of ideas and also practices, from an examination of tourist consumption and production practices, and the effects that these may have on the creation of policy for ecotourism development. The discussion focuses on how the process of 'labelling' has been used to create a set of conflicting signals concerning ecotourism, which must be addressed by policy makers in their attempts to shape constructive and meaningful policy.

We consider the context of ecotourism policies through a discussion of recent theory concerning what tourists do. Directly, this contests arguments concerning tourists as essentially exploitative and engages a discussion with prevailing interpretations of tourists as primarily sight-*seers*. Instead, tourists are identified as engaging environment, place and culture in more nuanced and complex ways, through an exploration of 'practice'. The discussion develops the contention that ecotourism is not only an activity enjoyed by tourists, but rather, tourists

© CAB *International* 2003. *Ecotourism Policy and Planning*
(eds D.A. Fennell and R.K. Dowling)

'buy into' ecotourism as an expression of personal identity. Secondly, it is argued that ecotourism as a label has been used by industry in a variety of ways that reflect the current consumer interest in such holiday 'products'. Familiar tourism communication, such as promotion, advertising and brochures, may lack acknowledgement of practice and turn instead to the representation of environments and cultures as 'products', objects of a particular kind of 'consumption'. This chapter problematizes 'consumption'. The discussion then goes on to argue that using labels in this way to package tourism offerings provides a dilemma for policy makers. The chapter seeks to offer a way out of this dilemma through re-thinking ecotourism as culture and consumption, and suggests ways of bringing such re-thinking into making policy.

Culture is a key factor in this discussion, both the cultural shifts in the tourism generating countries and also the cultures that are directly affected by policy, or the packaging of experiences from industry. It is suggested here that an overlooked part of the groundwork for ecotourism policy derives from the attitudes and practices that ordinary people hold, and how they are interpreted, or ignored, by the diverse agencies involved in tourism. Making sense of tourism policy requires insight into the cultures of the agencies operating in tourism, the cultures of the 'host' communities and the tourists themselves. This chapter considers the tourists' dimension of this and the potential for developing ecotourism policy responses that acknowledge more intelligently what tourists make of their touring. In particular, the potential for tourist practice as a resource for policy makers is identified because the 'everyday' knowledge that tourists have provides a largely untapped resource for making tourism sustainable in terms of both ecology and human rights of cultures that are the 'object' of what tourists do.

This discourse does not, of course, replace the measures available to policy in terms of controls and other management in pursuit of policy and responding to tourist numbers; juggling the pressures of development and sensitivity; and the rights of environment and 'other' cultures. The argument here does, however, point to the potential to engage the tourist positively in these parts of the process. One issue that threads through this discussion is that the way in which tourists are influenced in their regard and expectations of tourist places is the competing communication between tour operators and tourism promotion agencies in giving a particular kind of context to what happens in tourism. Policy for ecotourism has an important role to play here, and the way that role may be exercised is arguably influenced by the debate pursued here. However, the discussion begins with a consideration of these debates by making a critique of the way 'nature' and 'culture' have been engaged in both tourism promotion and in the literature of both tourism and academic debate.

'Nature': a Culture of Consumption

'Nature' is far from a new focus for leisure and tourism consumption. Taking the example of the development of touristic consumption of nature in the UK context, the point can be made that such consumption is socially constructed and therefore applicable in a more general sense. Although more often associated with cultural tourism, the Grand Tour of the 17th and 18th centuries often included visits to view the dramatic alpine landscapes of Switzerland, Italy and Austria (Borocz, 1996; Inglis, 2000). Later on in the Victorian era, nature became inextricably linked with issues of health and personal welfare. As the great industrial age created such desperate working and living conditions in the cities and towns, the industrial philanthropists realized that the development of both urban and country parks for recreation and health benefits could lead to better productivity in the factories (Clarke and Critcher, 1985).

Towards the end of the Victorian era, nature became linked to heritage as the focus of leisure and tourism consumption practice as bodies such as the National Trust were set up to preserve for posterity the landscapes and buildings created by the ancient aristocracy (Henry, 1993). This was a reaction to a perceived encroachment of urban sprawl in the UK that arose directly out of the wealth generated by industrialization. In discussing the social construction of the UK countryside, Urry (1990) describes the process by which nature has gradually been appropriated for touristic consumption by the development of an industry to appeal to a large market. In other words, as industrialization produced the wealth that enabled the masses to enjoy holidays and visits to the countryside (constructed as promoting health as opposed to the unhealthy conditions in the cities), a tourism industry rose to cater to that demand.

Later, government also provided support through legislation which reduced the working week, provided for paid holidays, and set up the general economic conditions that enabled the working classes to participate, through the increases in leisure time, in recreational pursuits and tourism activities (Clarke and Critcher, 1985). From 1900 to the outbreak of the Second World War, Henry notes that the state increasingly recognized leisure as a legitimate concern of government. The development of organized interest groups led to the introduction of the National Parks after 1951 (Glyptis, 1991). This development was of course much delayed in contrast with the American and Australian National Parks, the first of which were established in the late 19th century (Sharpley and Sharpley, 1997). However, the reasons for their creation were much the same: the diminishment of the rural, wilderness or countryside landscape in the face of industrial or other human encroachment. Alternatively the perceived need to enjoy nature as a reaction to the constrictions and the unhealthy lifestyle created by industrialization processes led to government initiatives to preserve nature for recreational, leisure or touristic uses (Glyptis, 1991). Leisure became increasingly tied into the welfare services, as part of community everyday needs, right up to the current era of free-market pluralism and the marketization of service provision (Henry, 1993).

Yet policy in the UK context has always remained (until recently) fairly ambiguous; the National Parks and Access to the Countryside Act of 1949, despite setting up the Parks, did not go far enough in empowering the authorities. In all but one case, despite national importance, the controlling interest in their administration was handed to local authorities whose role is ultimately to serve local interests. The spirit of the 1949 Act was reinforced by the Countryside Act 1968, which provided further guidelines for designating and pro-tecting more areas of countryside in AONBs (Areas of Outstanding Natural Beauty) or SSSIs (Sites of Special Scientific Interest). The Act also provided for the establishment of the Countryside Commission for England and Wales. The paradoxes still remain, despite this legis-lation, according to Seabrooke and Miles:

> ... all this legislation contains an unresolved paradox which leads to a land dilemma revolving around the distinction between recreation (implying public participation and enjoyment) and preservation or conservation (implying land use control even to the exclusion of recreation) and the extent to which the two are mutually exclusive. Official designations of outstandingly valuable countryside act as a powerful magnet to casual visitors as well as to those actively seeking the qualities which gave rise to the designation, thus subjecting the areas in question to greater pressure than they might otherwise attract. More recent legislation, the Wildlife and Countryside Act, 1981, perpetuates this dilemma.
>
> (Seabrooke and Miles, 1993: 4)

Glyptis (1991) argues that this dilemma is partially a result of the uniqueness of English and Welsh National Parks in that the designa-tion in international contexts usually applies to wilderness areas, owned, where possible, by the state. The English and Welsh National Parks, by contrast, are neither, being predominantly lived-in land-scapes with many competing uses, where ownership is spread between large institutions and smallholders alike.

Urry (1990) argues further that contemporary tourism is a socially constructed phenomenon. He sees tourism as a leisure activity that is distinct from, and presupposes its opposite, work. 'Modern' societies separate and regularize social practices, and tourism is one manifesta-tion of the fracturing of work and leisure (1990: 2). Tourist places and services are 'consumed' because they provide pleasurable experiences that are different from those encountered in everyday life. It is because of this difference from the normal environment that Urry claims modern tourism can be interpreted as a tourist 'gaze' (1990: 1):

> The tourist gaze is directed to features of landscape and townscape which separate them off from everyday experience. Such aspects are viewed because they are taken to be in some sense out of the ordinary. The viewing of such tourist sights often involves different forms of social patterning, with a much greater sensitivity to visual elements of landscape or townscape than is normally found in everyday life. People linger over such a gaze which is then normally visually objectified or captured through photographs, postcards, films, models and so on. These enable the gaze to be endlessly reproduced or recaptured.

Although Urry recognizes that English people tend to have a preoccupation with the countryside as a bucolic vision of a peaceful and deferential past, it is now becoming an increasingly popular object of the tourist gaze in the postmodern age of tourism (1990: 96). However, this romanticized notion of the English countryside is itself a construction and in the light of recent rapid changes in the rural, agricultural economy, the tension between the reality and the constructed myth of the countryside is heightened. The rise in demand for rural recreation is, according to Urry, powered by the service class, and is very much connected to particular notions of 'landscape'. Certain 'landscapes' are constructed in order to remove traces of their lived qualities, work and machinery, as well as other tourists; for the countryside is to be gazed upon and the idea of 'landscape' suggests separation from others. Urry claims that this is what distinguishes postmodern recreational users of the countryside from the right to roam campaigns of the inter-war period where access was fought for on the grounds of actual physical experience. The protestors wanted to walk and climb the land, as opposed to see it (1990: 98). In the postmodern sense, the countryside becomes a themed 'landscape' in contrast with modern emphases on its 'use', it has become a packaged, sanitized representation of rural life constructed and presented to visitors (Mcnaghten and Urry, 1998; Wilson, 1992).

As the 'cultural shift' to postmodernism, partially facilitated through international travel and a rise in awareness of global issues, brought about by a great surge in access to information, concern for the environment grew on a global scale (Holden, 2000). This led directly to the change in consumption patterns often referred to as the change from 'old' to 'new' tourism (Poon, 1993). Poon argued that tourists in the developed West became dissatisfied with the sun, sea and sand holidays that were developed for them by the travel industry in the boom of international travel from the 1960s until the late 1980s. Consumption trends moved towards holiday packages that featured the environment to a greater or lesser extent. Global 'green' issues became more focused in the minds of consumers after the publication of the Brundtland Report (WCED, 1987). The policies towards the environment of governments in the major tourism generating regions (reflecting both national and international concerns) then become transposed in an international context. The consumption trends and issues generated in the wealthy developed world mean that international tourists value the quality of destination environments that may not be available to them in their home countries.

Thus, the consumption of 'nature' becomes both validated and legitimized through the social construction process as a response to industrialization and by attempts of governments to facilitate and mediate access to and enjoyment of nature as something of social 'good'. Touristic consumers travel abroad in the hope of re-capturing the 'traditional' or pre-industrial ways of life and scenes of nature, both lived and landscape, that they now cannot obtain at home. This type of desire includes the cultures of tourism destinations. Therefore 'nature' becomes a core part of the 'product' of touristic consumption and provokes a response from industry that seeks to capture tourist spend. Thus a chain reaction is created by which policy makers seek to regularize, monitor and standardize practices to ensure that all concerned achieve the desired outcome.

Nature, Postmodern Ecotourism and the Practices of the Tourism Industry

Shifts in both the production and consumption of tourism to create labels for activity, places and tourism offerings, or 'products', as being ecotourism cannot therefore be treated outside of an exploration of the socially constructed nature of tourism as practised. Policy makers need to understand and relate to these changes in the creation of a label-driven culture of consumption and be aware of how they are used by tourists and the tourism industry when developing policy. However in the discussion above some direct consequences of this process of action and reaction were identified. In this section, some of the problematic issues in this production/consumption continuum are discussed.

Brown (1992) and Desforges (2000) argue that the consumption of tourism serves symbolic functions, including self-identity. This notion can lead to the creation of a hierarchy of tourism settings, where experiences of one or another place are seen as providing references for social groups. In a similar way, Brown goes on to consider the psychological bonds that we form with physical places. People can make attachments to and representations of self-image in tourism places. Destinations may come to embody shared meanings as symbols endowed with cultural significance to form a relationship between tourist and environment. One may develop psychological bonds with places that are spatially dispersed. This is opposed to the concept of place-identity associated with embeddedness in the home environment; identity may be related to greater geographical awareness.

Crouch (2000a) argues that the shift into postmodernity is characterized by a sense of rootlessness, the importance of the visual in tourist experiences, detachment, extreme sensations and fleeting movements through space, which are totally separated from everyday life. These features have resulted in a new geography of leisure in both the cities and the countryside. Nature has become more significant in the new geography of leisure, although it is often an adjusted, managed 'nature' (after Urry, 1995). This demand has spawned new places to see 'nature' such as wildlife parks, country zoos and open farms, as well as ecotourism. Central to an understanding of these concepts is an understanding of the lifestylization of leisure and tourism. The making of themed sites, or commercializing places under a name or brand such as ecotourism, depends on the re-imaging, or re-labelling of landscapes used in advertising. The design of new buildings and the copying of 'rural' types of landscape in new developments at the edges of cities blurs the traditional distinction between town and country, and furthermore in the context of heritage sites, creates an 'unreal', pastiche or false geography. This commercialization of places has changed the ways in which we think about and the meanings attached to places. Crouch argues that at the core of the consumption of leisure is an appreciation of 'culture' (2000a: 271):

> In terms of leisure we make and reproduce culture through what we do and use and the sense we make of places. The way we 'make sense' of leisure 'products', events and places is influenced by numerous contexts: advertising and other media, family, friendships and schooling. All of this helps us to make sense of places and our lives.

Urry (1994) argues that identity is formed through consumption of leisure goods, services and signs in the postmodern age, opposing the traditional notion that identity is formed through occupation and employment. Ecotourism practices by individuals in society then could be assumed to bestow a certain prestige upon those individuals who are able to partake. People may make representations about places labelled as ecotourist resorts. They may also create identity links between their own activities in ecotourism settings and make associations between themselves and those sharing their experiences, based upon those identifying markers. Issues of authenticity of experience in such circumstances become problematic. Tourists want to be assured that their ecotourism experiences are authentic, and as such the offerings of industry must have valid criteria on and through which their offers may be judged. However, in creating criteria, policy makers and industry alike contribute to the re-figuring of landscapes that may detract from their meaning and value to tourists and their desires to experience the culture of places. This must be a central concern of policy for ecotourism.

MacCannell (1976) argued that sightseeing is ritual and that tourists search for authenticity of experience. Mass tourism of this sort has become a central feature of consumer culture. Urry argues that the 'post-(mass) tourist' does not have to leave the house to experience all the 'framed' experiences of tourism, as he or she can watch them on TV. The post-tourist can move easily between 'high' culture and 'pleasure principle', the world is a stage and the post-tourist can play a multitude of games. One part of those games is that post-tourists 'know' that they are tourists, that there are multiple texts and no single authentic experience. This allows them to play many different games. But throughout these games, the tourist is concerned with space, what it represents and the cultures of places. Urry says that what are now consumed are not products but signs or images, and identities are constructed through the exchange of sign values (1988: 39).

However, some tourism is about 'gazing', and so Urry claims this is part of the commodification process, in the creation of signs or markers about tourism as the medium through which memories are created and stored (Urry, 1990; Crawshaw and Urry, 1997). However, some tourism is about sensing and experiencing bodily (Black, 1998), hedonically (Hyde, 2000). The tourism industry responds to these processes by creating and labelling places and experiences as being typical of a particular sort of tourism experience, thus allowing tourists to choose freely from the multitude of experiences available 'out there' in the mass of communications and images about places.

This type of labelling of places has been the subject of work by Dann (1999), who is concerned with the distinction between notions of 'traveller' and 'tourist' within travel writing. Being able to disregard aspects of space and time, the travel writer can connect with the 'anti-tourist' (Dann, 1999: 165, after MacCannell, 1989) in all of us. The distinction has sparked off a debate about the authenticity of tourist experiences, whether they seek contrived or genuine experience. It has helped to define tourism as, for example, 'sacred' journey (Graburn, 1989) or as play (Lett, 1983). The tension that is formed when tourists meet other tourists is called 'tourist angst' by MacCannell (1989), who assumes that tourists seek to distance themselves from their fellows. Tresidder (1999) has argued that the tourist landscapes of national parks represent a sacred space to tourists, creating a means of reference to allow us to find roots away from the 'horrors' of rootlessness in postmodern culture (Tresidder, 1999: 144). Dann argues that, although there have been studies of the messages of tourist brochures and their meanings (e.g. Selwyn, 1996; Dann, 1996a,b), there have not been any studies that investigate '...the sources of tourist angst, or the tensions they conceal or reveal' (Dann, 1999: 160).

Dann argues that travel writers, by banishing the tourist from the Eden of the undiscovered places of the world, unwittingly become mediators of the tourism industry, promoting the experiences that they write about as desirable places for others to visit. Dann further argues that writers are complicit in this. The travel book is still a marketing tool (Dann, 1996b). The surprising thing is that in the world of the mass-media 'gaze', the travel book is still so popular, Dann argues that it is precisely for these reasons that travel books appeal to the anti-tourist in us all. Dann uses what he describes as the universal human frames of experience, 'time' and 'space', which still must partially account for the popularity of such books. Roe (1992) writes of the social construction of space in travel writing as being composed of mobile, timed elements, which adds support to Dann's view.

The ways in which nature and culture have been constructed in policy discourse have worked from these particular perspectives: of tourist sites/destinations as product, cultures and natural sites being protected from tourists. As demonstrated here, tourism can be given a context of different sorts. This is likely to influence what tourists do. Much of this literature works from a perspective that privileges vision (the site-seer) and visual representations. In the following section this approach is unsettled. 'Culture' is regarded as important in giving context to the way the tourist encounters places (peoples, environments), although the discussion is not limited to 'culture' (i.e. as 'Western tourist culture') as the contexts delivered to people going on holiday. Instead, that limited version of culture is enlarged into a consideration of the tourist's own practices. In doing so, the discussion concentrates on processes. Policy makers for ecotourism can actively use and engage this wider debate in the development of policy. Policy must be flexible enough to enfold such a diversity of postmodern readings of ecotourism, and yet also be 'fixed' enough to be meaningful especially to those local cultures directly affected by policy. Therein lies a paradox in the development of ecotourism policy.

Tourist Practice and Ecotourism Cultures

There has been a considerable growth in recent years in the understanding of what tourists make of places, environments and cultures they visit. In this section some of that work is drawn upon in order to develop insight into the tourist and ecotourism, and the knowledge resource that the tourist constructs in 'doing' tourism. Tourism, like other practices in everyday life, may be considered in terms of 'the feeling of doing' (Harre, 1993: 30). The tourist is both multi-sensual and includes an awareness of things that exceeds manipulative sensual encounter (Merleau-Ponty, 1962). Space is all around the tourist. Tourists are engulfed, surrounded by space. This adds complexity to the capacities through which the world is encountered in tourism, complementary to the rational, linear, directed by objectivity, and complementary to mentally reflexive processes relying on the figuring of representations (Crouch and Toogood, 1999). These embodied practices go beyond 'sensation' and 'perception' to refigure meaning. The tourist encounters environments, cultures, in an embodied way.

Moreover, the tourist, like the not-tourist, never stands or thinks, or feels, or does things in a cultural vacuum, but in relation to it (Crossley, 1995). Travelling along complex cultural flows, in relation to places, cultures, other people, the tourist makes his or her own sense of what is there and what has been delivered in the tourist literature. Subjective embodied practice operates in a complex relation with contexts (Crouch, 2001). This interpretation provides fresh ground from which to consider both the work of ecotourism policy and the way that policy can engage and encounter tourists. Tourists may not be simply an object of policy but partners in policy development, communication and delivery.

Embodiment provides ground on and through which we can develop a grasp of how the tourist copes, manages, explores and negotiates his or her world. Tourism brochures, magazines and other media provide an influence, but each individual as a human being interacts materially and metaphorically with numerous other contexts for making sense of being a tourist (Game, 1991). If one considers cultures visited in tourism as felt bodily (Csordas, 1990), culture, which includes the policy of ecotourism, is lived, worked, made meaningful, developed, refigured, felt, laughed over, practised, through lived experience, not mere 'background' or 'setting' in a tourism *trip*. Tourists' subjectivities become less a separate reflection and more a practical involvement. Embodiment is a conduit through which one makes sense of heritage, culture, landscape; as tourists are involved in each of these, they feel them, work their emotions through them. These features become, even if for a moment, part of their lives, and those encounters become part of life when the tourist goes home and become transferred to the everyday, further to inform life in anticipation of the next trip.

While it is familiar to consider tourism as individualistic, sometimes solitary, even self-centred, this is only a partial story. Along the pier, in a campsite, across the beach, at Center Parcs and backpacking, tourism usually happens with, or at least among people. Encountering tourism is made among others who may or may not be known, intersubjectively and expressively. By our own presence we have an influence on others, on their space and on their practice of that space, and vice versa. This is often considered as negative, as a source of conflicts, but such a position overlooks its positive potential. We may be in this way open to each other, and so things we are doing, places we wander across, feel different (Crossley, 1995). The popular or lay geography of the tourist is one which is influenced through his or her relations with others, through which relation or encounter places and cultures make some sense and are given value (Shotter, 1993). Through the practice of shared body-space that space becomes transformed as being social, temporarily, although it may appear to be 'wilderness'. Our bodies are informed in more ways than are experienced alone, even among others we may not 'know' such as 'the crowd', and places become different (Nielsen, 1995). Sociality, then, includes 'closeness' of shared activity as a proxemic tribe (Maffesoli, 1996) and participation by tourists may be in a loose crowd. We submit that all of these aspects of 'practice' inform the way the tourist shapes and makes sense and value of environments, fragile and otherwise.

By developing interpretation through embodied non-representational geographies (that include encounters with representations) it is possible to pay closer attention to the apprehension of materiality and how that material world makes sense. Thus particular ecological areas, natural and cultural heritage, great views and intimate corners are brought into our lives and may not remain detached from our own identities. This means that artefacts, heritage features and landscapes are engaged in our own lives. Their materiality and metaphor are worked together, as Radley (1990) argues: 'Artefacts and the fabricated environment are also there as a tangible expression of the basis from which one remembers, the material aspect of the setting which justifies the memories so constructed'. Sharing body-space provides the means through which the value of being together in tourism in crowded beaches and ecotourism treks can be enacted and understood (Desforges, 2000). These places are 'performed' as enactments of identity, relations, the self, embodied and made sense (Crossley, 1995). This is important in terms of memory and the 'role' of artefacts, including other human subjects in shared body-space.

The body is active because it extends and connects among people and the material geography of places (Radley, 1995). Of course this can happen in terms of the inscription of culture on the body surface – the wearing of adornments in tourism experiences, hill climbing, surfing or at a Balinese festival – but also as a *means* through which to express oneself, as 'being', enjoying life, 'making fun', or not. Thereby the 'fun' of tourism may be a means of being in the world, of reaching and engaging the world, a medium through which it is enjoyed, and subjects declare themselves within that world. White-water rafting and other 'adventure tourism', and more mundane tourism such as camping and coach-touring provide exemplars that express not only systems of signs but expressions of feeling, subjectivity in the world, and our unique personality as we encounter them through and in our body (Cloke and Perkins, 1998). Spatial practices provide means through which the individual can express emotional relationships with others and construct a sense of what is there (Wearing and Wearing, 1996).

It is unsurprising that the tourist can be understood in terms of emotions and poetics, qualities long acknowledged by tour operators (Rojek, 1995), and this returns us not only to mental but embodied reflexivity. Walking is not merely mechanical purpose, objectivized in health, but subtle and expressive bodily practice, with both feet and the whole self, through which space can be felt intimately and imaginatively. As emotion is located within the body being a tourist may enable playfulness, an opportunity in which the world can be experienced as a child does. 'Being there' in playful practice can overflow boundaries of rationality and objectivity in an embodied rather than cerebral game, in *joie de vivre*, encountering deep feelings as well as surface play (de Certeau, 1984; Game, 1991). As tourists leave their traces of presence and practice on places and cultures they have utilized, partly non-exploitatively, nature and wilderness in the construction of their identities through the tourism practices discussed and the materiality that surrounds them. They do not merely combine the imagery of tourist promotion with an exploitative posture. They engage space, environment and its ecologies in ways that can enable a sensitivity, respect and positive value, partly on their own terms, through their own encounters.

Following this interpretation we next explore two empirical case studies, one in the UK and one in the Indian sub-continent, to consider how such a process works. Moreover, we develop insights from these cases guiding considerations to shape the way in which policy makers think about tourist–environment relations in ways that capitalize on the tourist's own knowledge and value. Thus, the chapter looks to harness the tourists' knowledge and value (rather than regarding them as antipathetic to ecotourism intentions) as valuable resources. Understanding more critically the tourist encounter with environment, and using their value and knowledge as resources enables policy makers to take 'tourists with them' in the objectives and operations of policy, and to fulfil these through the ways in which destinations are developed and managed.

Considering Cases

We narrate these dynamics of being a tourist and processes of an everyday 'making value' in environment with the evidence from one of the authors' investigations of caravan tourism in the UK (Crouch, 2001):

The 'outdoors' can be ambiguous and can animate particular 'rural' expectations.

> *Having your friends over is like entertaining at a big country house.*

> *No, it's not the countryside that matters so much. Last weekend we were in this horrible little field and it was nothing... It does help when you go somewhere and that's pretty. Sometimes there's fishing and we've been on nice long walks. The other week was really nice, deer walking around, forest all around us.*

Caravanning may happen in the 'outdoors' but this 'outdoors' combines things, people and what you are doing. The fields, the country, the outdoors can seem to offer endless opportunity for things to do, where, and how to feel. Going 'somewhere else' matters as much because it signifies getting away and seems still to evoke romance remembered from caravanning's past *and* yet also to be surrounded by familiar artefacts in fields and forests. Anticipation is marked by places en route: 'Once we're past Crook you feel different' (Crouch, 2001). Places signify numerous individual journeys of memory *and* of what you know happens when you arrive.

'The field', nature, becomes more than simply outdoors and a discovery of prefigured environment: '...in the middle of a field, in the middle of nowhere' (Crouch, 2001). 'Nowhere' is a space of escape and discovery. This can reveal very deep feelings in the way people engage themselves in an encounter with the spaces felt to be around them, significantly on their own terms:

> *Caravanning and camping, all of it makes me smile inside. I mean, everyone just comes down to the ford and just stands there and watches life go by. It's amazing how you can have pleasure from something like that. I just sit down and look and I get so much enjoyment out of sitting and looking and doing nothing. We wake up in the morning, open the bedroom door and you're like breathing air into your living.... We walk and talk.... I love to cycle and fish.*

Tim's son, who experiences disability, copes better when caravanning (Crouch, 2001).

This narrative suggests that people 'discover' 'the field' in different ways. The field emerges as a place to meet people, to manoeuvre the van, to experience escape, separation, freedom, to express oneself.

In another example, also from the UK, one of us explored the representations given well intentionedly by a regional tourist promotion of an area of country as wild as you get in England, the east side of the Pennines. Official tourism romantically and powerfully pulls on the long history of the area and labels it 'Land of the Prince Bishops'. While this satisfies on the grounds of exotic reference and at least partly constructed history, it places the locality in a time-capsule. The promotion of the area displays a centuries-gone landowning class. Another key dimension of what is, in the inappropriate language of 'products', 'on offer', is the representation of the area as a home of rare flowers depicted in beautiful photographs. Ironically, tourists are unlikely to see these species and in any case ecologists may prefer that they did not.

Working with a professional photographer one of us sought to construct an alternative story of the place and its distinctive, although hardly pre-modern culture in a series of stories of the place and its contemporary life. These stories were provided by people living and working in the area, in their own voices (although of course selected by us) with photographs of their surroundings, usually with our respondents in them in a way that, through discussing the images with those people, seems to depict the place as they know it (Grassick and Crouch, 1999; Crouch, 2000b). With an awareness of its limitations, we argue that the resulting narrative reinstates the environment in a relationship between past and present human activity and its environment, although the story emerges to entertain, to assist a mutual recognition if not identity, both with the people and the environment they experience.

The interpretation of space, the outdoors, 'countryside' and 'nature' suggest a further unsettling of taken-for-granted cultural capital, the power of context and pre-figured meanings in terms of spatial labelling (Hetherington, 1998; Macnaghten and Urry, 1998). There is instead negotiation between tourists in their making sense of space and in their making space. Human value and the physical encounter with flora and fauna provide unexpected outcomes.

Preferences for 'nature' locations are complex. People are not necessarily visiting outdoor places for rational observation of nature in isolation. Choosing a camping site near 'a pretty village', and to see deer, suggest familiar images of countryside and nature as visually aesthetic. On the other hand, 'the countryside does not matter'; 'all you need is a few cans and a field'. The language is familiar, but its constitution often unexpected: field, nature, forest, countryside, sanctuary, community, are often not coded in relation to context-explicit meanings. In a discussion of back-packers in the Andes, Desforges (1997) argues that people explore, visit, wander, spend time, because they are in a process of self (re)discovery, and the sites visited become valued through and as part of that process.

> It's a bit of a sanctuary. I don't think people started camping and caravanning for that sanctuary, *it probably comes out of what you're doing because of what you're doing and because of the values you treasure and how you respect other people's space.*

> (Denis, Yorkshire, 1998)

In the context of recent research work in Goa, India, McCabe and Stocks (1998) sought to assess the extent to which the natural environment could be developed as a feature of the tourism industry. Conducting observational research in National Parks in Goa, McCabe found an environmental resource that was partially neglected. Managed and yet severely under-resourced, the National Parks of Goa were not considered to be important for tourism, and yet were the focus of much touristic use. The dichotomy here was that the government underplayed the value of the natural resource for local people (including parties of school children, and visitors from other Indian States). As such, this touristic activity was not deemed to be worthwhile compared with the massive contribution played by Western international tourists who primarily came to Goa to experience the beach environment. A lack of physical infrastructure that could provide an appropriate level of services for international visitors was partially the problem, together with an inappropriate 'packaging' of the types of experiences and facilities that were presumed to be desired by international tourists. This meant that many international visitors only ventured into the National Parks to visit one or two key sites, mainly on excursions from the sanctuary of their beach resorts on organized tours. This type of activity limits the ability of local people to benefit financially from international tourism, creates greater stress on the natural environment and has the potential to create misunderstanding and conflict.

The National Parks of Goa provide a rich touristic resource for those interested in experiencing nature away from the crowded beaches and also for those wanting to experience something of the culture of Goa. Yet many international tourists are spatially concentrated around the beaches, and the five National Parks form a ring around the furthest hinterland of the Western Ghats away from the beaches. Governments in places such as Goa rely on the international community for their understanding of the desires of international visitors to their countries, as well as relying on their own research and policies. Once on holiday in a destination such as Goa, the tourist moves about and experiences places of interest, shopping, eating, drinking that are typical of the culture of the destination. Such a concentration on the beaches of Goa provides a point of contact for the tourist between self-identity, place and the culture of the destination, and it could be argued that the culture of Goa is essentially and inextricably linked to beach life. However, this is a social construction. The first international tourists who came to Goa were the hippies of the early 1960s in search of a 'rest' from the arduous treks around the harsher inland states of India (for example, Wollaston, 1997; Wilson, 1997). It is from that point onwards that the touristic and economic and social development of the state has focused on the beach and its culture. This type of activity was presumed to be what all tourists wanted from a visit to Goa. However, the hinterland of Goa provides for the tourist a richly diverse and authentic natural, cultural and economic landscape to explore. In terms of policy, the government and policy makers need to reflect not only upon what they think tourists want to get from their experiences but also on how they can use tourists' experiences as a part of the policy making process. Tourists want to show their concern for the environment and this aspect of concern is picked up by industry in the development of framed experiences as packaged, themed, trips.

Many Western tourists share a concern for the natural (and cultural) environment in Goa. This was evident in other research by McCabe and Stocks (1998). In research focusing on the beach resorts of Goa, McCabe and Stocks found that industry entrepreneurs, sensing the move towards environmental concern in the major tourism generating regions of the West, engaged in a 're-branding' process to capitalize on this trend, without tangible differences in 'product' offering. One five star hotel actually changed its name to add the prefix 'eco-' seemingly in a marketing oriented initiative rather than being based on any substantive ecotourism policy or strategy. Despite the fact that the hotel in question did have some environmental protection measures, these were nothing more or less than one would expect from recent standard waste management initiatives. The hotel had its own sewage treatment plant and its own groundwater borehole. However, these types of facilities not only undermine the natural environment in Goa (which suffers from a severe lack of water resources), but have a knock-on effect on the local population, who have limited access to water (Goa Foundation, 1993). The interactions between the tourism industry, local people in Goa and tourists themselves have been the focus of considerable attention (see, for example, Wilson, 1997). Pressure groups have been formed to raise awareness of the anomalies in the government system of land use planning and control in Goa, such is their lack of ability to control some forms of touristic development. The rapidity of development along Goa's coastal strip led to concern among local people and eventually to the rise of pressure groups and voluntary societies to voice concern about the effects of such rapidly expanding development. Such moves highlight the extent to which the tourism industry has opportunistically shown an overriding focus on profit maximization at the expense of concern for the protection of the environmental resource. Such voluntary and pressure groups should be encouraged to join in the debate on ecotourism policy development.

Nature as used in policy is given meaning through rational scientific purpose and the tourist is frequently 'provided' with a valuing of nature, and cultures, as 'products', objects to consume as a gaze

(Wilson, 1992). However, it emerges from our discussion in this chapter that nature itself, and cultures, are refigured and given meaning through what tourists do, into their own grasp of what nature is. Nature is both 'out there' and inside. Cultures, other people's lives and places are refigured similarly. This interpretation does not negate the value of promotional content but suggests a more dynamic encounter between nature and the tourist. This provides, it is argued here, imaginative means through which policy can engage the positive dimensions of tourist practice and the potential value constructed through being a tourist. This provides a big challenge through which to rethink policy approaches for ecotourism.

Ecotourism Policy and the Use of Lay Knowledge

There are some key points to discuss from the case studies described above. First of all, there is an issue about what can or cannot be called an ecotourism destination. Such issues of nomenclature should take into consideration the fact that tourists experience landscapes of all different sorts and types and levels. The field of the caravan site or the Equatorial rainforests or the beaches of Goa represent types of places as environment to tourists but all can be experienced imaginatively from an ecological perspective. Ecotourism policy does not have to be focused on the ecologically unique or threatened resource from the perspective of the tourist. Secondly we must also be mindful of the connotation that ecotourism must involve a 'solitary' tourist experience. The people with whom we share the trip and the people that we meet at the destination provide us with both a means of enjoyment and the making of the holiday, and also allow the tourist to perform important identity work. Policy makers must factor these ways of experiencing into their development of policy. Policy should further encourage governments and industry to review their attitude to ecotourism and the desires of tourists. It can often be the case that historical use and associations can mask the changing nature of tourism consumption. Industry must be responsible in the ways they create representations of places that reify the environment at the expense of the people of tourism destinations. These representations, while not wholly structuring the touristic experience, can exert an influence. The culture of tourism destinations is always a vital and enriching component of the tourist experience and policy must be directed towards understanding the embodied ways in which people realize other cultures through their interaction with places, environments and space.

The practice described here is not detached from cultural influences and contexts, but instead is relational to them; influenced, influencing, negotiating. People refigure the promoted and managerial contexts in their own terms. Nature areas, wild environments, peoples of other cultures are understood also through these practices. In that sense tourists encounter and engage. Promotion and product labelling can detract or distract from these meanings and the values therein, but may not disrupt them completely. Developing policy can pay attention to these nuanced encounters and think through how people frame what they do, and the places and so on that they encounter. Communicating the value of places can capture the character of these encounters and develop and enhance a positive attitude among tourists.

These dimensions of what the tourist does are potentially informing of tactics through which policy may be developed. Tourists' grasp of cultures and environments provides a resource for policy. Tourists constitute their 'lay' or everyday knowledge of environments and cultures. The way in which this is worked on by tourism promotion provides potential excitement but can also influence what the tourist makes of things. This is not a one-way influence, and our discussion suggests that the tourists instead practice a process of 'refiguring' environments and cultures through the complex dimensions of dynamic encounter with which they engage. Rather than see the tourist as necessarily a potential problem in, for example, fragile environments, environments everywhere and, for that matter, cultures, the tourist is a potential resource for the value of each of these. Tourists are partners in the process of a shift towards tourism development that is more sustainable, or at least in securing care, respect and attitudes of sustainability among themselves. Indeed, we argue that the groundwork for working with the knowledge resource of the tourist is already there. Policy can engage and work with this lay knowledge resource.

This combination of metaphor and materiality (practices and places) is crucial in making geographic knowledge. Crouch (1999a) refers to the embeddedness of the everyday practice and meaning of the use of space for leisure/tourist consumption, as 'lay' geographic knowledge, which he describes as:

> ... a process in which the subject actively plays an imaginative, reflexive role, not detached but semi-attached, socialised, crowded with contexts. The resulting knowledge resembles a patina and kaleidescope rather than a perspective with horizon, a series of mutually inflected and fluid images rather than a map....the subject bends, turns, lifts and moves in often awkward ways that do not participate in a framing of space, but in a complexity of multi-sensual surfaces that the embodied subject reaches or finds in proximity and makes sense imaginatively. This combination contains meanings of landscapes, fragments, spaces, whole and abstract places, abstractions of the city and the country, street, nation, gender, ethnicity, class, valley, arena and field, through which human feelings,

love, care and their opposites may be refracted. The subject mixes this
with recalled spaces of different temporality.

(Crouch, 1999a: 12)

It is important therefore that policy is mindful of the changing nature
of consumption and the interrelated nature of the effects of policy on
the development of new, labelled places for tourism, which in turn
has an effect on practices and experiences. That the culture of the people of tourism destinations should also be considered in the development of responsible and meaningful policy for ecotourism should be
axiomatic. It is too often the case though that the natural environment
becomes the focus for ecotourism development rather than a more
integrative view. The relationships between the natural environment,
the cultures of places and the practices and experiences of tourists
themselves as a totality should be included in the process of policy
development. We have argued here that the tourist can be actively
engaged in the policy development process; they have a great deal to
contribute to the debate. We recommend that action research be used
to develop better understanding of the meanings of tourism places for
tourists themselves.

Through such research we discover the ways in which the tourist's
self and social identities can be understood in relation to the development of policy for ecotourism initiatives. Intrinsic in this type of
research, tourists should be made fully aware of their role in their contribution towards development and in affecting the cultures of the
people of tourist destinations. Recent case studies have shown that the
people of wilderness, marginal, protected landscapes can care for and
manage the environments in which they live if they are empowered to
believe in the value of such landscapes to themselves and to tourists.
Tourists can be encouraged to be selective when choosing their holiday
destinations or packages if they are also more involved in and aware of
the effects of their trips on the environment and peoples that they visit.

References

Black, P. (1998) Walking on the beaches. Looking at the Bodies. Paper presented at the B.S.A. Annual Conference: *Making Sense of the Body.*

Borocz, J. (1996) *Leisure Migration: a Sociological Study on Tourism.*
Pergamon Press, Oxford.

Brown, G. (1992) Tourism and symbolic consumption. In: Johnson, P. and
Thomas, B. (eds) *Choice and Demand in Tourism.* Mansell, London.

Clarke, J. and Critcher, C. (1985) *The Devil Makes Work: Leisure in Capitalist
Britain.* MacMillan, Basingstoke.

Cloke, P. and Perkins, H.S. (1998) Cracking the Canyon with the Awesome
Foursome: representations of adventure in New Zealand. *Environment
and Planning D: Society and Space* 16 (1), 185–218.

Crawshaw, C. and Urry, J. (1997) Tourism and the photographic eye. In: Rojek, C. and Urry, J. (eds) *Touring Cultures: Transformations of Travel and Theory*. Routledge, London, pp. 176–195.

Crossley, N. (1995) Merleau-Ponty, the elusive body and carnal. *Sociology: Body and Society* 1, 43–61.

Crouch, D. (1999a) The intimacy and expansion of space. In: Crouch, D. (ed.) *Leisure/Tourism Geographies: Practices and Geographical Knowledge*. Routledge, London, pp. 257–276.

Crouch, D. (1999b) Introduction: encounters in leisure/tourism. In: Crouch, D. (ed.) *Leisure/Tourism Geographies: Practices and Geographical Knowledge*. Routledge, London, pp. 1–16.

Crouch, D. (2000a) Leisure and consumption. In: Matthews, H. and Gardner, V. (eds) *The Changing Geography of the UK*. Routledge, London.

Crouch, D. (2000b) Places around us: embodied lay geographies in leisure and tourism. *Leisure Studies* 19, 63–76.

Crouch, D. (2001) Spatialities and the feeling of doing. *Social and Cultural Geography* 2 (1), 61–75.

Crouch, D. and Toogood, M. (1999) Everyday abstraction in the art of Peter Lanyon. *Ecumene* 6 (1), 72–89.

Csordas, T.J. (1990) Embodiment as a paradigm for Anthropology. *Ethos* 18, 5–47.

Dann, G.M.S. (1996a) *The Language of Tourism: a Sociolinguistic Perspective*. CAB International, Wallingford, UK.

Dann, G.M.S. (1996b) The people of tourist brochures. In: Selwyn, T. (ed.) *Tourism Research: Myths and Structures*. John Wiley & Sons, Chichester, UK.

Dann, G.M.S. (1999) Writing out the tourist in space and time. *Annals of Tourism Research* 26 (1), 159–187.

de Certeau, M. (1984) *The Practice of Everyday Life*. University of California Press, Berkeley, California.

Desforges, L. (2000) Traveling the world: Identity and travel biography. *Annals of Tourism Research* 27 (4), 926–935.

Game, A. (1991) *Undoing the Social: Towards a Deconstructive Sociology*. Oxford University Press, Buckingham, UK.

Glyptis, S. (1991) *Countryside Recreation*. Longman, in association with the I.L.A.M. Graburn, Harlow, UK.

Goa Foundation (ECOFORUM), (1993) *Fish Curry and Rice: a Citizens Report on the Goan Environment*. Other India Press, Mapusa, India.

Graburn, N.H.H. (1989) Tourism: the sacred journey. In: Smith, V. (ed.) *Hosts and Guests. The Anthropology of Tourism*. University of Pennsylvania Press, Philadelphia, pp. 21–36.

Grassick, R. and Crouch, D. (1999) *People of the Hills*. Amber /Side Gallery, Newcastle upon Tyne.

Harre, R. (1993) *The Discursive Mind*. Basil Blackwell, Oxford.

Henry, I.P. (1993) *The Politics of Leisure Policy*. MacMillan, Basingstoke.

Hetherington, M. (1998) Vanloads of uproarious humanity: new age travellers and the utopias of the countryside. In: Skelton, T. and Valentine, G. (eds) *Cool Places*. Routledge, London, pp. 328–342.

Holden, A. (2000) *Environment and Tourism*. Routledge, London.

Hyde, K.F. (2000) A hedonic perspective on independent vacation planning, decision-making and behaviour. In: Woodside, A.G., Crouch, G.I., Mazanec, J.A., Oppermann, M. and Sakai, M.Y. (eds) *Consumer Psychology of Tourism, Hospitality and Leisure*. CAB International, Wallingford, UK, pp. 177–192.

Inglis, F. (2000) *The Delicious History of the Holiday*. Routledge, London.

Lett, J.W. (1983) Ludic and liminoid aspects of charter yacht tourism in the Caribbean. *Annals of Tourism Research* 10, 35–56.

MacCannell, D. (1989) *The Tourist: a New Theory of the Leisure Class*. Macmillan, London.

Maffesoli, M. (1996) *The Time of Tribes*. Sage, London.

McCabe, A.S. (1997) The future for national parks tourism in Goa. Paper presented at the IIDS International Seminar on Tourism and Economic Development, Bhubaneswar, India, 4–7 June.

McCabe, A.S. (2001) *'Worlds of Reason': the praxis of accounting for 'day visitor' behaviour in the Peak National Park. A qualitative investigation*. PhD thesis, University of Derby, UK.

McCabe, S. and Stocks, J. (1998) Issues in Social Impacts of tourism research with reference to the Indian state of Goa. In: Kartik, C. Roy and Tisdell, C.A. (eds) *Tourism in India*. Nova Science Publishers, New York, pp. 87–100.

Mcnaghten, P. and Urry, J. (1998) *Contested Natures*. Sage, London.

Merleau-Ponty, M. (1962) *The Phenomenology of Perception*. Routledge, London.

Nielsen, N.K. (1995) The stadium in the city. In: Bale, J. (ed.) *The Stadium and the City*. Keele University Press, Keele, UK, pp. 21–44.

Poon, A. (1993) *Tourism, Technology and Competitive Strategies*. CAB International, Wallingford, UK.

Radley, A. (1990) Artefacts, memory and a sense of the past. In: Middleton, D. and Edwards, D. (eds) *Collective Remembering*. Sage, London.

Radley, A. (1995) The elusory body and social constructionist theory. *Body and Society* 1(2), 3–23.

Roe, P. (1992) Textual tourism: negotiating the spaces of reading. *SPAN, the Journal of the South Pacific Association for Commonwealth Literature and Language Studies*, 33. www.mcc.murdoch-edu.au/ReadingRoom/litserv/SPAN33/Roe.html

Rojek, C. (1995) *Decentring Leisure: Rethinking Leisure Theory*. Routledge, London.

Seabrooke, W. and Miles, C.W.N. (1993) *Recreational Land Management*. E. and F.N. Spon, London.

Selwyn, T. (1996) Introduction. In: Selwyn, T. (ed.) *The Tourist Image: Myths and Myth Making in Tourism*. John Wiley & Sons, New York, pp. 1–31.

Sharpley, R. and Sharpley, J. (1997) *Rural Tourism: an Introduction*. Thompson International Business Press, London.

Shotter, J. (1993) *Cultural Politics in Everyday Life: Social Constructionism, Rhetoric and Knowing of the Third Kind*. Oxford University Press, Buckingham, UK.

Tresidder, R. (1999) Tourism and sacred landscapes. In: Crouch, D. (ed.) *Leisure/Tourism Geographies: Practices and Geographical Knowledge*. Routledge, London, pp. 137–148.

Urry, J. (1988) Cultural change in contemporary holiday making. *Theory, Culture and Society* 5, 35–55.

Urry, J. (1990) *The Tourist Gaze, Leisure and Travel in Contemporary Societies.* Sage, London.

Urry, J. (1992) The tourist gaze and the environment. *Theory, Culture and Society* 9, 1–26.

Urry, J. (1994) Cultural change and contemporary tourism. *Leisure Studies* 13, 233–238.

Urry, J. (1995) *Consuming Places.* Routledge, London.

WCED (1987) *Our Common Future.* Oxford University Press, Oxford.

Wearing, B. and Wearing, S. (1996) Refocusing the tourist experience: the flaneur and the choraster. *Leisure Studies* 15(3), 229–243.

Wilson, D. (1997) Paradoxes of Tourism in Goa. *Annals of Tourism Research* 24 (1), 52–75.

Wollaston, S. (1994) *The Guardian Weekend*, Easy Come, Easy Goa, 24 February.

Economic Instruments of Environmental Tourism Policy Derived from Environmental Theories

6

Tanja Mihalič

Faculty of Economics, University of Ljubljana, Kardeljeva pl. 17, 1000 Ljubljana, Slovenia

Introduction

Over the last two decades there has been growing awareness and concern about the relationship between tourism and the environment. The rapid growth of tourism has resulted in significant negative environmental impacts. Although a quality natural, cultural and social environment is the basis for most of the tourism business, in practice a paradoxical situation is produced when 'tourism destroys tourism'. Additionally, tourism also suffers from environmental degradation arising from other economic activities. Increasing environmental research, especially the development of a sustainable tourism development agenda, has significantly contributed to the awareness that the environment, the primary tourism resource, must be sustained.

In what follows, firstly the general theories on the creation, elimination and minimization of environmental damage as developed in the theory of environmental and welfare economics are applied to the field of tourism: behavioural theory, as well as growth and system theory. Economic policy instruments are derived from each of the above-mentioned theories and briefly explained. The main emphasis is laid on the market functioning of environmental economic instruments.

© CAB *International* 2003. *Ecotourism Policy and Planning*
(eds D.A. Fennell and R.K. Dowling)

Environmental Theories and Derived Instruments of Environmental Policy in Tourism

Today's environmental economic theories attribute ecological damage to various causes:

- system;
- growth;
- behaviour (Mihalič and Kaspar, 1996).

In the economic literature each of the mentioned theories is dealt with in more variants or theories, as also seen in Fig. 6.1. As those theories seek to explain the reasons for environmental damage they also dictate the instruments needed to eliminate it. Although some instruments are suggested by more than one theory, I shall present

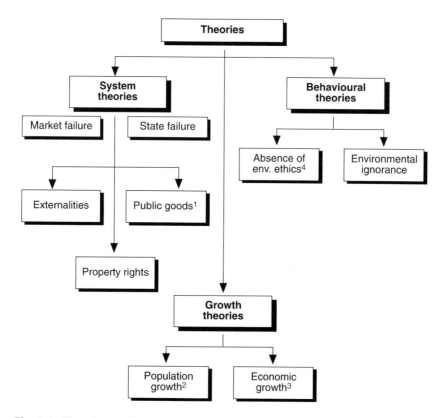

Fig. 6.1. Theories on the creation and elimination of environmental damage.
[1] Local public goods, [2] tourism growth (growth in number of tourists, tourism demand side), [3] tourism industry growth (growth in capacities, tourism supply side), [4] absence of tourism ethics (supply and demand side).

individual instruments along with the theory that suggests their usage to the greatest extent. Most instruments are aimed at the preservation of natural resources while only a few refer to any reduction of the negative effects on the cultural and social environments. The study focuses on instruments functioning through the tourism market; for example, by influencing supply and demand. In addition, some fiscal and administrative instruments, or even techniques that contribute to reducing the negative effects of tourism, are briefly mentioned.

Systems

The first theory implying the causes of environmental damage is the system theory (Fig. 6.1). Environmental damage is accelerated by the inefficient allocation of environmental resources as a result of: (i) failure of the market; and/or (ii) failure of the state. Allocation efficiency is defined as Pareto optimality, e.g. the impossibility of reallocating environmental goods to make one person in the economy better off without someone else becoming worse off. Environmental goods can be optimally allocated through the functioning of the market for environmental goods and/or by state intervention.

System theories

Three interacting system theories can be found in the economic literature:

- theory of externalities;
- theory of public goods;
- theory of property rights.

THEORY OF EXTERNALITIES. The main reason for environmental problems arising lies in the fact that the environment is cost-free, which leads to its excessive exploitation and degradation. Thus, the environment must become an economic good on which the users will be economizing; that is the environment must be given a price.

The theory differentiates between private and social costs and benefits. The total amount of private costs and benefits is not equal to the amount of social costs and benefits because quite often the firm does not realize its total product (positive external effects) and/or does not carry all of the (social) costs of its production (negative external effects). This means a loss of Pareto optimality as at least one economic subject is better/worse off at the expense of costs or benefits of others. A classic example of negative externality is a firm dumping organic waste into a river, thereby reducing the production possibilities of other firms as well as the recreational quality of the river for bathing and sports fishing (Hjalte *et al.*, 1977).

Pigou (1920: in Leipert, 1989: 7) argued that the divergence between private and social effects necessitates state intervention to achieve an optimal allocation of resources. He suggests taxes and subsidies. Subsidies refer to positive external effects and taxes to negative ones. If environmental taxes are introduced, the internal costs of a firm increase and the firm is consequently forced to reduce the quantity of its environmental use: by either cutting production or applying new, more environmentally friendly technologies. Hence, a better allocation of environmental resources is achieved through the impact of an increase in a firm's costs and prices on the market mechanism.

In tourism a divergence between private and social effects still exists but, at the same time, the environmental damage caused by a single firm also affects the tourism industry itself. Therefore, we distinguish two kinds of effects:

- external effects caused by non-tourism subjects but which affect tourism;
- external effects caused by tourism which affect:
  - other non-tourism subjects; or
  - tourism firms and tourists.

Firms discharge into the environment noise, exhaust gases, dust and other waste. In this way they affect the tourism business as tourism demand (numbers of visitors) is reduced or additional costs arise for the tourism firms trying to remove the consequences of such emissions. In addition, tourism firms bring about negative and positive external effects on the environment thus affecting other firms and/or local people. For instance, hotel sewage piped into the sea can reduce the number of fish caught; ski fields and ski lifts shrink the agricultural yields of the farming sector; the emission of noise from tourism activities affects local people. At the same time, the negative effects caused by tourism also affect tourism itself. For example, polluted sea, landscapes visually destroyed by tourism infrastructure, erosion caused by ski activities etc., reduce the quality of the tourism product and thereby the prices and revenues.

Economists call for the internalization of external effects: polluters should bear the social costs of any pollution they cause. Accordingly, for example, an industrial firm should cover the related costs if demand for accommodation facilities in a neighbourhood hotel falls because of negative effects caused by the firm (such as noise). Yet it is not only the loss caused by a fall in demand for existing facilities but also the loss caused by a fall in the expected tourism demand growth that is a point of issue. In this case, the damage stemming from the lower chances of capacity growth also arises. A mere tax on noise, depending on the production size of the noise maker, does not internalize all of the social costs (Tschurtschenthaller *et al.*, 1981).

THEORY OF PUBLIC GOODS.   Environmental goods such as clean air, clean water, diverse species or healthy forests are often public goods in the sense that they can be enjoyed (consumed) by many individuals simultaneously without affecting individual consumption (Lesser *et al.*, 1997: 8). A pure public good is a good whose consumption by one individual does not reduce the amount of it available for other consumers (non-rival consumption) and where no one is excluded from its provision (non-excludability). Another category of public goods, impure public goods, can be either non-excludable or non-rival but not both (Hanley *et al.*, 2001: 20). Air quality or biodiversity are examples of a pure public good. Common property and club goods like rivers, local parks and beaches are impure public goods because their benefits can be excluded from non-members of the group which owns the resource. In the case of open access to common goods, everyone has access, all have a right to the resource and scarcity value is ignored.

In contrast to private goods, which are excludable and rival, public goods are goods that can be used freely regardless of a user's participation in the related costs. Individuals do not wish to show their need for public goods and prefer to be 'free riders', therefore there is no demand for these goods. Accordingly, a market for public goods cannot be established. As public goods do not involve any price they are used heavily and can possibly be degraded (Stabler, 1997: 5).

In tourism, the theory of public goods is treated as the theory of local public goods. Tourism destinations offer natural goods like a beautiful countryside, a nice climate, clean air, pure water or unspoiled natural beaches. In this case we have in mind goods that are, generally open to use by all interested tourism users (firms), no matter whether provided by private or governmental bodies. Individual tourism firms do not have to pay for the quality of the environment. Nevertheless, they indirectly make profit from the presence of good environmental quality. Payment for all these (tourism rent) is included in the premium prices of tourism products that are set and formed as a result of greater demand because of the attractiveness of the natural goods concerned. Nature is, in this case, a local public good which benefits more or less all the parties of a given area. For known reasons, they are not willing to take over the costs of its preservation. In such paradoxical circumstances, the quality of the environment gets worse although everybody knows that only a well-preserved environment allows returns from tourism and its long-term economic efficiency. So long as tourism firms are not willing to admit their dependence and/or need for a quality natural environment and then participate in environmental preservation and protection costs, the conditions are constantly deteriorating. In tourism it is often believed that the costs of the environment, which is considered as a public good, are to be covered by a third party – that is by the state. This argument is primarily unjustifiable because the tourism industry is also an indirect seller of public goods. Yet, it is true that the environment is also being destroyed by other, non-tourism firms.

In certain cases, some of the environmental public goods mentioned may lose their public character by private ownership and exclusion, such as private beaches or parks. Another important environmental policy problem arises from congestible public goods. These goods are non-rivalrous, but only to the point where congestion begins. In tourism, many forms of congestible public goods are relevant. Too many visitors in a destination crowd the highways or streets, as well as beaches or parks.

THEORY OF PROPERTY RIGHTS.   Property rights theory derives from the theory of external effects. The market mechanism is not an optimal allocation mechanism if there are external effects that are not internalized. Nevertheless, this is not due to a market. The state has not succeeded in creating framework conditions to prevent cost-free utilization of environmental goods which in reality are relatively scarce. Where environmental goods appear as relatively scarce goods, property rights must be developed. Well defined property rights are a precondition for a market-oriented solution to the environmental problem (Hanley *et al.*, 1997).

Two variants of this theory are: (i) the polluter has a right to pollute; or (ii) the affected party has a right to non-pollution. In the former case, the costs of non-pollution are a burden on the affected party; in the latter case the costs of pollution are covered by the polluters (polluter pays principle). Coase suggests direct negotiations and compensation between polluters and the parties involved. Both Pigou and Coase base their theories on the fact that in a situation where the Pareto optimality is not achieved the total income of the affected party does not reach the maximum and thus both the affected party and the polluter are interested in negotiating. Additional income gained through the negotiations can be distributed so that not just one party benefits from the negative external effects.

Instruments derived from system theories

The system theories discussed above suggest the following instruments for doing away with and reducing environmental damage: taxes, subsidies and compensations that, through costs and prices, impact on the more optimal allocation of natural resources and/or reduce environmental utilization via the market mechanism. One of the system theories, the theory of local goods, suggests that the costs of environmental protection should be borne by a third party (state) and not by (the tourism) industry.

ENVIRONMENTAL TAXES.   Environmental taxes fall within the group of so-called fiscal market instruments. In the environmental area, with the help of fiscal policy instruments the state establishes framework conditions that enable the internalization of external effects through the market mechanism (see 'Theory of externalities', above).

It is possible to levy taxes on production factors, on harmful emissions or on tourism products themselves. The main problems here are setting the tax base and rate, as well as creating an information network for how to obtain proper data on tax base and/or for tax lifting.

In tourism a so-called 'overnight tax' has been suggested (Tschurtchenthaller *et al.*, 1981). Here, the tax base is the overnight stay. The suggestion does not take into consideration that the price elasticity of demand is relatively low as an overnight stay is a relatively necessary product. Accordingly in the case of higher accommodation prices a tourist will first cut other expenditures. The idea also does not consider the real possibility that a hotel manager would try to nullify the effect of the tax on the price through rationalization within the firm; in tourism mostly on the account of the employees and/or product quality. Another disadvantage is the fact that the tax base is inappropriate. Overnight stays themselves do not cause negative external effects (for example, overcrowded tourism places, overloaded infrastructure, etc.). A much greater burden on the environment stems from the special activities of tourists: skiing is much more harmful to the environment than walking; 1-day visitors cause greater environmental damage (for example, traffic chaos) than overnight guests. Therefore, Tschurtschenthaller *et al.* (1981) suggest that tax should be set on the basis of various tourism activities.

Further, the European Union is suggesting a 'package tax'. It proposes an environmental tax of 1% on the total cost of a tourism package (De Rivera Icaza, 1990). Such a tax base and uniform tax rate assume that all kinds of tourism packages cause the same level of environmental damage. Further, in this case the environmental tax would be paid only by tourists who travel on an all-inclusive basis and not those travelling on their own. Yet, it would be necessary to formulate tax rates for package tours according to the environmental damage caused by a certain package.

Environmental taxes on catering services have also been applied. Taxes on disposable packaging and on hotel cutlery thrown away after use would force catering managers to cut the production of waste. Another disadvantage of a tax as an instrument of preventing environmental damage lies in the fact that it is of no use to those who, due to the negative external effects, are the affected party. That deficiency could also be abolished through so-called tax transfers, for instance between tourism firms and farms. Farmers may even claim compensation due to lower agricultural yields because farming areas are used for skiing activities.

ENVIRONMENTAL SUBSIDIES.   Subsidies are transfer payments granted by the state to firms in order to change the relationship between costs and returns in their production. Compared with environmental taxes, the purpose of environmental subsidies is to lower the costs and prices of more nature-friendly production *vis-à-vis* more environmentally unfriendly production. As the prices of products polluting the environment remain unchanged, the demand for environmentally friendly products is increased through the market mechanism. Therefore, the production of these products increases and the price decreases. At the same time, demand for less environmentally friendly products is reduced and, consequently, their production. Tourism is also interested in subsidizing more environmentally friendly non-tourism activities as it is affected by the negative external effects of other activities.

In economic discussions subsidies are mainly criticized for taking over the social costs of environmental pollution and not forcing the polluter to further reduce their negative external effects. It is said that they do not contribute much to improving the allocation of natural resources. Of course, subsidies could be suitable means for diminishing the adjustable costs. Therefore, subsidies would be recommended when, for example, sewage discharge regulations were changed and adjusting to new regulations would lead to the bankruptcy of some (existing) tourism firms due to the high adjustment costs.

NEGOTIATIONS. Negotiation solutions discussed in the theory of property rights assume that the parties agree directly on a quality level of the natural environment and on compensation for its conservation and/or preservation. The solution through negotiations supposes a division into two groups: polluters and affected parties. Yet in tourism this division is impossible when tourists and/or the tourism industry are simultaneously both the polluter and the affected party. If we take the first principle in the theory of property rights by which the non-pollution costs are charged to the affected party, the tourism industry is the one charged for this. At the same time, it is also charged for pollution, whereas by the second principle this is the burden of the polluter. Negotiations between the affected parties and the polluter are impossible because there is one and the same party on both sides. An exception is the case where, for example, a tourism business in a destination intends to expand. As this would increase tourism supply and endanger the environmental quality of the existing supply and its competitiveness, prices are likely to fall. In this case, it is reasonable from the economic and environmental aspects for the existing tourism firms to pay compensation to those giving up their ideas of the planned development of tourism.

A division into two negotiating partners is, however, possible when negative external effects are caused by non-tourism activities on the account of tourism business or when the external effects are caused by the tourism industry at the expense of non-tourism industries or local inhabitants. The affected parties can seek compensation, yet the tourism industry refuses to negotiate if the environment is a public good whose quality must be protected and guaranteed by a third party. A similar viewpoint is also shared by local inhabitants if they are employed by tourism firms. In practice, there are cases of the internalization of positive external effects of farmers in favour of the tourism industry (for example, cultivated farm landscapes as a tourist attraction). In theory, internalization of negative effects on agriculture caused by tourism is also possible, for example payment of compensation because of poorer annual yields per acre on a piece of land used by the tourism industry for skiing purposes.

CONCESSIONS. Concession granting over natural goods sets out measures for environmental protection and enables the collection of funds for environmental protection and/or rehabilitation. The basis for concession granting is a concession agreement between the concession grantor and concessionaire. It should define:

- conditions regarding environmental protection, the method of management, use or exploitation of natural resources, and the concessionaire's obligations regarding rehabilitation, establishment of a new environment or restoration of the previous state of the environment;
- payment for the concession (concession value).

Regardless of the fact that concession granting can be market-based (competition among possible concessionaires), natural resources are protected due to the set environmental protection conditions. Thus, concessions are administrative instruments that have very similar effects to licences, prohibitions or enforced environmental standards. The implementation of concessions for natural goods faces many problems (Mihalič, 2000b). It requires the adoption of concession laws and natural goods must have their owner (e.g. the state). The model assumes that the concession grantor knows the true concession value or tourism rent (see 'Theory of public goods' above). Where tourism rent does not exist due to the low quality of a natural good (the example of environmental degradation), the concession value does not exist or is even negative.

PUBLIC INVESTMENTS. Public investments are not an instrument of internalization yet they do address external effects (for example, financing the building of a purifying plant, or waste disposal area). They can be financed entirely from tax resources or partly from fees in line with the principle that the costs of pollution and/or burdening the environment must be carried by the polluter.

In tourism, public investments are reasonable if, for instance, the state takes over some public investments like the building of a purifying plant for a lake. This act can enable tourism development in the area and prevent the emigration of local inhabitants. In this case, strict enforcement of the causer's principle could threaten the population policy. Shifting the costs of purifying the lake to tourism firms could threaten the further development of tourism and emigration would not be stopped. But we must be aware of the fact that without the introduction of fees (see below) in accordance with the polluter pays principle there would be no motive for cutting lake pollution.

FEES AND CONTRIBUTIONS.   Fees and contributions are, by definition, used as a replacement for the use of a state or public service. In tourism they are used in the same way as for other activities and households (for example, contributions for water usage, waste, etc.). The uniform use of fees and contributions does not allow any special ways to influence negative external effects. On the other hand, differentiation, for example of contributions dependent on the quantity of waste, would also motivate tourism firms to cut the quantity of waste.

Differentiation of parking fees through the system of parking zones of different distances from the centre of a tourism destination can significantly lift the burden off the place of tourism. High parking fees can cause a shift towards the use of environmentally friendlier public transport and thus noticeably increase the environmental quality of a tourism destination (polluted air, noise, overcrowding, etc.). Similarly, entrance fees in protected areas influence tourism demand, visitation use of the area and may be a way of collecting resources for environmental management and/or protection (see Lindberg and Huber, 1993; Lindberg, 1998).

PROHIBITIONS, LICENCES AND ENVIRONMENTAL STANDARDS.   In tourism there are some negative external effects that cannot be reduced or abolished easily if at all without administrative prohibitions or licences. This is mostly the case where the environment is a public good and property rights cannot be defined.

Administrative instruments do not present any (economic) motivation for private firms to economize on a relatively scarce good – the environment. As they are issued according to the current state of technology they do not stimulate the development of more nature-friendly technologies. Since they do not consider the costs of environmental protection, the environmental protection aims are not achieved with the minimum social costs. Their advantage lies in the fact that they work immediately and absolutely if there are sanctions against breaches. Accordingly, in many countries these are the predominant instruments in the field of environmental protection policy.

General prohibitions are usually related to harmful emissions in the atmosphere. Their purpose is to lower the emissions by a certain percentage or to even eradicate them entirely. Individual prohibitions refer to individual producers or users. One problem of the use of individual prohibitions lies in the fact that an enormous state bureaucracy is necessary and the related information problem is almost unresolvable. The same applies to licences.

Prohibitions and licences can be used in tourism. These instruments can relate to harmful emissions as well as rural protection. Accordingly, it would be reasonable to issue licences for covering ski fields with artificial snow, as known in, for example, Vorarlberger (Amt der Vorarlberger Landesregierung, 1990). In practice, prohibitions are also used to protect the cultural and social characteristics of certain places. For this reason some countries close down some areas almost entirely or partly to tourists to prevent the negative socio-cultural impacts of tourism development. Permanent licences (for instance, for pollution for a maximum period of 10 years) or the introduction of temporary prohibitions (moratoriums) are also possible.

Environmental standards also work in a similar way. Standards are legally defined regulatory instruments for limiting pollution. Several types of standards are possible: emission standards, technology standards, product or process standards, etc.

In tourism, standards for the use of space for tourism purposes are the most relevant in order to prevent congestion. One example is standards for skiing, expressed in terms of the number of skiers per hectare (Inskeep, 1991). These standards are not enforceable in court because they are not defined by law but are recommended and usually used when planning tourism capacities (see 'Carrying capacity and EIA techniques', below).

Growth

The second group of theories explaining the reasons for environmental damage and seeking out the possibilities of solving this problem are the so-called growth theories (Fig. 6.1).

Growth theories

According to the growth theories, constant economic growth and population growth are the most concrete and obvious reasons for a conflict arising between people's economic and natural environment and the indirect cause of worse living conditions on Earth.

POPULATION GROWTH/TOURISM GROWTH. The population growth theory implies that a growing population presents a burden on natural resources. Population growth causes environmental damage due to people's over-utilization of space, the building up of rural areas, which makes green areas disappear, and even climate change.

Since many believe that tourism development is one possible alternative to economic development in developing countries, there is also the so-called tourism argument for birth limitations. If developing countries specialized in the production of so-called environmental goods, this would be an additional argument for birth limitations on the population. For example, if in Africa future population growth were not so high, many natural reserves, representing tourism attractions, could be saved in the long term. A solution to the problem of population growth is thus sought for in birth limitations that should be regulated through so-called baby certificates.

The growth theory of tourism can be looked at as a version of population growth theory. Between 1950 and 1999 the total number of international travellers grew from 25 million to 664 million, corresponding to an average annual growth rate of 7% (WTO, 2000a: 1). A breakdown by country of origin shows that only rich countries participate in international tourism. 'The word is divided fairly sharply into the jet set who travel all around the world, ... and the poor set that hardly ever travel at all' (Boulding, 1985: 118). Since the right to enjoy the planet's resources is 'equally open to all the world's inhabitants' (WTO, 2000b) and not all of them travel, they are entitled to be compensated for not polluting the world's resources. On the other hand, those who do travel and thus pollute and use natural resources should pay for this, according to the polluter pays principle.

ECONOMIC GROWTH/TOURISM INDUSTRY GROWTH.   It is clear that quantitative economic growth causes environmental damage. Environmental damage is also caused by quantitative tourism industry growth. Space and the seasonal concentrations of tourism make the environmental problem even more pressing and serious. In the future, we can expect ongoing fast growth of tourism demand while strong space and seasonal concentration remain.

The question arises whether limitless development is at all possible. In the area of economic development, the idea of zero economic growth has already been replaced by the idea of quality and/or organic economic growth and, in the area of tourism, the concept of sustainable tourism has been put forward. Implementation of this concept in most cases refers to economically and environmentally acceptable tourism and, at the same time, to the satisfaction of guests and the quality of tourism products. But there remains the question of justice and the social equality of the participation of states and/or their inhabitants in incoming as well as outgoing tourism. Although equity is one of the fundamentals of sustainable development (Goodall and Stabler, 1999: 280; Wall, 2000: 567) it is often excluded from the debates on sustainable tourism.

The concentration of tourism supply and visitors is growing along with their congestion per space unit. According to the law of diminishing returns, additional quantities of variable factor input (new tourism capacities, numbers of beds and hence numbers of visitors and pollution quantity) to the fixed factor (natural attractiveness, space in the destination) influence the destination yield. The yield first increases at an increasing growth rate, and then raises ever more slowly at a falling rate, at the end reaching a peak of a maximum value. A further increase in the size of the tourism utilization of the space available results in a saturated destination, the total destination yield starts to fall, and the destination has to cut the prices and reorient itself to new market segments that are willing to market a lower quality environment at a lower price through the filtration process. The curve showing diminishing destination yield simultaneously presents the diminishing tourist satisfaction.

Instruments for regulating tourism growth

Instruments deriving from growth theories primarily have an impact of reducing or regulating tourism growth. Quality limitation through various types of certificates is of the greatest significance and market-based certificates or tradable permits are a guarantee that the price for the use of the environment for tourism purposes is set on the basis of demand and supply. At the same time, growth and the concentration of tourism can be limited by administrative instruments which are briefly introduced below.

CERTIFICATES.   A solution that involves certificates is quantity based and gives the owner of certificates a right to a certain quantity of pollution and/or use of the environment.
    In tourism we distinguish between:

- pollution certificates;
- certificates for use of the environment for tourism purposes;
- tourist certificates.

Pollution certificates. Pollution certificates involve reducing or eradicating all emissions of harmful or non-degradable material in the environment. In this case, emissions are measurable and there are also standards for permissible upper limits of pollution caused by individual pollutants. The most widely known practice model of tradable permit systems involves sulphur dioxide (SO_2) emissions in the USA. Within the EU, a carbon dioxide (CO_2) emissions trading programme has been suggested (IPE, 1998). The same system could also be implemented for noise or CO_2 emissions from tourist aeroplanes or tourist traffic.

Certificates for use of the environment for tourism purposes. The situation is different with certificates issued for use of the environment for certain tourism purposes. Such a model has been developed in Austria for restricting the expansion of ski resorts. Socher (1990) suggests the employment of certificates for the right to use the land for ski slopes. New ski-lift developers would have to buy permits. The quantity of certificates for additional ski resorts, which would be fixed by government, would be a mechanism by which the expansion of ski resorts was controlled. With just one fixed granting of certificates for land use for ski-lifts, year by year their price will grow with new investments. It is hard to think of a technological solution that would lead to a reduced need for such certificates due to reduced use of space. As seen in practice, the use of the environment in tourism is irreversible. Used certificates can no longer appear on the market. Because of the impact on costs and the quantity limited tourism supply, the price of this tourist activity would rise. This could result in, for example, the substitution of alpine skiing by cross-country skiing, which is more environmentally friendly.

A similar effect could also be achieved through other instruments like taxes. An advantage of certificates over taxes lies in the fact that they fix the planned size of use of the environment for new tourist attachments, while an exact response cannot be anticipated with taxes. New developments could also be limited with prohibitions and/or licences or environmental plans. Still, certificates seem to be better because they are market-oriented and their price is formed as a response to (a limited) quantity. They indicate how much the society values environmental preservation.

Tourist certificates. In the section on 'Population growth/tourism growth' (p. 109). we mentioned 'baby certificates' as a way of addressing the questions of population growth and social equity. The principle of justice would guarantee to all the right to have a child. This right could be expressed in the form of a baby certificate. To slow down population expansion, the total number of births per year could be allocated by an international organization to each state and their citizens. When parents wish to have more children, they would have to obtain additional certificates from parents with no children. The latter would get financial compensation for not having children. The initial model failed because the critics thought it was not ethical to regulate the number of births through the imposition of market forces (Frey, 1985: 101).

The idea of tourist certificates is close to the certificate model that addresses two questions: those of tourism growth and equity in travel (Mihalič, 1999). Every person would get a certain number of tourist certificates, and those consuming a larger amount of (international) travel would have to buy additional certificates on the market. On the other hand, those satisfied with lower degrees of tourist participation could sell their certificates. The total number of certificates should be determined on the basis of the desired growth rate in international tourism within a given year. Certificates could be distributed among countries according to the number of their citizens and could be tradable on the national and international markets. Introduction of financial compensation for those not travelling would bring money to the developing countries that could be used for further development programmes. At the same time, travel certificates would limit unsustainable tourism growth.

There are, however, some implementation problems (Mihalič, 1999: 130). The main problem relates to the unit of measure of certificates, which is similar to the problem of the tax base and could be an overnight stay or an activity. Distribution of certificates among countries and the ethical issue of charging for the 'freedom to travel' are also problematic.

CARRYING CAPACITY AND EIA TECHNIQUES.   Carrying capacity is a basic technique used in tourism planning to determine the upper limits of development and visitor use and the optimum utilization of tourism resources (Inskeep, 1991: 144). The law of diminishing destination yields explained above implies the setting of development limits. Limits refer to different kinds of capacities: physical, environmental, economic, social, cultural, etc.

Something similar can be applied to the Environmental Impact Assessment technique. The EIA procedure assesses the impact of proposed development projects on the society, economy and natural environment and is often required for administrative approval of a project. The EIA procedure is a very useful technique to ensure that the environmental impacts of proposed projects have been taken into consideration and preventive actions taken (Inskeep, 1991: 352).

Both techniques are widely used in tourism. They prevent certain environmental damage types, but fall into the category of administrative instruments and do not have a market effect on the more efficient allocation of environmental goods.

INSTRUMENTS FOR MAKING TOURIST TRAFFIC LESS SEASONAL.   While the concept of capacity planning refers to the environmental aspect of tourist concentration, by limiting the seasonality of tourism traffic we are trying to resolve or at least reduce the problems caused by the seasonal concentrations of tourism. The final aim of this kind of instrument is to prolong the tourism season and/or more regular and constant distribution of tourism demand throughout the whole year. This can be achieved by different planning of school holidays and paid leave, appropriate advertising activities and a policy of lower (subsidized) out-of-season prices and special travel arrangements in the low season.

Behaviour

The last group of environmental theories shown in Fig. 6.1 are behavioural theories. The theory of environmental ethics absence tries to explain the reasons for environmental damage from a philosophical aspect. From a historical point of view, it has its origins in Aristotle's practical philosophy (De Haas, 1989: 266), which implies the equality of three areas: politics, economics and ethics. As economics excluded itself from practical philosophy as an independent science, it developed a way of thinking based on rationality and considering only economic values. For this kind of reasoning ethical demands for justice, humanity and ecology are irrational. On this basis a theory was developed claiming that the absence of so-called social environmental ethics has caused the present negative attitude towards man's natural environment (Frey, 1985: 38).

Behavioural theories

Environmental behavioural theory explains the existence of environmental damage:

- through the absence of environmental social ethics;
- as a product of human ignorance.

ABSENCE OF SOCIAL/TOURISM ENVIRONMENTAL ETHICS.   According to the first explanation, the absence of environmental social ethics is the main reason for environmental degradation and damage.   The term environmental ethics refers to the 'standards and principles regulating the behaviour of individuals or groups of individuals' (Rue and Byars, 1986: 71) in relation to their environment. In general, ethics deals with questions such as 'what is right and what is wrong', and with moral obligations. In theory, it is assumed that humans possess environmental awareness and environmental ethics and will react in an environmentally friendly way if appropriate environmental information and know-how are available. According to some authors, environmental awareness includes the intention to act in an environmentally friendly way (Müller and Flügel, 1999: 53). A gap occurs because intentions are not necessarily transferred into actual behaviour.

In general we can draw a distinction between business ethics (the supply side) and consumer ethics (the demand side). A distinction on the demand and supply sides in tourism is also possible. Tourism ethics on the tourist side determine environmental principles regulating the behaviour of tourists, while ethical principles on the supply side regulate the attitude to the environment from the side of the state, destination or tourism firm.

ENVIRONMENTAL IGNORANCE.   The second variation of environmental behaviour theory involves human ignorance due to insufficient environmental research, education and information. The theory here says that environmental disasters occur over a long period. A direct link with specific actions is invisible, therefore a lack of understanding and information are the real reasons why disasters arise. If mankind had sufficient information about the consequences of its actions such disasters would not happen. In order to prevent manipulation by interested parties, research in this area must be intensified and the resulting information made public and easily accessible.

Although the said theory is specially treated in the economic literature (Frey, 1985: 39), it is quite justifiably criticized for being inappropriate. There is no doubt that sufficient information on environmental damage, together with knowledge about environmental behaviour is necessary, yet this is not the only condition needed to prevent damage. Prevention also depends on factors like the above-mentioned environmental ethics. Illustrated in the case of the tourism industry, we doubt that a seaside hotel owner would invest in an (expensive) sewage purifying plant for ethical reasons only. Thus, environmental information – in this case information on the absence of a purifying plant and information on the poor quality of bathing water – should be available to the public. It would create public opinion on the inappropriate behaviour of the hotel owner and a push for appropriate environmental behaviour. At the same time the information on the poor quality of bathing water would have a market effect: it would decrease tourism demand (and prices) and thus market mechanisms would push for appropriate environmental behaviour.

The two variations of the behavioural theory discussed complement each other. Environmental ethics can only be developed on the assumption that the reasons for environmental damage and methods (know-how) for improving and preserving the environment are known. Otherwise, knowledge about environmental disasters itself does not guarantee that behaviour regarding the environment will be friendlier.

Behaviour-based instruments

Instruments derived from behavioural theory assume that consumers are environmentally conscious and prefer environmentally friendlier products. In response to increased demand for environmentally friendlier products, demand for those products that mean a greater burden on the environment falls. Accordingly, through the market mechanism the production structure changes and the environmental burden thus decreases. In tourism it is also assumed that tourism demand is environmental quality sensitive and this makes tourism stakeholders behave towards the environment in a more friendly way and pay attention to the environmental quality of the destination.

ENVIRONMENTAL LABELLING. Environmental or eco-labelling for industrial products is well known and widely used in today's world. Eco-labelled industrial products communicate the message: 'lowered (negative) environmental impacts'. Eco-labels are awarded to products environmentally less harmful in comparison with other products from the same product group (Council Regulation, 1992).

Tourism products differ from industrial products. From the customer viewpoint, the quality of the natural, social and cultural environments forms part of the tourism product. Thus, for tourism customers it is not only the impact minimization but also the environmental quality of the destination that is the issue. In one study (Lübbert, 1998: 28), when asked to evaluate the importance of different labelling criteria, German travellers gave 60% to the 'environmental quality' criterion (poor water, clear air) and 26% to the 'lowering negative impacts' criterion (waste minimization and sorting, water and energy saving programmes, purifying plants, etc.).

Since environmental quality is the greatest concern of a tourism customer, the eco-label notion in tourism is often incorrectly restricted to the ecological quality of the tourism destination such as the cleanliness of bathing water, instead of negative impacts. In tourism, we must observe both aspects and so the term 'ecological labelling' has been introduced. The term encompasses both eco-labels as traditionally defined for industrial products the labels of the environmental quality of tourism places. Therefore, it is necessary to distinguish between:

- the environmental or eco-labels which refer to the impact of tourism products or tourism on the environment (as in the case of the EU's eco-labels for industrial products); and
- the environmental quality or eco-quality labels (labels of environmental quality) that refer to the tourism product's environmental attributes, e.g. to the environmental quality of the tourism destination.

Combined labels that simultaneously refer to the impact of the tourism product on the environment and to the environmental quality of the tourism product/destination are also possible. At the same time, many quasi tourism eco-labels can be found in the tourism market.

Eco-labels.   The eco-label in tourism identifies the (reduced) negative physical, visual, cultural and social influences of tourism. Very often eco labels refer only to some of the influences mentioned, most commonly to the influence on the natural environment.

We agree that eco-labels in tourism have a noticeable effect on tourism demand. The environmentally responsible tourist is clearly willing to buy eco-labelled tourism products in order to contribute to environmental protection.

Environmental quality labels.   The label of environmental quality refers to the degree of existing environmental (non-) degradation of a tourism destination, irrespective of the cause.

From the point of view of the tourism destination, the two kinds of environmental labelling are co-dependent. On the one hand, lowering the negative impacts of tourism preserves the environmental quality of the destination, yet, on the other hand, preserving environmental quality requires reducing the negative impacts of tourism activities at the destination. At the same time, from the standpoint of the consumer, there is an essential difference between the two. The environmentally responsible tourist would find information on environmental impacts essential to his or her choice of tourism package, hotel or carrier.

However, since we already know that destination choice is influenced by environmental attractiveness (e.g. the quality) of the destination in the first place (Tschurtschenthaler, 1986), the mere offer of low-impact tourism products is not sufficient. Customers look for eco-quality labels in the first place. But it is reasonable to believe that the impact minimization message given by the eco-label communicates an induced message: if the destination's product and organizations are environmentally responsible then the destination must also be environmentally responsible (induced message 1). Further, an environmentally responsible destination takes care of the environment and is environmentally sound (induced message 2). Since many potential customers are not sufficiently informed on how to distinguish between both aspects, eco-labels in tourism may have a similar market effect as eco-quality labels. Transmission of the wrong messages is also caused by the flood of (not necessarily authorized) environmental logos, the complexity and diversity of criteria and the lack of information on eco-labelling.

There are many signs and labels meeting the standards for eco- and eco-quality labels (Hamele, 1996; Viegas, 1998). Examples are the German Blue Angel and the European eco-label, the Blue Flag for beaches and marinas, and the Green Globe certification programme etc. (UNEP, 1996; Viegas, 1998; Green Globe, 2000).

Quasi eco-labels.   Quasi ecological labelling refers to those forms of environmental labelling that cannot be strictly called eco-labels or environmental quality labels because the criteria (for criteria see Mihalič, 2000a: 73) or proceedings for eco-labelling are not fulfilled. Many eco-logos are awarded only to the stakeholders within a local community, region or only to the awarding association's members. Very often the accreditation body is a tourism association or somebody from the tourism business, which raises the question of credibility. Such eco-logos that are not based on pre-determined expert criteria, where criteria fulfilment is not necessarily controlled and the awarding body is perhaps one-sided, fall into the category of quasi eco-labelling.

Another example of quasi labelling found in the tourism market involves eco-denominations for tourism such as green, ecological or eco-, natural, romantic, alternative, human or soft tourism. Nevertheless, for objective eco-labelling the difference between a self-appointed and an externally awarded eco-logo is crucial. Tourism companies often use the above-listed denominations on their own initiative and without any outside validation or control.

OTHER INSTRUMENTS RELATING TO ENVIRONMENTAL ETHICS. Increased demand for environmentally friendlier products can also be influenced by the environmental information on product declarations, published information provided by independent institutions on the basis of ecological product testing, and other information and advertising activities. It would be reasonable to also use the instruments mentioned in tourism. For instance, we can imagine the obligation of a tourism supplier to advise on conditions of the environment in its catalogue. Consumers' associations and some other organizations already issue maps showing the pollution rates of some parts of the sea.

Although environmental codes of ethics (Dowling, 2000: 86) are not real instruments of environmental policy they should be mentioned. These kinds of codes and declarations can have an important role in designing environmental ethics for the tourism industry, host communities, tourist countries and various associations and can offer know-how for the environmental treatment of concrete examples. Examples are WWF/Tourism Concern Principles for Sustainable Tourism, The Himalayan Tourist Code (UNEP, 1995) and Global Code of Ethics for Tourism (WTO, 2000b).

Conclusion

General theories on the creation, elimination and prevention of environmental damage as developed in the theory of environmental policy (behavioural theory, growth and system theory) can be applied in the field of tourism with only small modifications. Market, fiscal and administrative instruments as derived from the above-mentioned theories can be accommodated to be used in tourism in order to prevent or minimize environmental damage. They usually prevent and eliminate damage in the natural environment, only a few of them are appropriate to be used to protect the social or cultural environment, too.

Exclusively economic debate is rarely found in the literature on ecological, environmental or sustainable tourism. Most existing works broach the ecological issue primarily from the sociological point of view. Nevertheless, economic instruments of environmental policy or market-oriented instruments have a real chance of preventing or minimizing environmental destruction in tourism.

References

Amt der Vorarlberger Landesregierung (1990) *Richtlinien für Beschneiungsanalgen.* Amt der Vorarlberger Landesregierung, Bregenz, Austria, 11pp.

Boulding, K.E. (1985) *Human Betterment.* Sage, London, 224pp.

Council Regulation (EEC) No 880/92 of 23 March 1992 on a Community eco-label award scheme (1992) *Official Journal of the European Communities* 11.4.1992, 7pp.

Dowling, R.K. (2000) Code of ethics, environmental. In: Jafari, J. (ed.) *Encyclopedia of Tourism.* Routledge, London.

Frey, B.S. (1985) *Umweltoekonomie.* V&R, Göttingen, Germany, 164pp.

Goodall, B. and Stabler, M.J. (1999) Principles influencing the determination of environmental standards for sustainable tourism. In: Stabler, M.J. (ed.) *Tourism Sustainability. Principles to Practice.* CAB International, Wallingford, UK, pp. 1–25.

Green Globe (2000) *Welcome to Green Globe 21.* Green Globe, www.greenglobe.org

Hamele, H. (1996) *The Book of Environmental Seals & Ecolabels. Environmental Awards in Tourism. An International Overview of Current Developments.* Federal Ministry for Environment, Nature and Nuclear Safety, Berlin, 36pp.

Hanley, N., Shogren, J.F. and White, B. (1997) *Environmental Economics in Theory and Practice.* Macmillan Press, London, 464pp.

Hjalte, K., Lidgren, K. and Stahl, I. (1977) *Environmental Policy and Welfare Economics.* Cambridge University Press, Cambridge, 111pp.

De Haas, J.P. (1989) *Management-Philosophie im Spannungsfield zwischen Ökologie und Ökonomie.* Verlag Josef Eul, Köln, Germany, 368pp.

Inskeep, E. (1991) *Tourism Planning: an Integrated and Sustainable Development Approach.* Van Nostrand Reinhold, New York, 508pp.

IPE (International Petroleum Exchange) (1998) *A Proposal to Reduce CO_2 Emissions in the European Union through the Introduction of an Emissions Trading Program.* IPE, London, 16pp.

Leipert, C. (1989) *Die Aufnahme der Umweltproblematik in der Ökonomischen Theorie.* Sozial-ökologische Arbeitspapiere, IKO Verlag, Frankfurt am Main, 70pp.

Lesser, J.A., Dodds, D.E. and Zerbe, R.O., Jr. (1997) *Environmental Economics and Policy.* Addison Wesley, Reading.

Lindberg, G. (1998) Economic aspects of ecotourism. In: Lindberg, G., Wood, M.E. and Engeldrum, D. (eds) *Ecotourism. A Guide for Planners and Managers,* 2nd edn. The Ecotourism Society, North Bennington, Vermont, pp. 87–117.

Lindberg G. and Huber, R.M., Jr (1993) Economic issues in ecotourism management. In: Lindberg, G. and Hawkins, D.E. (eds) *Ecotourism. A Guide for Planners and Managers.* The Ecotourism Society, North Bennington, Vermont, pp. 82–115.

Lübbert, C. (1998) Umweltkennzeichnungen für touristische Angebote: Einstellungen deutscher Urlauber – Ergebnisse einer Pilotstudie. In: Lübbert, C., Feige, M. and Möller, A. (eds) *Fachtagung 'Umweltkennzeichnungen im Tourismus' am 29. Oktober 1998 an der Ludwig-Maximilians-Universität München (LMU).* DWIF, Deutsches Wirtschaftswissenschaftliches Institut für Fremdenverkehr e.V. an der Universität München, München, pp. 22–31.

Mihalič, T. (1999) Equity in outgoing tourism through tourist certificates. *International Journal of Contemporary Hospitality Management* 2/3, 128–131.

Mihalič, T. (2000a) Environmental management of tourist destinations. A factor of tourism competitiveness. *Tourism Management* 1, 65–78.

Mihalič, T. (2000b) Increasing tourism competitiveness through granting concessions to natural goods in transition countries. In: Robinson, M., Swarbroke, J., Evans, N., Long, P. and Sharpley, R. (eds) *Environmental Management and Pathways to Sustainable Tourism. Reflections on International Tourism.* Business Education Publishers, Sunderland, UK, pp. 133–152.

Mihalič, T. and Kaspar, C. (1996) *Umweltökonomie im Tourismus.* Paul Haupt, Bern.

Müller, H. and Flügel, M. (1999) *Tourismus und Ökologie.* Forschungsinstitut für Freizeit und Tourismus (FIF) an der Universität Bern, Bern, 310pp.

De Rivera Icaza, D.D. (1990) Report drawn up on behalf of the Committee on the Environment, Public Health and Consumer Protection on the measures needed to protect the environment from the potential damage caused by mass tourism, as part of the European Year of Tourism. PE DOC A 3–120/90. European Parliament, Luxembourg. 15pp.

Rue, L.W. and Byars, L.L. (1986) *Management: Theory and Application,* 4th edn. Irwin, Homewood, Alabama, 649pp.

Socher, K. (1990) Developing new instruments for restricting winter tourism in the Alps. In: *1990 – Year of the 40th AIEST-Congress and of the 25th Anniversary of the Tourist Research Centre TRC.* AIEST St Gallen, Switzerland, pp. 93–98.

Stabler, M.J. (1997) An overview of the sustainable tourism debate and the scope and content of the book. In: Stabler, M.J. (ed.) *Tourism Sustainability. Principles to Practice.* CAB International, Wallingford, UK, pp. 1–25.

Tschurtschenthaller, P. (1986) *Das Landschaftsproblem im Fremdenverkehr dargestellt anhand der Situation des Alpenraums.* Paul Haupt, Bern, 380pp.

Tschurtschenthaller, P., Socher, K. and Lukesch, D. (1981) Die Berücksichtigung externer Effekte in der Fremdenverkehrswirtschaft. *Jahbuch für Fremdenverkehr* 28/29, 93–135.

UNEP IE (United Nations Environmental Programme Industry and Environment (1995) *Environmental Codes of Conduct for Tourism.* UNEP IE, Paris, 69pp.

UNEP (United Nations Environmental Programme (1996) *Awards for Improving the Coastal Environment. The Example of the Blue Flag.* UNEP, Paris, 50pp.

Viegas, A. (1998) *Ökomanagement im Tourismus.* R. Oldengourg Verlag München, Wien, Austria, 193pp.

Wall, G. (2000) Sustainable development. In: Jafari, J. (ed.) *Encyclopedia of Tourism.* Routledge, London.

WTO (World Tourism Organisation) (2000a) *Global Code of Ethics for Tourism.* www.world-tourism.org/pressrel/CODEOFE.html

WTO (World Tourism Organisation) (2000b) Tourism Highlights 2000, 2nd edn. WTO, Madrid.

Local Government, World Heritage and Ecotourism: Policy and Strategy in Australia's Tropical Rainforests

7

Dianne Dredge[1] and Jeff Humphreys[2]

[1]*School of Environmental Planning, Griffith University, Nathan, QLD 4111, Australia;* [2]*Humphreys Reynolds Perkins Planning and Environment Consultants, Level 20, 344 Queen Street, Brisbane, QLD 4000, Australia*

Introduction

Across the Western capitalist world, local government is experimenting with new roles, responsibilities, structures and practices (e.g. Healey, 1997; Marshall, 1997). Local government is shedding its traditional emphasis as a provider and administrator of local services and is becoming an important strategic partner in sustainable development and economic reform processes (e.g. Clarke and Stewart, 1993; Mayer, 1995). In this context, tourism is emerging as an important conduit for local economic development, and is of particular interest to rural local governments where the declining importance of agriculture and out-migration are destabilizing local economies (e.g. Wahab and Pigram, 1997; Butler *et al.*, 1998). However, rural local governments, often less well resourced than their urban counterparts, are frequently faced with vexed policy problems where tourism growth occurs in sensitive natural environments. On the one hand, tourism offers opportunities to sustain and even enhance local economic activity. On the other, significant social and environmental impacts can be difficult to manage with limited resources. The policy approach adopted by a particular local government depends on the interplay of interests and the flow of inter-agency relations and resources occurring within complex institutional environments. It is the exploration of these aspects in the local government policy making framework that is the focus of this chapter.

Douglas Shire, located in North Queensland, Australia, is one rural local government that has become a 'hot spot' for tourism planning and policy making (Fig. 7.1). Approximately 80% of the Shire is located in the Wet Tropics World Heritage Area (WTWHA), which is marketed internationally as 'Australia's Tropical Rainforests'. Characterized by environmentally significant lowland rainforests, fringing reefs and a spectacular mountain backdrop, Douglas Shire has become an important international ecotourism destination. However, the Shire has had to confront some vexed issues and has had to make some difficult policy decisions. This chapter examines the range of factors that have influenced the role and nature of Douglas Shire's involvement in tourism policy making, and, in particular, the Shire's role in ecotourism management north of the Daintree River ('the Daintree'). It discusses the range of environmental, social and economic pressures arising from ecotourism and the policy responses of one particular local government. In addressing these issues, this chapter goes beyond description of the particular policy approach, to demonstrate the complexity of local government's role in ecotourism management, especially where overlapping jurisdictions and responsibilities give rise to complex policy making environments.

World Heritage, Land Tenure and Ecotourism

World Heritage Areas (WHAs) are cultural and natural heritage sites of universal value that have been inscribed on the Register of the World Heritage under the UNESCO Convention for the Protection of World Cultural and Natural Heritage (1972). The relationship between World Heritage Listing and growth in visitation is difficult to establish since growth can be attributed to a range of factors, not just listing (Drost, 1996). The listing process does, however, highlight the outstanding values of a site and the symbolism associated with elevating a site's natural, ecological, scenic and cultural importance can attract global attention (e.g. Shackley, 1998; Haigh, 2000). As a result, the tourist gaze can focus on listed sites, and potential tourists begin to construct and attach additional meanings to these locations. Over time, increased visitation is a natural consequence, although Shackley (1998) observes that rates of growth can also be influenced by a range of other factors including how the site is managed and marketed.

Far from being an honours list, inscription of a site on the World Heritage List implies a significant international obligation for those governments signing the Convention (see Haigh, 2000). Signatories have an international obligation to identify, conserve and present sig-

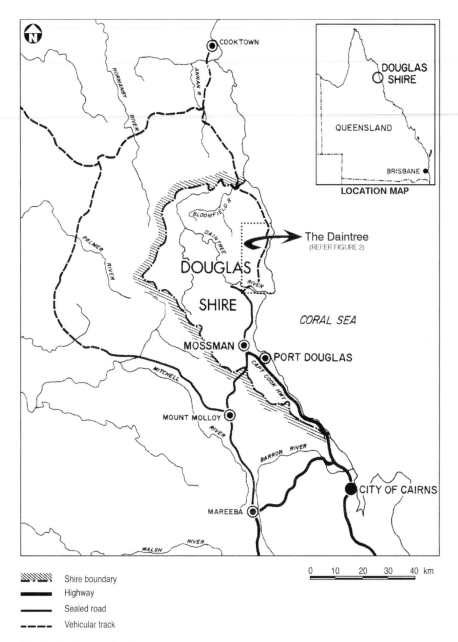

Shire boundary
Highway
Sealed road
Vehicular track

Fig. 7.1. The regional context.

nificant sites, and to transmit these sites to future generations for the purposes of education and enjoyment. Member countries are required to set up and implement a management framework, and periodic monitoring is conducted by the World Conservation Monitoring Centre to ensure that the site is adequately managed (World Conservation Monitoring Centre, 1995). The signing of the international convention also has a range of policy consequences that filter down to agencies at different levels of government, albeit indirectly. International obligations are not easily translated into policy and difficulties are compounded where more than one agency, and more than one level of government, is involved. Put simply, many policy sectors are interrelated and the actions of one policy agent can have intended and unintended impacts on a range of other agents (e.g. Davis, 1989).

The World Heritage Committee, which oversees the Convention, requires boundaries to be precisely delineated but does not have the authority to prescribe them. Nor can the Committee prescribe a management framework. The identification of a site's boundaries and the design and implementation of a management framework are domestic matters. The site's boundaries are usually based on scientific advice regarding the site's bioregional integrity and can include lands in a range of tenures, not just lands in public ownership. Models of natural protected areas that include a mosaic of land tenures have an important advantage in that they allow the impact of tourism to be carried by different tenures whereby pressures on national parks and reserves can be reduced (Sharpley and Sharpley, 1997; Wahab and Pigram, 1997; Figgis, 2000). This line of reasoning assumes that the land of highest environmental value is situated within national parks. However, this is not always the case.

As will be seen in this case study, sometimes land of extreme environmental significance can be excluded from protected area designations because the political, social and economic costs of its inclusion are deemed by decision makers to be undesirable. Inclusion of private lands in protected area designations can give rise to a range of political, legal, social, economic and administrative difficulties that make the adoption of multi-tenured models less attractive on a political level. For example, legal difficulties may arise where existing use rights have been limited by the listing. The vexed issue of compensation inevitably arises. Moreover, different tenures may require the involvement of an increased number of management agencies, with implications for the development of an effective and efficient integrated management network. As a result, land in freehold tenure can be excluded on the basis of legal, political or administrative difficulties despite the fact that it may be environmentally significant. The following case study illustrates that, where land of environmental significance in freehold title is excluded from WHAs, considerable responsibility for the planning and management of ecotourism can fall to local government. How local government deals with this responsibility is derived not only from the local political discourse, but also from the complex policy making environment which spans different levels of government.

Institutional Context

In 1974, the Australian government signed the Convention for the Protection of World Cultural and Natural Heritage, thus accepting an international obligation to identify, protect, present and put into place a management framework for sites of universal value. However, under the Australian constitution, the States have traditionally had control over resource exploitation and management and environmental protection. The signing of the Convention by the Commonwealth government represented an incursion into these State responsibilities. Two States, Tasmania and Queensland, resented the Commonwealth's move into policy areas traditionally considered sovereign, and a number of bitter and protracted legal battles ensued (e.g. see Davis, 1989; Christie, 1990; Hall, 1992; Lane, 1997).

The WTWHA was inscribed on the World Heritage Register in 1988, following unilateral action by the Commonwealth government (e.g. see Davis, 1989; Richardson, 1990; Hall, 1992). Given the difficult, even hostile, intergovernmental relations that existed at the time, the Commonwealth government was unable to obtain a comprehensive list of properties from the State. This was because the division of powers in the Australian federal system is such that the States have responsibility for the registration of property titles, and the State's refusal to cooperate meant that a comprehensive list of properties could not be prepared. The listing proceeded based on the best available information, but resulted in the inclusion of some freehold land and the exclusion of other parcels. Significantly, because of this WHA listing process, many freehold properties of high ecological significance in the Daintree were not included in the WTWHA (Brannock Humphreys, 1994; Rainforest CRC, 2000). In the Daintree, at least 40 rare or endangered species of flora and fauna are estimated to occur outside the WHA on freehold allotments, and four species occur only in these areas (Humphreys, 1994a, b). Some relatively intact forests on private land have outstanding conservation value, including rare and/or threatened species, narrow endemics, species with widely disjunctive ranges and poorly known and/or unidentified species (Rainforest CRC, 2000). Responsibility for the environmental management and planning of these areas falls predominantly within the administrative domain of local government.

After inscription, the Queensland State government continued to pursue legal action claiming that the Commonwealth had acted unconstitutionally. However, after a change in government from a National/Conservative government to a Labor government in 1989, this action was withdrawn. Following this, a new phase of cooperation emerged, albeit reserved. In 1990, State and federal governments agreed to jointly manage the area, but it was not until 1993 that the

Queensland government passed legislation to set up a management framework for the WTWHA (Queensland State Government, 1993). The following year the Commonwealth also enacted complementary legislation for the purpose of developing a cooperative management arrangement (Commonwealth Government of Australia, 1994). The legislation led to the creation of a statutory corporation, the Wet Tropics Management Authority (WTMA), which is vested with certain powers to prepare management plans, enter into Cooperative Management Agreements with landowners and to make regulations. However, these arrangements only apply to land within the WTWHA.

The exclusion of environmentally significant land from the WTWHA resulted in a situation where considerable opportunity for ecotourism development lay outside the WHA boundaries, and outside the direct management responsibilities of the WTWHA. As a result, in this instance, local government (which in Australia operates under delegation from State powers) has had the opportunity to play an important role in ecotourism management. Of particular relevance, local government has a legislated responsibility to undertake land use planning, infrastructure provision and servicing and environmental management activities, and through the exercise of these powers has been able to influence the development of tourism infrastructure (e.g. accommodation, attractions and services), the nature and location of tourist activity (e.g. by manipulating access and urban infrastructure availability for tourism development), and the presentation of the WTWHA (through infrastructure development and land use controls in the adjacent areas). Douglas Shire is one local government that has attempted to address tourism proactively and deal with the range of vexed policy issues that has emerged.

Ecotourism Growth and Emerging Pressures

The Daintree–Cape Tribulation area is a small section of the much larger WTWHA (Fig. 7.2). The total WHA covers approximately 900,000 ha including a range of vegetation types and geological units and extends for more than 450 km along the Queensland coast. It straddles three distinct geomorphic regions (i.e. tablelands, lower coastal belt and intermediate escarpment) and is home to diverse vegetation types and avifaunal communities (World Conservation Monitoring Centre, 1995). The area contains more than 3000 plant species, 700 of which are endemic to the region and 390 are classified as rare or threatened. The avifauna of the rainforests is regarded as the most diverse in Australia, containing 30% of Australian marsupial species, 50% of Australian bird species, 60% of Australian butterfly species, 26% of Australian frog species and 17% of Australian reptile species. Over 370

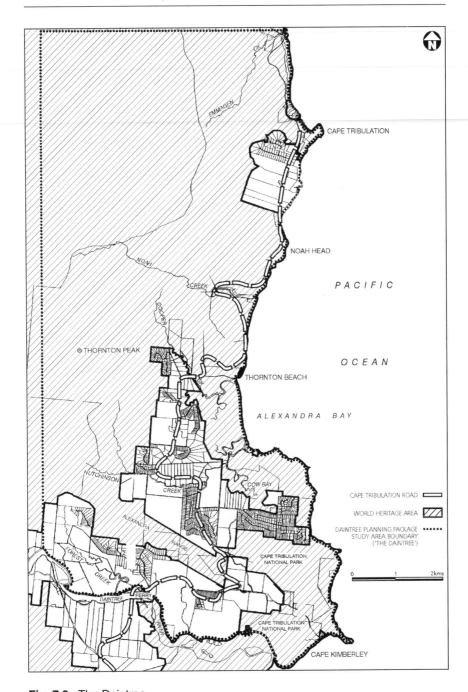

Fig. 7.2. The Daintree.

rare or threatened avifaunal species are recorded, and the rainforests are refuges for many species of flora and fauna regarded as relics of ancient times (Wet Tropics Management Authority, 1995). There are also many features of outstanding scenic beauty including lowland coastal rainforests interfaced with extensively developed fringing reefs, an association that is unrecorded elsewhere in the world (World Conservation Monitoring Centre, 1995). Aboriginal occupation of the area is thought to date back 40,000 years, possibly making the current indigenous occupants the oldest rainforest culture in the world (World Conservation Monitoring Centre, 1995).

The Daintree Section is one of the most diverse and ecologically sensitive sections of the WTWHA. It comprises a small coastal strip extending 37 km from the Daintree River in the south to Cape Tribulation in the north, and is located 100 km north of the international gateway of Cairns. Access to the area is limited to a Council-managed vehicular ferry at the Daintree River (Fig. 7.2). Despite this limited access, the Daintree has become an important focus of tourism activity within the entire WHA (Brannock Humphreys, 1994; Jenkins and McArthur, 1996; Rainforest CRC, 2000). Estimates reveal that total visitation (day visitors and overnight visitors) in 1986 was less than 100,000 persons. Over the period 1991–1999, visitation increased from 223,000 to 426,000 persons per annum, representing an average annual growth of 8%. In 1999, tourism was estimated to inject between Aus$80 and 100 million into the local economy (Rainforest CRC, 2000).

This growth in tourism activity can be attributed to the confluence of a number of domestic and international factors. On the domestic level, political conflict and heightened media coverage associated with the construction of the unsealed coastal road between Cape Tribulation and Bloomfield River during the 1980s, and the hostile relations between the Commonwealth and Queensland State governments during the WHA listing process, have stimulated public interest in the natural values of the area (Jenkins and McArthur, 1996). In addition, proximity to the Cairns International Airport (opened 1984), and development of mass tourism infrastructure at Cairns and Port Douglas, within day-tripping distance of the Daintree, have increased the area's accessibility to a larger pool of tourists. At an international level, the worldwide growth in nature-based tourism, and increasing consumer awareness of environmental issues, has encouraged visitation. Interest has also been stimulated by aggressive marketing activities conducted by regional and State tourism agencies seeking to exploit the area's unique location between two WHA's, the Wet Tropics and the Great Barrier Reef. A list of international celebrities visiting the Shire, celebrity property purchases and the use of the area as a film set have all attracted attention (MacDermott, 1999). In addition, Jenkins and McArthur (1996) suggest that geographically imprecise marketing of the area as the 'Daintree World Heritage Rainforests' may have also contributed to concentrated levels of visitation in this small section of the much larger WHA.

While these factors have served to focus the tourist gaze on the Daintree, it is somewhat ironic that much of the ecotourism experience is associated with lands that are outside the WTWHA boundary. In particular, the road between the Daintree River ferry crossing and Cape Tribulation travels for the most part through privately owned land outside the WTWHA (Fig. 7.2). This road is the spine that provides access to all of the visitor sites in the Daintree, within and outside the WTWHA. In so far as visitor impressions of the Daintree are determined from the road, the management of these freehold areas adjoining the main access route is of critical concern, and is the primary responsibility of the local government (Humphreys, 1996).

The first major study of tourism planning and management issues in the Daintree commenced as part of Douglas Shire Council's town plan review in 1992 (Brannock Humphreys, 1994). Substantial development pressures requiring policy attention were identified in this review process and have been consistently reaffirmed in later studies (e.g. Brannock Humphreys, 1994; Beeton and Bell, 1998; Douglas Shire Council, 1998; Rainforest CRC, 2000). These include the following:

1. Environmental values in freehold properties were threatened by inadequate controls over occupation, development and land use. Weed infestation and feral domestic animals have also emerged as significant problems.

2. Uncontrolled growth in visitation had the potential to damage the environment and erode the quality of the visitor experience.

3. The only road into the area was unsealed, and on a tortuous horizontal and vertical alignment. In dry weather the road was very dusty, impacting on environmental and experiential values, and in wet weather, it was hazardous. Increasing visitor traffic made the road even more dangerous.

4. There were social and environmental concerns over the impact of further accommodation development on visitor wilderness experiences. That is, increased spatial and temporal presence of visitors could have negative consequences for wilderness experiences.

5. There was pressure to provide more tourist accommodation in the area, but basic infrastructure and servicing was lacking. Since significant local government financial commitment was required to improve levels of servicing, a carefully conceived strategic approach to infrastructure provision was required.

6. One thousand freehold allotments had been subdivided in the Daintree in the 1980s, and were being progressively settled. Community issues were emerging with respect to the lack of local employment and community facilities, environmental destruction associated with settlement, absence of reticulated power and demands for its installation. Put simply, population growth resulting from historical subdivision was beginning to create development pressures that had the potential to perpetuate the urban growth cycle.

7. There were concerns over the competition between residents and tourists for access to the ferry service across the Daintree River. This ferry service provided the only means of vehicular access to the area north of the Daintree River and, while it was at times a bottleneck, it was also the mechanism by which visitor flows to the area could be regulated.

8. In some aspects, ecotourism conflicted with aboriginal interests. Sites of potential interest to visitors were, in some cases, culturally inappropriate or unacceptable to the local Aboriginal population. The Wujal Wujal aboriginal community on the Bloomfield River in the north of the area requested improved vehicular access for social and economic reasons. However, better access for conventional vehicles conflicted with ecotourism management strategies aimed at limiting access to more remote parts of the Daintree. Such strategies were favoured as a means of reducing environmental impacts of ecotourism on some remote and pristine sections of the area.

These concerns present significant challenges for environmental conservation and ecotourism management and have been the subject of ongoing policy development by Douglas Shire Council.

Politics and Ecotourism Policy

Policy and politics are inextricably linked (e.g. Davis *et al.*, 1993; Boehmer-Christiansen, 1994). The content and direction of policy is as much a result of the local political discourse as it is a product of the particular world-views of the tourism planners and resource managers involved in its production. A mix of social, cultural, economic and environmental factors occurring over time and at different spatial scales influence the political discourse associated with ecotourism management. Policy responses are negotiated within, and woven around, this context and cannot be separated from the flow of interests and resources in the particular community and its place in the broader social, economic and administrative contexts (e.g. Considine, 1994; Hall, 1994). As a result, ecotourism policy is subject to a dynamic ebb and flow of commitment and resource availability.

Douglas Shire Council has been acknowledged as a rare example of a local government authority with a strong ongoing commitment to ecological sustainability, and its approach to tourism planning is mooted as innovative (Dredge, 1998; Roughley, 2000). Douglas Shire differs from most other municipalities in that tourism growth is not a primary objective of the Council. The Council is one of the first local governments in Australia to try to impose limits on tourism activity and is adopting strategies to shape the nature of ecotourism development (Thomas, 2000). Proper planning and management of tourism activity to ensure the maintenance of ecological quality and integrity and protect visitors' wilderness experiences underpin its approach to the management of ecotourism (Humphreys, 1994a). It is an approach that has been secured, strengthened and consolidated incrementally over more than 10 years using a variety of policy instruments and taking advantage of diverse opportunities that have emerged from the flow of governance in Australia's federal system.

There are a number of reasons for the positive, even activist response to the challenges of planning and managing ecotourism in the Daintree. Firstly, there is a coincidence of interest between locally driven green interest groups and commercial interests primarily based in the resort town of Port Douglas. Together, these political interests now outweigh the historically important rural sector of the Shire's population, which has traditionally been less disposed towards environmental management and the adoption of strict land use controls necessary to achieve outcomes that support ecotourism. Secondly, from 1992, Douglas Shire has had a mayor who is a committed green activist with a well-developed political savvy. Ironically, perhaps, he only emerged from his hippy rainforest retreat to participate in the protests over the Cape Tribulation to Bloomfield Road in the 1980s. There he honed his political skills and galvanized his interest in local political issues, before moving on to edit the local newspaper and thence to the mayoralty. He has succeeded in three re-election campaigns, often without obtaining majority support within the elected councillors. However, he has managed to garner sufficient local political support to promote effective local government strategies, including ecotourism and environmental management. At the same time, he has applied considerable energy and skill to promoting the interests of Douglas Shire in issues facing the management of the Daintree, and in developing effective collaboration with relevant leaders and agencies to obtain funding and support for planning initiatives. Thirdly, Douglas Shire industry stakeholders could see that the location could fill a market niche as the less developed, more sophisticated and green destination area outside Cairns.

Douglas Shire Council's Policy Approach

Douglas Shire Council's approach to the management of tourism has been progressively refined and consolidated over more than 10 years.

The Council expresses its commitment to planning and managing tourism through statutory mechanisms, including a strategic land use plan and development control plans (Douglas Shire Council, 1996), and through non-statutory mechanisms such as a local tourism strategy (Douglas Shire Council, 1998). Douglas Shire Council also manifests ongoing commitment in its day-to-day decision making and management practices in areas such as environmental management, development assessment, roads and transport and community services. It has also collaborated effectively with State and Commonwealth agencies to pursue improvements to its research base and to its policy framework, revealing a preparedness to build partnerships to achieve better planning and environmental outcomes. These instruments and studies are outlined below.

Land use planning instruments

The town planning scheme, prepared in 1990–1992 and gazetted in 1995, articulated a vision for tourism development of the Shire. It sought to retain and enhance environmental quality, destination identity, protect visitor experiences and ensure the compatibility of resident and visitor activities (Humphreys, 1992). The strategy identified nodes where different styles of tourism were to be promoted, and identified land uses considered consistent and inconsistent with the overall vision. Notably, the plan restricted development north of the Daintree River to that which would 'facilitate the exploration and appreciation of the natural environment' (Douglas Shire Council, 1996). Development of tourist accommodation beyond the level of a considerable suite of existing approvals would not be permitted until those approvals had lapsed. Instead, Port Douglas, half an hour south of the Daintree River, would be the main centre within the Shire for visitor accommodation and other support services. Other areas south of the Daintree River, which could be developed to relieve pressure on the Daintree, were identified.

   Secondly, in critical aspects, the plan embodied a move away from traditionally prescriptive planning approaches, whereby land use and development of land is tightly controlled in terms of detailed planning guidelines and criteria. Prescriptive approaches are thought to stifle innovative development, where it is easier to gain approval for a mediocre development because it conforms to prescriptive standards rather than one which attempts to achieve superior planning outcomes (Weir, 1995). The Douglas Shire planning scheme embodied a balance between normative and prescriptive approaches, where broad goals and objectives are complemented by detailed description of the desired nature of tourism activity, desired styles of development and acceptable levels of impact. This is particularly important in relation to tourism, where constantly evolving market demands and the emergence of new niches stimulate innovative product development. Strictly defined prescriptive standards can quickly become outdated and are less able to accommodate innovative development.

Thirdly, the plan recommended policy approaches in other areas of Council responsibility in order to achieve the desired planning outcomes. For example, vehicular access to the Daintree is via a small Council-operated ferry across the Daintree River. As part of the strategy to maintain the wilderness experience north of the river, it was recommended that the ferry capacity should not be upgraded. It was also recommended that the Council consider not extending urban servicing beyond existing urban nodes, and upgrading of roads within the Daintree was to be minimized. As will be discussed later, as conditions have changed, visitation has increased and political debates have been played out, these strategies have been further refined, often in unexpected ways.

Fourthly, this broad strategy was backed up by a development control plan that provided specific guidance on planning and development issues in the Daintree. In particular, protection of visual quality and landscape integrity were considered essential, and a Development Control Plan provided the vehicle through which detailed planning intent could be expressed. The desired nature of tourism development in specific nodes and areas was fully described, and flexible site, design and landscaping criteria were articulated. Since the adoption of the planning scheme, partnerships between Douglas Shire Council, the WTMA and other public and private sector agencies have resulted in the preparation of other planning and policy initiatives which complement and reinforce these original intentions.

Daintree planning strategy

In 1993, Douglas Shire Council obtained a Commonwealth government grant and commissioned the Daintree Rescue Strategy. The Rescue Strategy was intended to be a 'road map' for environmental protection and development in the area, and was supported by an implementation strategy containing detailed proposals for, among other things, the public purchase of allotments, town planning controls, road and traffic management and environmental rehabilitation (Beeton and Bell, 1998). Following the preparation of this Strategy, the Commonwealth and Queensland State government agreed to jointly fund the Daintree Rescue Programme in 1995 (Beeton and Bell, 1998). The Programme was aimed at enhancing the environmental integrity of the area and reducing the negative social and economic impacts of world heritage listing. Specifically, Aus$23 million was allocated for the Rainforest Protection and Visitor Facilities and Infrastructure Programme Sub-programmes. The funding flowing from the Daintree Planning Package was managed by the Daintree Coordination Group, a collaboration of WTMA, the Queensland Department of Environment and Douglas Shire Council, with other stakeholder representation. The creation of this organization has had a significant long-term impact on developing a cooperative and collaborative culture among stakeholders and administrative agencies charged with the management of the Daintree.

A review of the Programme was undertaken in 1998 (Beeton and Bell, 1998). At that time, 83 allotments had been purchased and only 10 cooperative management agreements (out of 120 expressions of interest received) between landowners and the WTMA had been finalized. In addition, the lack of detailed planning and feasibility studies for the particular projects contained in the Visitor Facilities and Infrastructure Sub-programme meant that this particular component of the Strategy had stalled and over 75% of the funding allocated to the Sub-programme had not been spent.

Douglas Shire Tourism Strategy

In 1997–1998, the Douglas Shire Tourism Strategy was prepared (Douglas Shire Council, 1998). The Strategy was jointly funded by Douglas Shire Council and the Port Douglas Daintree Tourism Association. While the preparation of the Strategy involved further stakeholder consultation, it took as its basis the direction contained in the planning scheme; it built a marketing/visitor management framework for the Shire. It acknowledged the strong interrelation between marketing and visitor flows and used marketing as a tool for visitor management. While it does not have the same legal status as strategic and development control plans, it represents a further stage in the development of an integrated framework for tourism management.

Daintree Futures Study

In 1999, the WTMA commissioned the Daintree Futures Study in an effort to further understand policy issues and priorities, and to identify and evaluate alternative planning and land use strategies (Rainforest CRC, 2000). This report re-examined the main issues identified in previous reports, including complex property rights and conservation management issues associated with privately owned land, roads and ferry access, community servicing and facilities development and tourism management. Based on this improved information, the strategy for management of complex ecotourism management, servicing and community development issues was further refined and improved. Preparation of the Strategy was aimed at building a coordinated policy approach between the WTMA, the Shire Council and other stakeholders, and can be regarded as the first collaborative attempt at developing a strategic management approach which reconciles ecotourism management with conservation, community development and infrastructure provision and servicing.

Improvements to institutional arrangements for collaboratively managing the Daintree have been put forward for discussion as a result of the Daintree Futures Study. However, these proposals do not negate the legislated responsibilities of the local council in land use planning, environmental management and community servicing. Accordingly, Douglas Shire Council undertook to prepare a new land use planning scheme in 2001, which will build on previous studies, analyses and preferred strategies in order to refine the Council's constantly evolving policy approach.

Discussion

Douglas Shire Council's policy approach to managing ecotourism in the Daintree takes place in the context of constantly changing political, institutional, social and economic conditions. Evolving interpretations of and attitudes towards the natural environment in general, and ecotourism experiences in particular, have underpinned ecotourism demand, visitor expectations and perceptions of this World Heritage Area. In this dynamic context, Douglas Shire Council's ecotourism policy approach has been developed incrementally over more than a decade. It has been developed through engagement in regional, state and national policy spheres, and has been refined and implemented at the local level through debate and negotiation with a range of stakeholders. Over this time, the Council has been able to evaluate and progressively strengthen its policy directions and has made use of opportunities to further its research and information base. The core policy issues have not changed significantly and policy solutions have not been modified to any great extent, but each study has brought with it an updated and improved understanding of policy problems and the interrelated nature of policy issues.

While the Council's broad policy direction has been maintained, strong growth in visitation has placed considerable pressure on the Council's management framework. Overlapping roles and responsibilities between governments have also exacerbated the difficulty of adhering to this framework. For example, 10 years ago the town planning scheme recommended that the Daintree to Cape Tribulation road should not be sealed, with a view to limiting uncontrolled access by conventional vehicle. However, increased visitation and growth in the local population saw demand for the ferry service escalate. For predominantly financial reasons, the Council increased the capacity of the vehicular ferry, which led to increased traffic on the unsealed Daintree to Cape Tribulation Road. After this, the Queensland State government, having responsibility for the development and management of main roads (or roads considered to be of State significance), became involved in debates over whether or not the road was a main road and should be sealed. The Council decided to seal the road in an effort to head off a situation whereby the State government widened and sealed the road in accordance with the standards set for main roads. The view adopted by the Council was that, in sealing the road themselves, they were able to retain tree canopies and vegetation closer to the road than would have been possible under the Queensland government standards for main roads. The Council considered that the environmental and presentation benefits of sealing the road outweighed the problems that would be caused by additional visitor pressure due to improved accessibility. Reinforcing this position was the notion that the ferry could operate as a valve, not only to limit visitor numbers for experiential reasons, but also to limit traffic volumes on the access spine.

However, sealing the road from the Daintree River to Cape Tribulation, and widening it in some locations, has led to an increase in self-drive independent visitors, the impacts of which are more difficult to manage than those of visitors on guided tours. The shift in balance between tours and self-drive visitors and the growth in overnight visitors relative to day visitors has meant that there is an extended temporal and spatial presence of visitors in the area (Rainforest CRC, 2000). As a result, operators have observed decreased satisfaction with the wilderness experience among tourists.

The Council's policy directions have also come under pressure from development proposals for accommodation and other tourist-related facilities on the remaining freehold land. Innovation occurring with the tourism sector has been acutely demonstrated in the study area, and has also been testing the Council's policy approach. For example, one requirement of the strategy was that accommodation development should be of small scale so that it could be integrated into the natural setting, minimize noise and visual impact and have a negligible impact on visitor wilderness experiences. Douglas Shire Council in the period 1993–1996 successfully fought a proposal for a resort to accommodate 1000 overnight guests at Cape Tribulation, based on the town planning scheme's cap on accommodation development, the unsuitable size of the proposal, servicing problems (human and infrastructure) and inappropriate focus, that is, extending beyond the nature-focused theme recommended for north of the Daintree River.

These emerging issues illustrate that tourism is not a discrete policy domain, much less, ecotourism. Ecotourism shares a complicated set of relations with other policy areas, where decisions in one policy domain influence other policy arenas. Moreover, ecotourism policy making cannot be conceptualized as occurring in a rational, linear process, where planners move from problem definition through stages of analysis, identification of alternatives, production of alternative policy solutions and implementation. Tourism policy, as with other areas of policy, is dynamic and recursive (e.g. Considine, 1994). Tourism policy is derived out of the contest of values, interests and ideas between policy agents, where issues are simultaneously identified, discussed and evaluated and solutions are negotiated, implemented, evaluated and modified (Fischer and Forester, 1993). In the case of the Daintree, the policy approach for managing ecotourism is being progressively developed and refined, but this work is subject to local political commitment, lobbying and networking skills, partnerships with other policy agents and taking advantage of opportunities created at other levels of government.

Conclusions

This chapter has sought to discuss local government involvement in ecotourism policy making and to identify policy problems and issues associated with managing ecotourism. The case of Douglas Shire in North Queensland, Australia, illustrates that a range of factors are at play and the confluence of these provide opportunities and constraints for the development of ecotourism policy responses. These factors include:

- the institutional arrangements that structure the roles and responsibilities of local government;
- policy approaches and actions adopted at other levels of government;
- local government planning and environmental management practices and policy making processes;
- local political conditions, including the skills, interest, drive, commitment and continuity of elected representatives;
- interest structures within local communities, including alliances between groups that elevate tourism to an important and legitimate item on the political agenda;
- financial and other forms of resource commitment.

In conclusion, while the case of Douglas Shire represents the ecotourism policy approach of one particular local government, this
chapter illustrates the importance of recognizing and taking advantage
of the interconnections between different policy issues and developing a well-conceived policy approach. In the case of the Daintree, the
elaboration of a tourism strategy that considered the land use implications of ecotourism activity has been the cornerstone of a policy
approach that has been successful in managing many aspects of the
ecotourism experience. Most notably, the Council has been able to
maintain environmental quality and protect visitors' wilderness experiences and has resisted development that is not consistent with the
appreciation and enjoyment of the natural environment.

References

Beeton, B. and Bell, L. (1998) *Evaluation of the Daintree Rescue Package.*
    Report prepared for the Queensland Department of Environment and
    Heritage, Brisbane.
Boehmer-Christiansen, S. (1994) Politics and environmental management.
    Journal of Environmental Planning and Management 37, 1–69.
Brannock Humphreys (1994) *Daintree Planning Package.* Report prepared for
    Commonwealth Department of Environment and Douglas Shire Council,
    Brisbane.
Butler, R.W., Hall, C.M. and Jenkins, J.M. (eds) (1998) *Tourism and Recreation
    in Rural Areas.* John Wiley & Sons, Chichester, UK.
Christie, E. (1990) The Daintree rainforest decision and its implications:
    Comment. *Queensland Law Society Journal* 20, 223–228.
Clarke, M. and Stewart, J. (1993) From traditional management to the new man-
    agement in British local government. *Policy Studies Journal* 21(1), 82–93.
Commonwealth Government of Australia (1994) *Wet Tropics of Queensland
    World Heritage Area Conservation Act.* Canberra.
Considine, M. (1994) *Public Policy: a Critical Approach.* Macmillan,
    Melbourne.
Davis, B. (1989) Federal–state tensions in Australian environmental manage-
    ment: the world heritage issue. *Environmental Planning and Law Journal*
    6(2), 66–78.
Davis, G., Wanna, J., Warhurst, J. and Weller, P. (1993) *Public Policy in
    Australia,* 2nd edn. Allen and Unwin, Sydney.
Douglas Shire Council (1996) *Douglas Shire Town Planning Scheme.* Douglas
    Shire Council, Mossman, Queensland.
Douglas Shire Council (1998) *Douglas Shire Council Tourism Strategy.* Report
    prepared by Gutteridge Haskin and Davey for Douglas Shire Council and
    the Port Douglas Daintree Tourism Association. Douglas Shire Council,
    Mossman, Queensland.
Dredge, D. (1998) Land use planning policy: a tool for destination place man-
    agement. *Australian Journal of Hospitality Management* 5(1), 41–49.

Drost, A. (1996) Developing sustainable tourism for world heritage sites. *Annals of Tourism Research* 23(2), 479–492.

Figgis, P. (2000) The double-edged sword: tourism and national parks. *Habitat Australia* 28(5), 24.

Fischer, F. and Forester, J. (1993) *The Argumentative Turn in Policy Analysis and Planning.* Sage, Newbury Park, California.

Haigh, D.J. (2000) World heritage – principles and practice: a case for change. *Environmental and Planning Law Journal* 17(3), 199–213.

Hall, C.M. (1992) *Wasteland to World Heritage: Preserving Australia's Wilderness.* Melbourne University Press, Melbourne.

Hall, C.M. (1994) *Tourism and Politics: Policy, Power and Place.* John Wiley & Sons, Chichester, UK.

Healey, P. (1997) *Collaborative Planning: Shaping Places in Fragmented Societies.* Macmillan Press, London.

Humphreys, J. (1992) Planning for tourism in Douglas Shire. Paper presented at the *Royal Australian Planning Institute Biennial Congress 1992, Canberra.* www.hum-plan.com (20 August 2002).

Humphreys, J. (1994a) The planning approach north of the Daintree. Paper presented at the *Queensland Environmental Law Association Conference, Fraser Island,* pp. 9.1.1–9.1.9. www.hum-plan.com (20 August 2002).

Humphreys, J. (1994b) Planning for sense of place in a tourism destination – North of the Daintree River. Paper presented at the *Combined Royal Australian Planning Institute and Local Government Planners Conference, Hobart, Tasmania.* www.hum-plan.com (20 August 2002).

Humphreys, J. (1996) Implementing the vision in a world heritage area: Conservation and tourism in the Daintree. Paper presented at the *Royal Australian Planning Institute National Conference, Perth, Western Australia.* www.hum-plan.com (20 August 2002).

Jenkins, O. and McArthur, S. (1996) Marketing protected areas. *Australian Parks and Recreation Journal* 32(1), 10–15.

Lane, M. (1997) The HORSCERA inquiry into world heritage area management. *Australian Parks and Recreation* 33(1), 12–16.

MacDermott, K. (1999) Port Douglas looks forward to a boom year. *Australian Financial Review,* 29 April, p. 15.

Marshall, N. (1997) Introduction: Themes and issues in Australian local government. In: Dollery, B. and Marshall, N. (eds) *Australian Local Government: Reform and Renewal.* Macmillan Education Australia, Melbourne, pp. 1–14.

Mayer, M. (1995) Urban governance in the post-Fordist city. In: Healey, P., Cameron, S., Davoudi, S., Graham, S. and Madani-Pour, A. (eds) *Managing Cities: The New Urban Context.* John Wiley & Sons, Chichester, pp. 231–249.

Queensland State Government (1993) *Wet Tropics World Heritage Protection and Management Act.* Queensland State Government, Brisbane.

Rainforest CRC (2000) *Daintree Futures Study: a Report to the Wet Tropics Ministerial Council.* Rainforest Cooperative Research Centre, Cairns, Queensland.

Richardson, B.J. (1990) A study of Australian practice pursuant to the World Heritage Convention. *Environmental Policy and Law* 20(4/5), 143–154.

Roughley, A. (2000) Rainforests, reefs and rising rents. *Australian Planner* 37(1), 20–27.

Shackley, M. (1998) Introduction – World Cultural Heritage Sites. In: *Visitor Management: Case Studies from World Heritage Sites.* Butterworth Heinemann, Oxford, pp. 1–9.

Sharpley, R. and Sharpley, J. (1997) *Rural Tourism: an Introduction.* International Thomson Business Press, London.

Thomas, H. (2000) Northern overexposure. *The Courier Mail,* 28 September, p. 15.

Wahab, S. and Pigram, J.J. (1997) *Tourism Development and Growth: the Challenge of Sustainability.* Routledge, London.

Weir, M. (1995) Performance-based zoning and strategic planning: a new era for Queensland. *Queensland Environmental Practice Reporter* 1(5), 143–152.

Wet Tropics Management Authority (1995) *Draft Wet Tropics Plan.* Wet Tropics Management Authority, Cairns, Queensland.

World Conservation Monitoring Centre (1995) *Protected Areas Programme World Heritage Areas.* UNEP World Conservation Monitoring Centre. www.wcmc.org.uk/protected_areas/data/wh/wettropi.html

Processes in Formulating an Ecotourism Policy for Nature Reserves in Yunnan Province, China

8

Trevor H.B. Sofield and Fung Mei Sarah Li

Tourism Programme, University of Tasmania, Locked Bag 1–340G, Launceston, TAS 7250, Tasmania

Introduction

This chapter explores the issues and complexities of ecotourism policy formulation in China and is intended to cast a little light on some of the conceptual and practical issues integral to such an exercise given the particularities of the Chinese cultural context and Chinese values associated with nature and wilderness. The process of planning for tourism in developing, non-Western countries has in a large number of cases been carried out by foreign experts under a wide range of development assistance programmes. Countries of the European Economic Community, the USA, Canada and, to a lesser extent, Australia and New Zealand have been actively involved in underwriting such planning programmes in Africa, Asia, Central America, the Caribbean and the Pacific and Indian Ocean regions. Multilateral agencies such as the United Nations Development Programme and the World Tourism Organization have also been active in sponsoring national tourism planning exercises. Despite the undoubted expertise of the consultants employed under formal agreements with the governments of the recipient countries, many of their plans fail to be implemented. A lack of fit with existing government priorities, the politics of the situation and the power exercised by vested (often competing) interests in the recipient society, and local cultural values which conflict with or negate the 'imported' values espoused in the plans – rather than economic capacity or lack of local expertise – have been identified as some of the major causative factors (Sofield, 2000).

The process of formulating an ecotourism strategy for five newly designated nature reserves in Yunnan Province, China, included all of these issues and others as well. It juxtaposed the Western paradigms of environmental conservation, wilderness and sustainability, upon which ecotourism is based, with centuries-old Chinese values and views about the natural environment and the role of humans interacting with nature. The latter are essentially anthropocentric, and the exercise thus produced a discourse of difference that was grounded in a strong contrast with the diametrically opposed biocentric Western approach. The forging of an appropriate ecotourism strategy needed to reconcile these differing, culturally determined values.

The strategy also needed to contend with consumer demand for access to natural resources on a scale unmatched in most other parts of the world. As the Chinese population of 1.3 billion grows wealthier and disposable incomes increase to the point where recreational travel and tourism become attainable, literally millions of domestic visitors stream into the countryside. Existing sites are often overwhelmed with thousands and thousands of daily visitors and management regimes are often inadequate to deal with the numbers. Since ecotourism as defined in the Western sense tends to be 'small tourism' based on a strict regulation of numbers through application of the concepts of carrying capacity and limits of acceptable change, a recreational land-planning and management scheme for the nature reserves of the Province had to be incorporated as an umbrella, with ecotourism presented as only one part of a much larger mosaic and for which planning could not occur in isolation. The concept of the Recreational Opportunity Spectrum (Clark and Stankey, 1979) was utilized for this umbrella purpose.

At the same time the proposed ecotourism strategy had to be consistent with a raft of existing Chinese policies concerning, *inter alia,* rural development, conservation of forest resources and poverty alleviation of Minorities Nationalities communities. The Minorities Nationalities were in fact required by the Yunnan Provincial Government to be major beneficiaries of an ecotourism strategy because under recent, stringently regulated nature reserves policy more than 800 such communities had been denied much of their previous access to traditional forms of exploitation of forest resources in the nature reserves. Those activities, such as swidden agriculture, timber cutting, fuelwood gathering and charcoal making, bamboo harvesting, and hunting and gathering, had been the mainstay of their economic survival for centuries. Many of these activities were extractive and destructive while ecotourism was perceived to be a benign, sustainable, non-exploitative alternative use of forest resources. Because of the requirement to utilize ecotourism for poverty alleviation the strategy thus had to incorporate an understanding of ethnic cultural systems as well as conservation imperatives.

This chapter is based on field work carried out in Yunnan Province, China, in 1999 by the authors as part of a consultancy team sponsored by the Sino-Dutch Forest Conservation and Community Development Programme, supplemented by participation in two conferences/workshops; one on 'Conservancy and the Environment in Yunnan' (Kunming, September 1999, sponsored by the Center for US–China Arts Exchange, Columbia University) and the other on 'Anthropology, Chinese Society and Tourism' (Kunming, October 1999, sponsored by the University of Yunnan and the Chinese University of Hong Kong). The Yunnan fieldwork was also backed by eight previous research projects undertaken by the authors in China during the past 6 years. The new ecotourism policy was to be developed for five nature reserves: Caiyanghe, Gaoligongshan, Tongbiguan, Wuliangshan and Xiaoheishan.

A 6-week study was undertaken which included:

- comprehensive field trips and site inspections throughout each of the reserves;
- meetings at Prefecture, County and Township levels with senior officials from the Chinese Communist Party, Government dignitaries, senior staff from the Forest Department, and representatives of tourism authorities in the capital Kunming, and eight other cities/towns;
- an assessment of previous research, particularly RRAs (Rapid Rural Appraisals) of rural communities located within the buffer zones of the five nature reserves;
- inspections of a wide range of tourism attractions, products and developments – nature based, cultural and 'contrived' (e.g. theme parks) – in areas adjacent to the nature reserves.

The Setting

Yunnan Province in the south-west of China is richly endowed with both natural and social capital. It is situated on a plateau that borders Tibet and the Himalayan ranges in the north, Myanmar (Burma) in the west, and Laos and Vietnam in the south. ('Yunnan' in Chinese means 'South of the clouds'). The Province covers a north–south distance of 900 km, encompassing permanent snow-covered peaks in the north (Kagebo Peak is the highest point at 6740 m) and tropical lowland rainforests in the south. More than 95% of the Province is classified as mountainous. Three of Asia's mightiest rivers rise in its mountains and form impressive gorges and valleys: the Yangtze, the Irrawaddy and the Mekong. Tengchong County has more than 90 volcanoes and numerous thermal springs in a scenic area of less than 100 km². Shilin has karst formations, pinnacles and limestone caves extending for an area of more than 350 km², now gazetted as a national nature reserve. Forests still cover 65% of the province, the most extensive such tracts in China.

Although the area of Yunnan accounts for only 4.1% of China's total land mass, the variety of its landforms, climatic zones and environments form habitats which have provided ecological niches for a biodiversity unmatched in China (Zhang Baosan, 1998). Fifty-five per cent of China's vertebrates (1704 species of a total of 3099) have been recorded in the Province. Some 200 of these have been classified as endangered and/or rare and include the Asian elephant, the Asian leopard, the Yunnan golden monkey, gibbons, wild ox, hornbills and other birds. Tigers, the Asian rhinoceros and the giant panda are no longer found in Yunnan. More than 90,000 insect species of a national total of 130,000 inhabit the Province. Yunnan is also home to the greatest number of plant species in China: some 18,000 (62.5%) of a total of 30,000 (Zhang Baosan, 1998). Since 1983 the Province of Yunnan has been active in conserving its environments and it now has a network of more than 20 reserves covering more than 10% of the Province. Seven of these are classified as Level A (national importance) nature reserves because of their outstanding natural features and biodiversity. (In the Chinese classification system nature reserves rank higher than national parks and connote inclusion of major stands of old growth virgin forests.) The increase of nature reserves in Yunnan mirrors a similar increase nationally. In 1978 (the strategically important year of the 'Open Door' policy which introduced 'capitalism with a socialist face' and opened China's borders to tourism) there were only 34 nature reserves covering 0.13% of the nation's territory, but by 1999 the number of reserves had expanded to 1146 covering 8.8% of the country (Han, 2000). There are 136 Level A national nature reserves, 16 of which have been listed under UNESCO's global Man and the Biosphere Programme (UNESCO, 1999).

Culturally, the Province is also among the most diverse in China. According to a 1995 survey (China State Council Information Office, 1999) for the whole country, there were 108.46 million people registered as belonging to minority ethnic groups. They accounted for 9% of the total Chinese population of 1.3 billion. Yunnan, which has 26 different Minority Nationalities totalling more than 25 million people, has the greatest and most diverse numbers of Minorities Nationalities (PRC State Council Information Office, 1999).

Yunnan's border position, incorporating 4060 km of China's external boundary, constitutes a highly strategic political and military region that was closed to outsiders until a decade ago, when tensions between its neighbouring states eased. As a consequence of Yunnan's imposed isolation, its mountainous terrain and its distance from centres of power and development, it is one of the least economically advanced provinces in contemporary China. A Province Government statement issued in January 1999 identified 73 of its total of 128 counties as 'poor' and 3.5 million of its people out of a total population of 44 million as 'below the poverty line' (Yunnan Province, 1999: 2). Tourism has been accepted as one of the major means for changing this situation. Thus every prefecture, county and even some townships now have their own tourism bureaux and a variety of local plans and policies for tourism-led development. The Yunnan Province Tourism Policy is predicated on utilization of its natural environment and the cultures of the 26 Minority Nationalities who make up almost 60% of the total provincial population of 44 million (Yunnan Province, 1999).

Chinese Values about Nature and Ecotourism

The context of ecotourism policy formulation in Yunnan Province was revealed as one where Chinese values concerning 'nature' were often diametrically different from those associated with the Western paradigm of ecotourism. The Chinese word for 'nature', *da-jiran*, may be translated literally as 'everything coming into being' and expresses the totality of mountains, rivers, plants, animals, humans, all bound up in their five elements, fire, water, earth, wood and metals (Tellenbach and Kimura, 1989). 'Man is based on earth, earth is based on heaven, heaven is based on the Way (*Tao*) and the Way is based on *da-jiran* (nature): all modalities of being are organically connected' (Tu Wei-Ming, 1989: 67). In ancient China there was the archetypical ideal of Confucian thought, 'a sentiment of consanguinity between persons and nature ... an awareness of active participation [by humans in] the well-balanced and harmonious processes that are the cosmos itself' (Shaner, 1989: 164). It is an anthropocentric perspective with a sociological definition in which man[1] lives and works in harmony with nature, where, because nature is imperfect, man has a responsibility to improve on nature (Chan, 1969; Elvin, 1973). It is thus distinct from a Western perspective that separates nature and civilization (humans), which views nature ideally as free from artificiality and human intervention. This attitude encompassing humans and nature as indivisible

[1] The term 'man' is used here to reflect accurately the Mandarin usage and should not be interpreted as unthinking sexist language on the part of the authors.

continues in modern China. In effect, any venture which is set in the Chinese countryside and utilizes natural resources and attractions tends to be classified as 'ecotourism' when a Western definition would define it as nature-based tourism or simply a tourism facility located outside the urban area (Sofield and Li, 1996). For example, the 'Ecotourism Plan' for the rural surrounds of East Lake 5 km from Huangzhou township in Hubei consists of five 'development zones': a conference centre, a hotels and resorts area, fishponds (aquaculture for both commerce and recreation), water sports (including water skiing) and a visitor reception area with amusement park (Huangzhou County Government, 2001).

The biocentric orientation of Western ecotourism encompasses five generally accepted components which distinguish it from nature-based tourism:

1. Conservation of nature is its fundamental criterion.
2. Education about biodiversity, habitats and the need for conservation is an integral component for both host communities and tourists.
3. Any income generated from ecotourism has a significant proportion ploughed back into maintaining the quality of the resource and its conservation.
4. Local communities when they are associated with a development must be able to share equitably in the benefits of ecotourism.
5. Ecotourism ventures or activities must be designed to be sustainable ecologically, economically and socio-culturally.

Ecotourism is thus defined as *a holistic system of management* of a natural resource for sustainable tourism, with the principle of conservation taking primacy over economic profit making and human comfort (anthropocentrism). In China nature-based tourism ventures are generally characterized by an almost total absence of any conservation message for both visitors and hosts. The economic imperative drives its development (Han, 2000) so that invariably the outcome is mass tourism, again in contrast to ecotourism which tends to be developed around relatively small visitation levels.

In addition to differences over basic definitions of what constituted nature and ecotourism, the policy process in Yunnan had to contend with contrary values concerning 'wilderness'. Perceptions of wilderness vary greatly and across cultures. What may be wilderness to one observer (e.g. the Australian Outback to a Caucasian Australian) may be another person's home (e.g. the Australian Aborigine who is familiar with every topographical detail and its biota, and for whom it will be the 'Inback'). Hendee *et al.* (1990) note that etymologically the English language word 'wilderness' 'is derived from the Old English *wild-deor-ness,* the 'place of untamed beasts', and that civilization by contrast is 'an environment under human control.' In their view, 'the only wilderness true to the etymological roots of the word is that which humans do not influence in any way whatsoever' (Hendee *et al.*, 1990: 27). There is no similar Chinese word for wilderness, however, the closest probably being *huangyie* meaning 'uninhabited countryside' which does not carry the same connotations of pristine, unsullied isolation. Rather its connotations are negative, in the sense that the land is 'bad', or 'poor' or 'not fertile'. Since man is always a part of nature in the Chinese perception there is an absence of the paradox described by Nash (1982) which exists in the western concept of management of wilderness; that is, if wilderness is an area not under the influence of human agency, its management in fact requires human control of nature. Hendee *et al.* (1990: 28) refer to this as 'the intellectual dilemma' posed by the concept of 'managed wilderness': for some 'just the knowledge that they visit an area by the grace of, and under conditions established by, civilization is devastating to a wilderness experience.'

There is a millennia-old tradition behind China's construct of nature (*da-jiran*). Under Confucian values scholars and mandarins were exhorted 'to seek ultimate wisdom in Nature' (Chan, 1969; Overmyer, 1986). The current slogan 'Man and Nature Marching in Harmony Towards the Twenty-first Century' is drawn from both Confucian thought and Taoist philosophy on the need for man and nature to bring opposing forces into a symbiotic relationship. This is a perception of the world in which 'harmony' rather than 'difference' or 'opposites' is dominant, where things do not occupy their own separate space but rather 'a seamless web of unbroken movement and change, filled with undulations, waves, patterns of ripples and temporary "standing waves" like a river, [in which] every observer is himself an integral function of the web' (Rawson and Legeza, 1973: 10). The essence of life itself, the cosmic force called *ch'i*, was the major determinant in the growth of all things, whether trees and crops would thrive, to what height a mountain reached, how fast a river flowed. Taoism perceived opposites flowing towards one another and being mutually dependent rather than being drawn into conflict. Summer flowed into winter and back to summer again; the sun rose, sank and was replaced by the moon that rose and sank and in turn was replaced by the sun; hot could not exist without cold, light without dark, male without female; death could not exist without life. Nature without man and man without nature were incomplete. Tao united everything, exemplifying the need of man and nature to bring opposing forces into a fluctuating harmony (Rawson and Legeza, 1973).

Mountains were particularly venerated and the complementary force fields of man and nature came together most powerfully in the Taoist concept of *yin–yang*. Like a magnet with its different force fields, both are needed for the magnet to function, and man is seen as indivisible from nature (Ropp, 1992; Spence, 1992). Under the religious belief system that evolved over centuries, there were nine revered sites of particular significance, five sacred *shan* or mountains and four rivers or *shui*. (These were the eastern Tai Shan, the southern Heng Shan ('Balanced/harmony mountain'), the western Hua Shan, the northern Heng Shan ('Eternal Mountain') and the central Song Shan; and the Chang Jian (Yangtze), the Huang He (Yellow River), the Huai Shui and the Ji Shui (*Illustrated History of China's Five Thousand Years*, 1994.)) It was a fundamental responsibility of Chinese emperors to visit these sacred mountains on a regular basis to propitiate the spirits, gods and ancestors. Failure to do so could place the entire prosperity and well-being of the empire at risk. Grand roads were constructed for the emperor to approach the sacred mountains (the imperial way). Steps, termed 'staircases to heaven' since the emperor was revered as the son of heaven, were carved into their slopes for his ascent to the summit. The Taoist religious philosophy developed its own sacred mountain centres, one of which is Mt Wudang, in Hubei Province, where a series of Taoist temples and monasteries perch on peaks and cliff-tops. Wudang has additional fame because it was the only site selected by the Ming dynasty to construct an imperial Taoist temple outside the Forbidden Palace in Beijing (Purple Cloud Temple, built in the 15th century) and its monks developed the Wudang form of martial arts most recently popularized by the Chinese award winning film *Crouching Tiger Hidden Dragon*. All of these sites have become immortalized in Chinese poetry, essays and art over the centuries. They feature in such lists as the China State Council's 1998 *State Level Scenic Wonders and Historical Sites* and a survey undertaken in 1998 by the China Travel Service of the 'Top Forty' most favoured tourist spots (China Travel Service, 1998). Wudang Mountain was designated in 1996 as a World Heritage Site for both its biodiversity and its outstanding cultural values.

From being a virtually non-existent activity in China only 20 years ago, tourism in China is rapidly becoming one of its largest sectors (Zhang Guangrui, 1995). Following the 'open door' policy introduced by the Leader of the Communist Party, Deng Xiaoping in 1978, tourism's growth has been nothing short of spectacular. In 1978, a total of 1.81 million tourists visited China, 1.6 million of them from Hong Kong. In 1979 that figure expanded to 4.2 million (3.8 million from Hong Kong), an increase of 232 per cent. By 2000, the China National Tourism Administration (CNTA) recorded more than 750 million domestic tourists and more than 60 million overseas visitors. Total tourism revenue (domestic and overseas) was estimated at US$41.4 billion in 1998, accounting for 4.3% of gross national product (China National Tourism Administration, 1999, *Yearbook of China*

Tourism Statistics, 1998; China Daily, 1999). Allowing for the fact that these figures include individuals from Macao and Hong Kong who will be crossing the border every day to work (and who thus will each be counted more than 300 times per year as overseas Chinese visitors), the numbers still indicate trends which are significant and which constitute one of the most dynamic tourism industries in the world.

A probe into rural tourist visitation in China reveals that the motivation for many domestic travellers lies in the traditions established over 5000 years and now firmly entrenched in the Chinese psyche. As Ying Yang Petersen (1995: 149) stated:

> To the Chinese people, visiting a scenic spot or a historical place is always attached to symbolic expectations. In order to watch the sun rising over Mt Tai (Tai Shan), for example, many old and young Chinese wait in the chilly darkness for hours. What they are really looking for is not simply the scene of the sun rising from the clouds but the experiences and reflections which have been memorialized again and again in Chinese poetry over centuries.

In examining how Chinese values about landscape and wilderness are translated into tourism attractions, the anthropocentric position accepts (indeed encourages and facilitates) programmes to alter the physical and biological environment in order to produce desired 'improvements' (Sofield and Li, 1998). These may include landscaped parks, facilities for recreation and tourism, roads for ease of access, observation towers and so on. Increasing direct human use is the objective of management and the character of the wilderness will be changed to reflect the desires of humans and contemporary standards of 'comfort in nature'. Styles of recreation and tourism will be tuned to the convenience of humans, so trails will be concreted, resorts permitted inside reserves, cable cars approved and so forth. Even the very centre of the ideal of Taoist reverence for nature, Mount Wudang, is now adorned with a cable car that takes 3000 tourists a day to a restaurant on the highest peak next to the famed Golden Hall Temple. The integrity of the pilgrimage experience manifested in the tough climb to the Temple along the ancient sacred way through three 1000-year-old 'gateways to Heaven' has been aborted. The anthropocentric approach, taken to its extreme, means the loss – in Western eyes – of an essential wilderness quality: naturalness (Hendee *et al.,* 1990: 19).

The biocentric approach, in contrast, emphasizes the maintenance or enhancement of natural systems, if necessary at the expense of recreational and other human uses (Hendee and Stankey, 1973; cited in Hendee *et al.,* 1990). 'The goal of the biocentric philosophy is to permit natural ecological processes to operate as freely as possible, because [in the Western system of values] wilderness [integrity] for society ultimately depends on the retention of naturalness' (Hendee *et al.,* 1990: 18). It requires controlling the flow of external, especially human-made, pressures on ecosystems by restricting excessive recreational or touristic use of the bio-geophysical resources. The recreational use of wilderness is tolerated with this position only to the degree that it does not change the energy balance inordinately. Thus ecotourism is in general an acceptable part of a biocentric approach to nature reserve management. A biocentric philosophy requires recreational users to take wilderness on its own terms rather than manipulate it to serve human needs. Like the anthropocentric approach the biocentric approach also focuses on human benefits, but the important distinction between them is that biocentrically the benefits are viewed over a longer term and as being dependent upon retaining the naturalness of the wilderness ecosystems (Hendee *et al.,* 1990: 19).

Both the Western and Chinese values attributed to 'wilderness'/*huangyie/da-jiran* are similar, but they find very different expression in use, management and acceptable behaviour. If we take the three main values of wilderness identified by Hendee *et al.* (1990) – experiential, scientific, symbolic/spiritual – a cursory examination is enough to highlight the differences between the Western and Chinese perspectives:

Experiential

The experience of feeling close to nature, of experiencing the mystical forces which shape the universe. The wilderness experience is seen as valuable in its own right. For Westerners this may be translated into a form of ecotourism which allows them to experience the solitude and freedom of nature with no sight or sound of humans anywhere, camping out under the stars. For Chinese, it may be sufficient simply to visit a forest resort and, surrounded by the forests, enjoy playing cards, mahjong or karaoke in the air-conditioned comfort of built facilities. Three resorts located inside the boundaries of Caiyanghe Nature Reserve in Yunnan, for example, exhibit this form of tourism. They are representative of many similar facilities throughout most of China's nature reserves including those with World Heritage Site status such as Yellow Mountain (Huangshan) in Anhui Province, Shennongjia in Hubei Province and Jiuzhaigou in Sichuan Province. The first regime will apply stringent conservation management. The second will place the comfort of visitors first: it could at a stretch be termed nature-based tourism but not ecotourism in the Western lexicon.

Scientific

Wilderness areas are seen:

as valuable assets; as natural baselines that reveal the extent of impacts elsewhere; as sites where scientists can study natural processes; as gene pools maintaining the diversity of nature and providing a gene reservoir we are only now learning how to use; and as sanctuaries for [rare or endangered flora and fauna].

(Hendee *et al.*, 1990: 9)

The Chinese accept and support this concept in principle through the creation of biosphere reserves under UNESCO's Man and Biosphere Programme, which designates 'no-go' core areas reserved for scientific investigation and conservation. However, purpose built facilities are tolerated (e.g. forest research centres in Gaoligongshan Nature Reserve, Yunnan; the Wolong captive panda breeding programme in Sichuan Province).

Symbolic/spiritual

Wilderness symbolizes both simplicity and stability in a fast-changing world where individuals have little or no capacity to exercise control over the pace and stress of modern life. The Western world appears to have 're-discovered' this virtue of wilderness only in the 20th century: for the Chinese it has been philosophically a guiding tenet of their society for several thousand years, as noted in the brief outline above of Taoist and Confucian values. For many people (e.g. some of the Minorities Nationalities of Yunnan) a belief in animism will inhabit the wilderness with spirits which protect and safeguard their communities now and into the future. Their 'management' of wilderness resources will be foregrounded in their cosmology. For the majority of Chinese their Taoist/Confucian heritage invests nature with a very strong spirituality and symbolism abounds in Chinese metaphors and similes drawn from nature. In the Long Jun Xi tributary entering the lowest of the Yangtse River's Three Gorges, for example, two waterfalls are rich in symbolism. One, the Thunder Dragon waterfall, has five outflows roaring forth from a tunnel in the cliff: it is described by the guides as tumultuous, expanding therefore powerful, ambitious, virile, masculine. On the opposite side of the valley is the Phoenix Piano Waterfall which has a fine lace-like flow that tinkles gently down the cliff-face: it is described as delicate, beautiful, demure, coquettish, feminine. The dragon symbolizes the emperor, *yang;* the phoenix symbolizes the empress, *yin.* The peaks above are *yang,* male; the valley is *yin,* female. The waterfalls and their names derived from the animal kingdom, combined with the human characteristics attributed to them, are manifestations of the Taoist forces of *yin* and *yang* and the Chinese tenet of 'man in harmony with nature'. The 2-km trek along the uninhabited valley floor to the waterfalls is as much a cultural experience as it is a wilderness experience. There are literally thousands of similar examples all over China[2].

In Western countries the management of wilderness, its protection, conservation and rehabilitation (where degraded), will be based on the three main values outlined above, and the establishment of nature reserves in Yunnan signifies to some degree a manifestation of them. Nevertheless the societal context in Yunnan is not necessarily accepting or understanding of these values; the Chinese view is succinctly made by a senior parks administrator from Jiuzhaigou Biosphere Reserve: 'Without man there is no wilderness only nothingness because wilderness needs man to appreciate it' (Zhang, personal correspondence, October 2000). However, Chinese legislation for the establishment of nature reserves (Regulations of the People's Republic of China on Nature Reserves 1994) follows mainstream Western thought and incorporates reserves with core areas based on biological values to which public entry is prohibited and all economic activity is to be excluded. Article 26 states: 'It is prohibited to carry out such activities as cutting, grazing, hunting, fishing, gathering medicinal herbs, reclaiming, burning, mining, stone quarrying and sand dredging, etc.' And Article 28 prohibits 'tourism, production and trading activities' from the core areas (PRC, 1994).

A small group of Chinese environmentalists, notably those associated with China's MAB Programme, with the Bureau of Natural and Ecological Protection in the State Environment Protection Administration, and some academics understand and promote the biocentric approach. But management regimes in all cases aside from a few fall far short of the legislative rhetoric. In this context Xue Dayuan (2000: 61) noted that 'Though resource development activities in nature reserves are not allowed ... in fact almost every reserve now practices some form of such activities within their prohibited zones.' Li Wenjun (2000: 70) noted that while 'the ideology of strict protection' for nature reserves was incorporated in the 1994 legislation, there existed no national policy on ecotourism for reserves and 'tourism in nature reserves is largely uncontrolled'. A study of 83 reserves (54 of them Level A national nature reserves) in 1998 by Zhuge Ren (2000), revealed that 68 (82%) had at least one of the prohibited activities occurring inside their boundaries, 54 of them had three or four activities, and 14 of them had five to eight activities. Of all reserves he surveyed 40% had forms of tourism activity within their boundaries, including within their core areas. In the 3 years since Zhuge's survey, tourism development has been vigorously pro-

---

[2] Authors' note: at the time of our field visit in 2001, the valley's pristine nature was under threat – that is, according to Western values – from a developer with plans to construct a cable car along the length of the valley.

moted by the state authorities and anecdotal evidence by Sofield and Li (field notes: 1999, 2000, 2001) suggests that tourism activity in many different forms is occurring unchecked at a significantly higher level than Zhuge's 40%.

Field trips to a number of existing nature-based tourism sites in Yunnan illustrated the cultural differences between the Chinese approach to utilization of natural resources for tourism and the Western notion of ecotourism. In Tengchong County, for example, the 'Big Hollow Mountain' volcano has been developed in a way which is the antithesis of Western notions of empathy with nature, the aesthetics of wilderness and conservation. This mountain is only small, perhaps 300 m high, but its perfectly symmetrical cone shape has made it an object of veneration for centuries. From a concrete expanse of 5 acres (2 ha) built for tour buses, the way to the mountain is entered via a replica of the Imperial Gate that stands in front of the Forbidden Palace in Beijing. A concrete boulevard, 50 m wide and 1 km long, has been constructed from the gate to the foot of the mountain. The undulations of the lava flows have been levelled, all natural vegetation removed and replaced with flower-beds along both sides of the straight boulevard. Thousands of plastic flags are strung across the boulevard, which ends in a 2-m-high raised oval platform about 100 m in diameter at the foot of the volcano. Set in the centre of this concrete platform is a ceramic tiled mosaic of the *yin–yang* symbol 10 m in diameter. On either side of the platform two broad sets of concrete steps with steel handrails ascend in straight lines up the slope of the volcano, to converge at the summit. Two huge concrete platforms jut out from the steps as observation points on the ascent. The overall impression is of grand, linear, concrete forms visible from miles away which dominate the landscape and subjugate the volcano to human construction. The sense of wilderness and naturalness are absent.

In Chinese eyes, however, this development enhances rather than detracts from their appreciation of the site. Big Hollow Mountain accurately captures all of the key points of Taoism: the regal entrance gate, the imperial way, the staircase to heaven, the *yin–yang* forces of man and nature. The plastic flags may look to Western eyes like a used car lot, but they are in fact akin to Buddhist prayer flags and walking beneath them to the mountain one is blessed 10,000 times as they ripple in the breeze. In effect Chinese visitors can 'play' at being emperor as they climb the concrete steps to the summit since all the different 'developments' combine to provide a highly visible and experiential link through which they can symbolically span several millennia of their heritage. For them, the site is an example of 'man in harmony with nature'.

A second site provides another graphic example of the difference in approach which highlights the cultural differences inherent in the formulation of ecotourism policy. This is the development of Moli Forest Scenic Reserve, famous for 2000 years in Chinese history for two interlinked features: a waterfall deep in a subtropical rainforest where the oxygen levels are reputed to be so high that to walk to the waterfall is believed to re-invigorate the body and turn old age back into youth; and a hot spring pool in a valley at the foot of the waterfall where an imperial princess went daily to bathe and where a Buddhist temple 2000 years old venerates the spot. For many years a track from the Ruili River led up to the temple and hot spring pool and thence for 2 km through the old-growth forest to the waterfall.

In May 1998 a ferry trip down the Ruili River was introduced. It includes a 2-hour stop for passengers to walk to the waterfall. Some 12,000 visitors had undertaken the walk in the first 12 months, free of charge. Shortly thereafter a Taiwanese investor was granted approval to 'develop' the site. The track from the river bank to the temple was widened and concreted for tour buses, with a huge traditional Chinese 'Imperial' gate erected at the entrance. Two hotel blocks were constructed beside the hot spring pool and temple, both rectangular concrete chunks of no particular architectural merit, owing nothing to their surroundings in either shape, form or colour, clad in white ceramic tiles and roofed in orange-red tiles. The stream leading to the pool was bricked and the pool itself tiled in the manner of a Western, suburban, backyard swimming pool. A large restaurant-cum-karaoke hall was built opposite the temple. Five acres of forest at the entrance to the waterfall walk were cleared and landscaped with formal gardens, lawns, two huge aviaries and a lily pond complete with the mandatory carp and a bronzed ('traditional' wooden) waterwheel fountain. The aviaries were built, so the Taiwanese developer informed us, to display the birds and animals of the forest. The 2-km walk to the waterfall was to be 'developed' with concrete paths and bridges so losing the pristine state experienced at the time of our field visit. In its post-development state, Moli Scenic Reserve is expected to attract upwards of 50,000 visitors per year, with a Renminbi 20 (US$2.50) entrance fee.

When set against the Western concept of ecotourism, the Moli Forest development is seen as incongruous, as destructive of nature, as contradictory to wilderness management principles, the idea of catching forest birds and animals for the aviaries as inimical to sound conservation, the architecture of the hotel lacking all empathy with its environment. Yet to the Chinese, the development is an excellent example of 'man working in harmony with nature', of 'man improving on nature'. After all, the forests are so thick one cannot see the birds and animals and the aviaries are considered perfect for displaying them. The water of the thermal pool would often become cloudy, its bottom slimy with composting leaves and mud, and now it is filtered and clean and able to be used every day of the year. The landscaped gardens are much 'nicer' than untidy forest growth and the hotel provides facilities previously lacking. For Chinese the experience of a walk to the waterfall has been significantly enhanced (Sofield *et al.*, 1999).

There are hundreds of developments like Big Hollow Mountain and Moli Forest Scenic Reserve throughout Yunnan and China. Both of these are relatively small, involving capital expenditure of perhaps US$2 million. There are many much larger developments (e.g. Wudang Mountain) involving many millions of dollars.

Scale of Nature Visitation in China

Yunnan has a reputation in China as a 'green', forested province and while Yunnan's mountains may not be as famous as some others in China, its 'Three Rivers' – the Irrawaddy (Salween), the Mekong and the Yangtze – are very famous. They all rise in Yunnan and flow through impressive gorges between mountain ranges, where many of Yunnan's nature reserves are located. The Kunming branch office of the CTS (China Travel Service) organized tours to Yunnan's nature-based attractions for 40,000 visitors in 1998, before any of the five new nature reserves had been opened for tourism. The problem for the Forestry Department will be how to manage the millions who will want to see and experience Yunnan's nature reserves once they are opened to the public – and for that it requires a recreational management regime for tourism-specific land use planning for each reserve. As noted, ecotourism is basically small tourism not mass tourism. Ecotourism has a vital role to play in the conservation of the nature reserves but it cannot provide a tool for the Forest Department to handle the sheer scale of visitation waiting to access the mountainous forests. As general living standards rise, incomes increase, leisure time becomes an accepted part of daily life and the crowded cities grow ever more congested, millions of China's urban dwellers seek respite and recreation in the countryside and the forests. In 10 years, visitation to Yellow Mountain rose from 50,000 to more than 3 million in 1999, for example. Providing only low-key, low numbers access would not meet this demand. It would run the risk of ad hoc decision making which could destroy the integrity and conservation values of the reserves. A compromise was necessary: the strategy for Yunnan could not be restricted only to ecotourism but had to incorporate aspects of a broader planning regime for visitation to the reserves (Sofield *et al.*, 1999).

One possible answer lay in an application of the Recreational Opportunity Spectrum (ROS). The ROS is a tool for the management of nature reserves which is based on determining the best possible balance between the biodiversity values of the area in question, its geophysical features and their possible use for recreation. It was developed in the late 1970s to assist nature reserve managers to regulate recreational use of the areas under their control (Clark and Stankey, 1979). It incorporates a range of categories of land use inside and outside the boundaries of a nature reserve, it takes full cognizance of the biodiversity values of the environment, the level of degradation or otherwise of the geophysical features of the landscape, and then attempts to match appropriate recreational opportunities to those two defining characteristics. It is usually applied at the macro scale but may also be used for more specific detailed zoning and planning.

The ROS has six main categories of land use (called 'opportunity classes'): Primitive; Semi-primitive non-motorized; Semi-primitive motorized; Roaded natural; Rural; and Urban (Clark and Stankey, 1979). Primitive can be defined as:

- an area characterized by an essentially unmodified natural environment;
- fairly large in size;
- interaction between users is very low;
- evidence of other users is minimal;
- the area is managed to appear essentially free from evidence of human-induced restrictions and controls (although they may be strict in fact);
- motorized use within the area is prohibited (Clark and Stankey, 1979).

At the other end of the scale, 'Urban' is characterized by an entirely man-made or 'built' environment in which nature has been replaced and subjugated to human habitation.

Between the extremes, a range of conditions exist for which managers are able to develop planning to maintain diversity of recreational opportunities according to wilderness values and demands for access. The ROS combines the biocentric with the anthropocentric although its management philosophy is clearly grounded in the biocentric position. The highest values for conservation and protection will be located at one end of the spectrum, and lowest wilderness values (e.g. urban areas) will be located at the other end of the spectrum. Thus a reserve may contain a degree of endemism and the rarity, scarcity and/or vulnerability of flora and fauna may require their complete protection from any external intrusion or influence if their survival is to be assured. In such a case, with the exception of scientific study under carefully determined regulations, no access would be permitted. There may be areas of pristine wilderness within the reserve where any permanent human presence would undermine its wilderness qualities. Therefore no constructions (buildings, towers, communications facilities) of any kind would be permitted. In such a case there may be only one trail to provide access and a management regime may restrict numbers of tourists to a pre-determined maximum per day (e.g. 100, in groups of a maximum size of ten, each group departing from a staging spot at 1 hour intervals so that the possibility of the groups merging is limited and the wilderness experience for each group may be maximized). At the other end of the scale there may be a river or natural swimming pool within the reserve where there are few significant biological values, and where forms of mass tourism may be appropriate (e.g. a major picnic spot beside the pool or along the banks of the river, with paths, fences and other such devices to ensure a modicum of control, to prevent erosion, etc.). A completely degraded area of no inherent biological value (e.g. a clear-felled eroded area, a former quarry) may provide an appropriate site for a car park, toilet block, small food outlet, park rangers' quarters, etc.

In effect the ROS was deemed to be able to deliver a middle way (*zong yong ji dao*[3]), consistent with the values of both Chinese and Western concepts of wilderness/*huangyie*. It provided a sound basis on which to plan for mass tourism that would be based on the principles of sustainable development of nature reserves for visitation, with conservation as the core element of interpretation. Given the anticipated scale of tourism it had to be accepted that those few sites selected for mass tourism would inevitably undergo significant adaptation. They would have to have high touristic values (good scenic qualities, spectacular landscapes or formations, high quality forests, etc.) but would not incorporate environmentally significant areas which required full protection. They would become sacrificial areas to keep most visitors out of the sensitive core areas. Some degradation of the sites would be accepted: for example, the large numbers of tourists would drive some birds and animals out of the vicinity, and a car park and other facilities such as a shop and toilets would replace some forest cover. But hardened trails, fences, and other management measures would restrict adverse impacts. This is an application of the concept of Limits of Acceptable Change, LAC. This is a process that requires 'managers to identify where and to what extent varying degrees of change are appropriate and acceptable' in assessing visitor impacts on a site (Hendee *et al.*, 1990: 221). If the tourism industry were free to develop the reserves without Forest Department jurisdiction and the application of the ROS, inevitably sites of high environmental values would be the focus of their attention and the profit

---

[3] Confucian thought extolled 'the middle way': avoidance of extremes combined with a blending of both ends, and one of his four famous texts known to all Chinese is titled *Zong Yong* (The Middle Way).

motive would drive their involvement. Any conservation measure which either costs money or restricts earning potential (such as limits on visitation numbers) would be opposed. The ROS in its macro land use planning and its micro applications was thus judged able to provide a tool for restraining undesirable business interests, channelling desired investment into selected areas, and maintaining conservation as the key principle for nature reserve access and their utilization for tourism. A balance between the anthropocentric and biocentric approaches could be reached (Sofield *et al.*, 1999).

The following example is provided to illustrate the way in which an application of the ROS was designed by the authors for training purposes to develop a greater understanding by Forest Department staff of conservation-based ecotourism planning within a nature reserve. The theoretical reserve is mapped and areas classified according to the six ROS classes: Primitive (Wilderness); Semi-primitive non-motorized wilderness; Semi-primitive motorized; Roaded natural; Rural; and Urban. There is a small town (Urban) just outside the north-west boundary of the reserve.

In examining the natural resources of the reserve that could be utilized for tourism, five waterfalls have been identified: two in the Primitive (Wilderness) zone; one in the Semi-primitive non-motorized

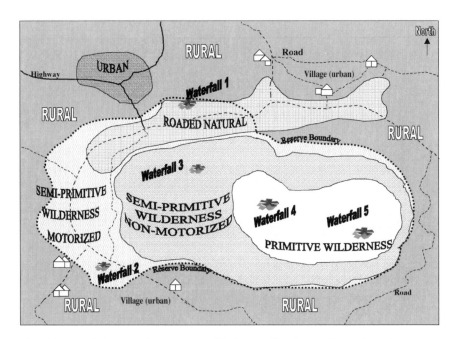

Fig. 8.1. Spatial areas of a park classified according to the Recreational Opportunity Spectrum.

(Wilderness) zone; one in the Semi-primitive motorized zone; and one in the Roaded natural zone.

The five waterfalls may then be classified according to their relative environmental values on a scale of 0–100 and their relative touristic values assessed, also on a scale of 0–100. Waterfalls 4 and 5 are both in virgin forest and provide a habitat for rare and endangered plant species and endangered birds and animals, with respective environmental values of 95 and 85. Their height, volume of water, and scenic value of the gorges and cliffs over which they plummet are high in touristic values: 92 and 80 respectively. Waterfall 3 is in a wilderness area which may have been selectively logged in the past 50 years (hence, the classification 'semi-primitive' because it has limited old-growth forest), but it retains relatively high biodiversity values (70). It also has high scenic qualities (70). Waterfall 2 is in a semi-primitive area which is bisected by a road, its surrounding forests have been logged and degraded, and its scenic qualities are not very high (environmental value, 30; touristic value, 40). Waterfall 1 is in a Roaded natural area, which was once impacted by swidden agriculture but has now reverted to natural vegetation; its biodiversity values are even lower (20). Its forests have been severely degraded, the walls of its gorge are eroded, and scenically it has a low attraction rating (28). The waterfalls may be graded according to Fig. 8.2.

When the ROS is utilized for the waterfalls, taking account of the underlying principle of conservation, a range of different tourist applications to satisfy a range of different visitor types could be devised.

Waterfall 5 would be unavailable for general tourist visitation. Its high environmental values require maximum protection and would be

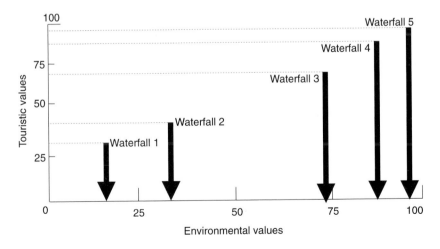

Fig. 8.2. Classification of waterfalls.

of significant scientific interest, hence entry into the area would be restricted for scientific study.

Waterfall 3 also has some limited scientific interest but it could be selected to satisfy the general tourist demand for visitation to the reserve. The road could be extended to a point where a car park and lookout could be constructed, with short trails to the waterfall itself. This waterfall would be designated for mass tourism and in effect become a sacrificial area to keep most visitors out of the Primitive (Wilderness) zone. The relatively high tourist qualities of the waterfall would satisfy most tourists. The decision to make this waterfall available for mass tourism would mean that the land use zone would need to be reclassified to Semi-primitive motorized, to enable the road, car park and tour bus facilities to be provided.

Waterfall 4 could become an **ecotourism** site. Access would be limited and designed along stringent conservation management guidelines. The trail from the car park of Waterfall 3 would be deliberately designed for a certain degree of difficulty rather than easy for comfortable walking; it would be long (perhaps 5 km or more); and only small numbers would be permitted to walk along it, each group mandatorily accompanied by a trained guide. Its scientific and environmental values would be mostly retained.

Waterfall 1 could also become an ecotourism site. This would appear paradoxical, given its low environmental values. However, there are examples of successful ecotourism to degraded sites which actively involve the tourists in helping to rehabilitate the environment (e.g. montaine reserves in Costa Rica). This type of ecotourist product takes advantage of the desire of many people to be active conservationists willing to make a personal contribution to improving degraded sites. Thus visitors to Waterfall 1 could participate in tree planting, erosion control and eradication of undesirable exotic plant species. This approach would 'fit' with the Chinese Taoist value of 'man improving on nature', but it must be noted that few Chinese *tourists* adhere to this principle; as *tourists* they demand to be entertained, not to be put to 'work'.

Waterfall 2 might be left for the occasional visitor to find his or her way there but no active attempt would be made to develop it for tourism. The Forest Department would utilize its own resources (park rangers) to carry out site upgrading through tree planting, erosion controls, and so on, over an extended period of time. Its low biodiversity value and its relatively low tourist value would accord it the lowest priority.

This exercise, in Chinese fashion, was quickly christened by workshop participants as 'The Five Waterfalls Model'. It has been used by the authors in a series of seminars in other parts of China because of its success in disseminating conceptual complexities through a practical, concrete process which helps to bridge the cultural differences between Western and Chinese paradigms of development.

The ecotourism strategy for Yunnan's five new nature reserves was thus constructed with:

- the ROS as its overarching framework;
- a management philosophy for recreation and tourism based on the natural resources of the reserves which placed conservation above short-term economic exploitation and was therefore more biocentric than anthropocentric;
- the provision nevertheless of a select few sites in each reserve for mass tourism; where
- ecotourism was just one part of the wider nature-based tourism system, with an emphasis on an integrated, holistic approach; and a series of ideal types (archetypes or modules) which could be applied, with specific modifications, to match each particular site and targeted communities located near different nature reserves.

In considering ecotourism ventures that could involve Minorities communities living around the nature reserves, archetypes included village homestays, guided ecotourism trails, interpretation centres, water activities (canoeing/rafting), forest platforms/hides, and a mix-and-match model incorporating various components of any or all of the five archetypes mentioned above (Sofield *et al.*, 1999). Each of these archetypes was designed to take advantage of the intimate knowledge (both biological and cultural) of the forests held by the Minorities people in order to enhance interpretation and permit benign use of the forests as a resource for ecotourism. An extensive series of workshops with government officials, forestry staff and other stakeholders and decision makers was a necessary component of the strategy to promote understanding of the ROS and its application, before embarking on specific ecotourism developments. With reference to the latter, some ten different sites/ventures were identified as model projects for full feasibility, environmental and socio-cultural impact studies. These workshops commenced in late 2000 and continued throughout 2001.

Summary

The entire process of introducing an ecotourism strategy for Yunnan's forest reserves may at a philosophical level be interpreted as an engagement between Said's Orientalism and its counterpart Westernism (Said, 1978). This issue arises because of the nature of the consultancy that assumed an unquestioning 'Western/modern' authority accompanying the ownership of the term 'ecotourism', evident in the very terminology used to describe the team leader – 'foreign expert'. Cultural diversity, as revealed on the one hand in the Chinese anthropocentric approach to nature and on the other hand in the Western paradigm of ecologically sustainable development and conservation, is

> an epistemological object – culture as an object of empirical knowledge – whereas cultural difference is the process of enunciation of culture as knowledgeable, authoritative ... If cultural diversity is a category of ethnology, cultural difference is a process of signification through which statements of culture or on culture differentiate, discriminate and authorize the production of fields of force, reference, applicability and capacity.

> (Bhabha, 1994: 34)

The formulation of an ecotourism strategy for Yunnan involved a problem of cultural interaction which emerged at the significatory boundaries of culturally derived different paradigms (Chinese and Western), where meanings and values were initially misunderstood or misread. The terms of reference on ecotourism, being a product of the Yunnan Forest Department, led to an initial belief (by us as the foreign experts) that the terminology of ecology and the concept of Western ecotourism were both accepted and understood; and all that was required was an exercise in implementation, i.e. to complete the consultation according to the given guidelines. The deeper we went into the process of policy formulation, however, the more the cultural diversity turned into cultural difference; therefore, at those points where there was a loss of meaning, culture emerged as problematic.

This contestation centred on 'the problem of the ambivalence of cultural authority', as Bhabha (1994: 34) has described it: the formulation of a policy based on external values in the name of 'expertness', i.e. the authority of culture as a knowledge of referential truth. 'International experts' were considered to have that authority. There was thus an acceptance by the Chinese authorities that a biocentric approach should be adopted in preference to any anthropocentric approach as the foundations of an ecotourism strategy. This was so, largely because the Forest Department had already adopted a policy that had resulted in the establishment of a network of nature reserves based on international best practice. Reserves were accepted as fundamental to the conservation of the biodiversity of the Province and were set up with two zones: a core zone which is designed to protect key biota and is a zone of almost total exclusion (only forest rangers and accredited scientists allowed entry); and an experimental zone (buffer) around the core zone where a range of activities might be permitted.

At another level, the formulation of an ecotourism policy for Yunnan introduces elements of globalization because it links local understandings with worldwide concepts that transcend national boundaries. In this context ecotourism in Yunnan may be seen as a vehicle for modernization (somewhat paradoxically given ecotourism's 'normal' connotation with untrammelled wilderness in the Western philosophical paradigm), and as an agency for transnational flows of ideas about environmental conservation and ways to manage nature reserves. It thus raises the thorny issue of the ethics of an etic-driven approach being introduced by an outside agency and superimposed on existing value systems. (An etic approach is one based on 'outsider' values which makes judgements according to those foreign values, rather than an emic, 'insider' or actor-oriented approach derived from the value system of the culture/society in question.) The ethics of imposing a policy based on external, foreign values never arose in a formal way throughout the consultancy, although it could be argued that in fact the formulation of an ecotourism policy for Yunnan had this profound issue at its core. The acceptance of such a policy by the Yunnan authorities, however, demonstrated the cooptation of the concepts of the western paradigm of ecotourism into mainstream Chinese policy for nature reserves. This in turn leads to the creation of a degree of homogenization and thus an extension of the process of globalization.

At yet another level it may be argued that the Western approach to parks and nature reserves is also anthropocentric since they are themselves cultural constructs. As Hendee *et al.* (1990: 28) acknowledged, 'in the final analysis, wilderness is a state of mind ... defined by human perception'. However the major differences in management, where Western biocentric concerns override human comfort argue for the validity of drawing such a distinction between the Western and Chinese paradigms and the very different meanings attributed to the term 'wilderness'. This is not pedantic because implementation of management policies will result in the conservation of biodiversity at the expense of human comfort in Western parks whereas in China biodiversity may be sacrificed for human comfort.

This issue also raises the question of what has been termed 'the implementation gap' (Dunsire, 1978), when a break occurs between policy intention and actual result. The 1994 Chinese legislation on nature reserves has clearly defined goals for the protection of natural resources, but the rhetoric bears little relationship to management practices. There is widespread tolerance of – indeed, sometimes vigorous official encouragement of and support for – prohibited activities in core areas, a lack of environmental assessment, a total absence of monitoring of impacts and a paucity of conservation education. Tourism development inevitably means in Chinese terms the construction of something for economic gain; the tenet of minimizing human intervention/presence that underlies the Western concept of ecotourism is lacking. Instead, the Chinese concept of 'man improving on nature' is evident in many of the activities undertaken inside the boundaries of nature reserves; the tension between Western and Oriental values about ecotourism is not confined to Yunnan Province but is present throughout reserves in all provinces.

One of the answers to the conundrum faced in Yunnan was to adopt the Chinese *zong yong ji dao* ('the middle way') and develop a strategy which incorporated both the anthropocentric and biocentric approaches. Thus the application of the ROS provided an environmentally sound approach for management of the reserves allowing for 'development' of sites able to handle very large numbers of tourists (upwards of 5000 per day). This was considered absolutely essential. The pressures for large-scale, mass tourism with a range of facilities accepted by Chinese domestic visitors as necessary for a quality experience (pavilions, restaurants, etc.), though distant from Western values of conservation and wilderness considered equally necessary to conserve and protect the environment, could not be ignored. Planning had to take account of the Chinese desire to visit reserves in extremely large numbers and any attempt to pursue rigid Western ideology and restrict planning to small-scale ecotourism ventures could have put the entire reserves at risk. Two or three 'sacrificial' sites per reserve able to satisfy the anthropocentric approach (man improving on nature) carry the probability that the remaining 98% of the reserves can be protected.

Acknowledgements

The authors are indebted to members of the Sino-Dutch Forest Conservation and Community Development Programme, especially the Chinese and Dutch team members who gave so generously of their time and expertise; and the many Chinese County and Prefecture officials and Forestry field staff who supported our fieldwork. Special thanks are due to the Australian Cooperative Research Centre for Sustainable Tourism who facilitated the authors' participation in the Sino-Dutch Programme.

References

Bhabha, Homi (1994) *The Location of Culture.* Routledge, London.
Callicott, J.B. and Ames, R.T. (eds) (1989) *Nature in Asian Traditions of Thought: Essays in Environmental Philosophy.* State University of New York Press, New York.
Chan, Wing-tsit (1969) *A Source Book of Chinese Philosophy.* Colombia University Press, New York.

China Daily, (1999) China records great increase in tourism. *China Daily,* 8 December 1999, p.4.

China National Tourism Administration (1999) *Yearbook of China Tourism Statistics, 1998.* CNTA, Beijing.

China State Council (1998) *State Level Scenic Wonders and Historical Sites.* Information Office, Beijing.

China State Council Information Office (1999) *National Minorities Policy and its Practice in China.* A White Paper issued 27 September 1999. Information Office, Beijing.

China Travel Service (1998) The 'Top Forty' most favoured tourist spots. *Special Zones Herald,* Haikou, 28 January 1998.

Chinese National Committee for MAB (2000) *Study on Sustainable Management Policy for China's Nature Reserves.* Chinese National Committee for Man and the Biosphere Programme, Beijing.

Clark, R.N. and Stankey, G.H. (1979) *The Recreation Opportunity Spectrum: a Framework for Planning, Management and Research.* General Technical Report PNW-98. US Department of Agriculture and Forest Service, Portland, Oregon.

Dunsire, A. (1978) *Implementation in a Bureaucracy.* Martin Robertson, Oxford.

Elvin, M. (1973) *The Pattern of the Chinese Past.* Stamford University Press, Stamford, California.

Han, Nianyong (2000) Analysis and Suggestions on Management Policies for China's Nature Reserves. In: Chinese National Committee for MAB *Study on Sustainable Management Policy for China's Nature Reserves.* Chinese National Committee for Man and the Biosphere Programme, Beijing, pp. 3–32.

Hendee, J.C. and Stankey, G.H. (1973) Biocentricity in Wilderness Management. *BioScience* 23(9), 535–538.

Hendee, J.C., Stankey, G.H. and Lucas, R. (1990) *Wilderness Management.* 2nd edn. North America Press, Golden, British Columbia.

Huangzhou County Government (2001) *Ecotourism Plan for the Development of East Lake.* Department of Planning, Huangzhou.

Illustrated History of China's Five Thousand Years (1994) [No place of publication, no publisher. In Chinese.]

Lew, A.A. and Yu, L. (eds) (1995) *Tourism in China. Geographical, Political and Economic Perspectives.* Westview Press, Boulder, Colorado.

Li, Wenjun (2000) Study on ecotourism management for China's nature reserves. In: Chinese National Committee for MAB *Study on Sustainable Management Policy for China's Nature Reserve,* Chinese National Committee for Man and the Biosphere Programme, Beijing, pp. 68–75.

Nash, R. (1982) *Wilderness and the American Mind.* Yale University Press, New Haven, Connecticut.

Overmyer, D.L. (1986) *Religions of China: the World as a Living System.* Harper and Row, San Francisco.

Petersen, Ying Yang (1995) The Chinese landscape as a tourist attraction: image and reality. In: Lew, A.A. and Yu, L. (eds) *Tourism in China. Geographical, Political and Economic Perspective.* Westview Press, Boulder, Colorado, pp. 141–154.

PRC (1994) *Regulations of the Peoples Republic of China on Nature Reserves 1994*. Information Office, Beijing.

Rawson, P. and Legeza, L. (1973) *Tao. The Chinese Philosophy of Time and Change*. Thames and Hudson, London.

Ropp, P.S. (ed.) (1992) *Heritage of China: Contemporary Perspectives on China*. University of California Press, Berkeley.

Said, E. (1978) *Orientalism. Western Conceptions of the Orient*. Penguin, Harmondsworth, UK.

Shaner, D.E. (1989) The Japanese experience of nature. In: Callicott, J.B. and Ames, R.T. (eds) *Nature in Asian Traditions of Thought: Essays in Environmental Philosophy*. State University of New York Press, New York, pp. 163–181.

Sofield, T.H.B. (2000) Regional tourism planning and development through international cooperation. Keynote paper delivered at the First Korean International Symposium on Regional Tourism, Chokso, Korea, September 2000.

Sofield, T.H.B. and Li, F.M.S. (1996) Rural tourism in China. In: Page, S. and Getz, D. (eds) *The Business of Rural Tourism*. International Thomson Business Press, London, pp. 120–142.

Sofield, T.H.B. and Li, F.M.S. (1998) China: tourism development and cultural policies. *Annals of Tourism Research* 25(2), 323–353.

Sofield, T.H.B. and Bhandari, S.P. (1998) *An Independent Evaluation of the Partnership for Quality Tourism Programme, UNDP, Nepal*. UNDP, Kathmandu.

Sofield, T.H.B., Li, F.M.S. and Khang, Yunhai (1999) *An Ecotourism Strategy for Nature Reserves in Yunnan Province, China. A Report for the Sino-Dutch Forest Conservation and Community Development Programme*. CRC Tourism, Brisbane.

Spence, J. (1992) Western perceptions of China from the late sixteenth century to the present. In: Ropp, P.S. (ed.) *Heritage of China: Contemporary Perspectives on China*. University of California Press, Berkeley, pp. 1–14.

Swain, M.B. (1993) Women producers of ethnic arts. *Annals of Tourism Research* 20(1), 32–51.

Tellenbach Hubertus and Bin Kimura (1989) The Japanese concept of 'Nature'. In: Callicott, J.B. and Ames, R.T. (eds) *Nature in Asian Traditions of Thought: Essays in Environmental Philosophy*. State University of New York Press, New York, pp. 153–162.

Tu Wei-Ming (1989) The continuity of being: Chinese visions of nature. In: Callicott, J.B. and Ames, R.T. (eds) *Nature in Asian Traditions of Thought: Essays in Environmental Philosophy*. State University of New York Press, New York, pp. 67–78.

UNESCO (1999) List of biosphere reserves, March 1999. Available from www.unesco.org/mab/brlist.htm (accessed 10 February 2000).

Xue Dayuan (2000) Study on utilization and management of natural resources in China's nature reserves. In: Chinese National Committee for MAB *Study on Sustainable Management Policy for China's Nature Reserves*. Chinese National Committee for Man and the Biosphere Programme, Beijing, pp. 59–67.

Yunnan Province (1999) *General Policy Statement on the Construction of*

Yunnan Province into a Great Province of Nationalities Cultures. Yunnan Province Government Information Office, Kunming.

Zhang Baosan (1998) *The Biology of Yunnan Province.* Forestry Department, Kunming.

Zhang Guangrui (1995) China's tourism development since 1978: policies, experiences and lessons learned. In: Lew, A.A. and Yu, L. (eds) *Tourism in China. Geographical, Political and Economic Perspective.* Westview Press, Boulder, Colorado, pp. 3–16.

Zhuge Ren (2000) Questionnaire survey on participation of local communities in nature reserve management. In: Chinese National Committee for MAB *Study on Sustainable Management Policy for China's Nature Reserves.* Chinese National Committee for Man and the Biosphere Programme, Beijing, pp. 76–85.

Ecotourism Development and Government Policy in Kyrgyzstan

9

Karen Thompson[1] and Nicola Foster[2]

[1]School of Leisure, Hospitality and Food Management, University of Salford, Frederick Road, Salford M6 6PU, UK; [2]Centre for Tourism, School of Sport and Leisure Management, Sheffield Hallam University, 1127 Owen Building, City Campus, Sheffield S1 1WB, UK

Introduction

During the Soviet era, the Ysyk-Köl region of northern Kyrgyzstan was one of the Union of Soviet Socialist Republics' (USSR) most important recreational resorts. With the demise of the Soviet Union, the Great Silk Road became the major focus for international tourism in Central Asia. The few historical and cultural remnants of the Silk Road which remain in Kyrgyzstan cannot compete with the outstanding architectural heritage of Samarkand, Bukhara and Khiva and, despite assistance from the World Tourism Organization (WTO), Kyrgyzstan profits little from Silk Road tourism. None the less, international tourism has been identified as an important area for economic growth by the government of Kyrgyzstan. Its remote location and mountainous topography have led to destination marketing in Kyrgyzstan being focused on adventure, nature and ecotourism. Furthermore, the close relationship of the formerly nomadic Kyrgyz people with their land, heritage and culture facilitates the adoption of ecotourism principles within the host community.

Geographical Situation of Kyrgyzstan

The former Soviet state of Kyrgyzstan covers an area of 198,500 km^2, landlocked by Kazakhstan, China, Tajikistan and Uzbekistan (Fig. 9.1).

© CAB International 2003. Ecotourism Policy and Planning
(eds D.A. Fennell and R.K. Dowling)

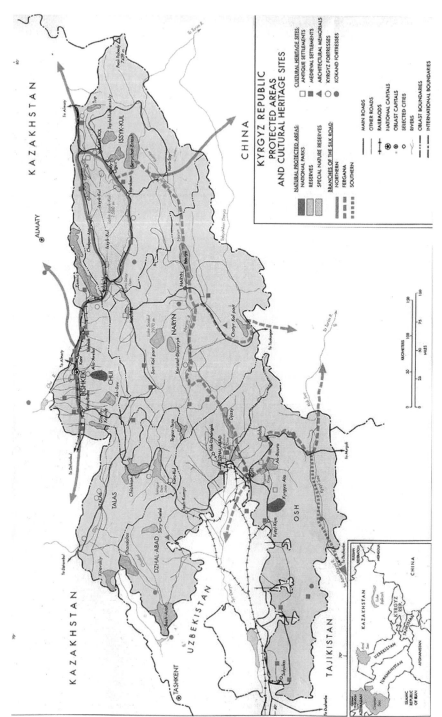

Fig. 9.1. Kyrgyz Republic: protected areas and cultural heritage sites.

Mountains cover 95% of Kyrgyzstan's territory. Pobedy (Victory) Peak is the highest at 7439 m, but almost half of Kyrgyzstan is above 3000 m elevation. The majority of Kyrgyzstan's 4.9 million population live in two flat and fertile valleys; the Chui valley in the north, where the capital Bishkek is situated, and the Ferghana valley in the south. Large areas of the country are remote and accessible only with difficulty and protect a wealth of rare wildlife and alpine plants. Crucially, Kyrgyzstan has not suffered from the problems of pollution, deforestation and litter that have affected other mountain tourism destinations like Nepal.

Tourism in Kyrgyzstan

While data on tourist arrivals in Kyrgyzstan are not available before 1995, it is clear that during the Soviet era, Kyrgyzstan had been an important leisure tourism destination for visitors from all over the Soviet Union. Kyrgyzstan was able to offer a variety of tourism products ranging from sports camps and sanatoria to hunting and skiing. Its unique flora and fauna and the dry, sunny climate ranked Kyrgyzstan alongside the Crimea and Black Sea for Soviet holiday-makers. The centre of tourism in Kyrgyzstan was Lake Ysyk-Köl, a large mountain lake, 182 km long, situated in the Alatau Mountains. The waters of Lake Ysyk-Köl have a high mineral content, and the northern shores of the lake were developed as a health resort where trade union sanatoria accommodated visitors from all over the USSR. The area was also popular with Soviet officials.

> By all accounts, during the pre-democratic period this area [Lake Ysyk-Köl region] was alive with tourism business from within the Soviet Union. Next to the Black Sea it was one of the most coveted vacation destinations in the union.
>
> (Eckford, 1997: 3)

The dissolution of the Soviet Union largely removed Kyrgyzstan's leisure tourism market and led to a situation of over-supply. In 1996, a total of 52,865 bed spaces existed in state-funded accommodation (Khanna, 1996). However, in the same year the Kyrgyz State Agency for Tourism and Sport (KSATS) recorded only 41,650 international visitors, of which around 80% were estimated to be business visits and just under 30,000 were from former Commonwealth of Independent States (CIS) countries (KSATS, Statistics and documentation provided to authors).

Prior to the disintegration of the Soviet Union, Kyrgyzstan was already among the poorest of the Soviet regions with a gross national product per capita of US$1550 in 1991 compared with US$2470 in Kazakhstan (Anderson, 1999). The creation of the new Kyrgyz Republic, combined with the state of emergency declared in 1990 as a result of fighting on the Kyrgyz–Uzbek border, led to the onset of economic crisis in Kyrgyzstan. The level of inflation jumped from 200% in 1991 to 900% in 1992 (Anderson, 1999). As economic reform progressed and external investment and assistance increased, tourism was identified as an important industry sector and potential means of attracting revenue from the developed world. A programme was set in place to privatize existing tourism resources and encourage the development of new products and services. Kyrgyzstan became a member of the WTO in 1993.

Existing Tourist Facilities

While the natural and cultural elements of tourism supply in Kyrgyzstan remain a valuable resource, the facilities and service mentality (lack of a service ethic) bequeathed by the Soviet era place constraints on future tourism development. These constraints were highlighted in a report commissioned by the State Committee for Tourism and Sport of the Kyrgyz Republic (forerunner to the KSATS) in 1996 (CBO Consulting, 1996). International access to Kyrgyzstan is normally by road from Almaty in Kazakhstan due to the lack of direct flights. While it was hoped that the reconstruction of Manas Airport, funded by the Japanese Government in 2000, would attract new international routes, this has yet to be realized. Due to a lack of competition, travel costs are high. Travel within Kyrgyzstan is also fraught with difficulties due to limited road and rail links which are poorly maintained. These difficulties are exacerbated by the topography of the country. Accommodation facilities in Bishkek dating from the Soviet era, and the sanitoria on the northern shore of Lake Ysyk-Köl, are badly designed and constructed and environmentally unsympathetic to their surroundings, while other areas remain entirely undeveloped. Communications are poor outside Bishkek, and few workers in the tourism and hospitality industries speak foreign languages.

None the less, the push to transfer ownership within the tourism industry from the state to the private sector has resulted in an overall improvement in the quality of tourism products and services, led by foreign investment and indigenous entrepreneurship. By 1999, only 4 of 249 accommodation providers (constituting 350 bedspaces of a total 15,121) in Kyrgyzstan were state owned (KSATS, statistics and documentation provided to the authors). In Bishkek, two Western hotel chains have constructed quality hotels for business travellers. In towns outside the capital, however, accommodation is mainly provided in small, family run hotels or in private homes. Several tour operators have established themselves in Bishkek and are making use of the Internet to market their products.

Existing Tourism Markets

In 1994 the WTO launched a project to market the tourism product of the Central Asian republics as a linear attraction by capitalizing on interest in the ancient Silk Road. By 1997 this scheme involved 18 countries with Silk Road heritage (WTO, 1997). This scheme may have been partially responsible for increased arrivals to Kyrgyzstan from 1995 onwards, however, Table 9.1 shows that the majority of visitors have been attracted from former CIS states, whereas WTO marketing efforts were concentrated in Germany and the UK. None the less, as can be seen from Table 9.1, while arrivals from former CIS countries peaked in 1997, arrivals from other countries appear to remain on the increase.

The tourism market in Kyrgyzstan is still relatively insignificant (Table 9.1). By way of contrast, Nepal received 418,000 international visitors in 1997 (WTO, 1998). Furthermore, Table 9.2 shows that 57%

Table 9.1. International arrivals to Kyrgyzstan by area of residence 1995 to 1999. (Source: Kyrgyz State Agency for Tourism and Sport, statistics and documentation provided to the authors.)

Area of residence	1995	1996	1997	1998	1999
Commonwealth of Independent States	25,710	28,625	72,202	46,287	31,229
Other	10,713	13,025	15,184	13,076	17,051
Total	36,423	41,650	87,386	59,363	48,280

Table 9.2. Ten main markets for inbound tourism to Kyrgyzstan by purpose of visit 1999. (Source: Kyrgyz Office of National Statistics, statistics and documentation provided to the authors.)

Country of origin	Visitor arrivals	Visitor nights	Leisure	VFR	Business	Health/ study
Kazakhstan	13,671	84,758	10,427	7	3,236	1
Russia	10,518	33,505	1,125	1	9,390	2
Uzbekistan	6,249	46,116	3,721	0	2,526	2
USA	2,868	8,301	628	13	2,224	3
Japan	1,934	7,560	446	0	1,488	0
France	1,696	5,658	854	0	842	0
Germany	1,695	5,955	478	0	1,217	0
Turkey	1,689	5,120	0	43	1,646	0
China	1,416	4,051	29	0	1,387	0
United Kingdom	1,398	4,747	385	0	1,011	2
'Other'	5,147	23,633	2,535	62	2,544	5
All countries	48,280	229,404	20,628	126	27,511	15

VFR, visiting friends and relatives.

of visitors to Kyrgyzstan from the ten main geographic markets in 1999 were travelling for business purposes. France was the only non-CIS country from which leisure visitors outnumbered business visitors in 1999.

Existing Tourist Activities

The traditional tourist activities, centred on health and sports, hold little attraction for the markets of the developed world and Kyrgyzstan is now marketing itself as a destination for nature tourism, adventure tourism and cultural tourism. The major tour operators offer mountaineering, trekking, spelunking (cave exploration), climbing, skiing, hunting, botanical and ornithological tours and cultural and historical tours. Activities are concentrated in the mountains and along the route of the ancient Silk Road. Figure 9.1 shows the location of natural protected areas and cultural heritage sites in Kyrgyzstan.

Weaver (1998) distinguishes between active and passive ecotourism. Active ecotourism occurs where there is a change in attitude or lifestyle of the tourist as a result of the experience, and where there is evidence of positive environmental impacts at the destination (Weaver, 1998). The environmental and social impacts of ecotourism in Kyrgyzstan will be discussed below, in connection with the Issyk-Kul Biosphere Reserve Project. Changes in attitude and lifestyle are more difficult to measure, but may result from educational activities. There is some evidence of an educational focus in nature tourism throughout Kyrgyzstan. Botanical and ornithological tours take place in alpine nature parks and the nut forests of southern Kyrgyzstan. In the north, the Biosphere Reserve Issyk-Kul promotes two-way education on environmental awareness between local people and tourists (Biosphere Reserve Issyk-Kul, 1997: 38). However, while the larger tour companies are keen to stress their use of local expertise, educating tour guides in methods of interpreting Kyrgyz culture and heritage for the visitor and in the more general principles of ecotourism has been identified as an area for improvement (Touche Ross & Co/International School of Mountaineering, 1995: 28; KSATS, 2000).

Passive ecotourism requires only that tourism has minimal negative impacts on the physical environment (Weaver, 1998). Environmental impacts of nature and adventure tourism in Kyrgyzstan are limited both naturally and by government intervention. The inaccessibility of the destination and the cost of the journey limit the number of tourists willing to travel. At the same time, the underdeveloped infrastructure and minimal facilities attract 'elite' tourists, who use pre-arranged, native facilities, are small in numbers and adapt easily into surrounding environments (Smith, 1989). Eckford (1997: 3) notes that in the Ysyk-Köl region '…aside from a few wealthy group tourists, elite mountaineers, diplomats and resident expatriates from nearby republic capitals, few outsiders have an opportunity to visit.' Thus the impact on the destination remains limited.

Kusler (1991) identifies the phenomenon of 'ecotourists on tour'. Highly organized tours are common in Kyrgyzstan, as they are in many remote and underdeveloped destinations. Tour operators liaise with local communities and families to provide for the basic needs of the tourist. The Kyrgyz are a formerly nomadic people and attempts at collectivization have only partially succeeded. Organized tours often travel on horseback and are accommodated in yurta (felt tents) with family groups who have relocated to higher ground for the duration of the summer. This practice is mutually beneficial to host and guest. It is precisely this type of activity that the Kyrgyz government is keen to encourage, as it is felt that the host population can be relied upon to confer their strong sense of environmental and cultural heritage to the tourist.

While the impact of tours organized by local operators remains limited (almost by default) there have been concerns that the need for outside investment and training will result in ownership within the tourism industry being removed from the host population. In addition, there is a need to regulate other types of tourist activity; in particular, hunting, skiing and high altitude mountaineering. On 6 November 1995, a 'Conception of development of tourism in the Kyrgyz Republic through 2000' was adopted by the government of Kyrgyzstan (Lozovskaya, 1996). While one of the primary objectives of this policy was the consolidation of tourism as a key sector of the economy, environmental and social concerns were given equal priority.

The Political Context of Kyrgyzstan

In essence, Kyrgyzstan may be described as a 're-developing country'. The legacy of the USSR remains, whereby development, albeit restricted and 'embryonic' by Western standards, took place. The transition from a centrally planned economy to democracy remains a challenge, hindered by political and economic instability.

The parliament of Kyrgyzstan (Jogorku Kenesh) has been established based on a presidential system of government, a system that became the focus of national dissatisfaction during the collapse of the Kyrgyz economy in the mid-1990s. Rationalization has recently taken place. It has been reported that the number of government bodies has been reduced by approximately 30% (TCA, 2001). At a regional level, seven administrative provinces, termed 'oblasts', exist: the capital city Bishkek; Naryn; Ysyk-Köl; Chu; Talas; Osh and Jalal-Abad. The last two oblasts are situated in the south of Kyrgyzstan. Critics have challenged the process by which oblast governors have been appointed. In theory the governors possess considerable autonomy from central government in terms of economic and social functions such as tax collection and pensions. However, in practice, proportional representation appears weak. Notably, the appointment of oblast governors, decided by the Kyrgyz president, Askar Akayev (a northerner), favours northern rather than southern citizens.

Strong evidence of a north–south divide exists, both politically and economically. Southern Kyrgyzstan was historically in favour of colonization by Britain rather than Russia on the basis that 'the British would be easier to get rid of later' (Alymbek-datka cited in Osorov, 2000), while the north was historically in favour of colonization by Russia. With respect to the path that history took it is perhaps no coincidence that the citizens of northern Kyrgyzstan have dominated positions of power in government for the majority of the last two centuries. Inter-regional tension within the country is a major factor affecting Kyrgyzstan's relations with neighbouring countries (the north is sympathetic to Kazakhstan and China while the south, incorporating a heavy concentration of the ethnic population of Uzbeks in the city of Osh, sympathizes with Uzbekistan). Significantly, Uzbekistan has been reported to be emerging as the strongest state in post-Soviet Central Asia, thus the potential for inter-regional conflict remains.

The promise of increased government powers for southern Kyrgyzstan in the pre-election addresses of President Akayev in 1990 has still to be honoured. The country's northernmost town, Pishpek (renamed Bishkek), was established as the capital of the country. Received standard Kyrgyz, formed on the basis of the northern dialect, was established as the official national language and the political elite was, and still is, composed of 80% northerners. Despite such anomalies, Kyrgyzstan has received international praise for its democratic principles and is generally recognized to be the most democratic, albeit the poorest, of all former USSR countries.

However, the extent to which Kyrgyzstan is successfully achieving democracy may be questioned. Allegations have been made that, '...the parliament in Kyrgyzstan, like in any other CIS country, is practically inaccessible to the citizens and public organizations if they do not have any personal ties in the parliament' (Tayanova, 2000: 8). Although the actual structure of the Jogorku Kenesh facilitates democratic governance, in practice, accusations of corruption impede the full acceptance of democracy by Kyrgyz citizens.

Economic Development

Historically, a heavy dependence on and excessive specialization through trade with the USSR has characterized the Kyrgyz economy.

To provide an idea of the degree of insularity, during the period 1989–1991 Kyrgyzstan's exports to the rest of the world (i.e. countries outside the USSR) amounted to less than 2% of total exports (Consortium for International Earth Science Information Network, 1993).

Following independence from the USSR, Kyrgyzstan has been proactive in terms of developing international attention and integrating into the world market (Guttman, 1999). By early 1993, only 2 years after gaining independence from the Soviet Union, it was reported that Kyrgyzstan had been recognized by 120 nations and had established diplomatic relations with 61 of them (The Library of Congress, 1996, lcweb2.loc.gov/cgi-bin/query/r?frd/cstdy@field(DOCID+kg0016)). Notably, Kyrgyzstan works in association with a range of international and supranational organizations including the WTO, UNESCO, the United Nations (UN), The World Conservation Union (IUCN), the European Union (EU) and the American Soros foundation. The country is also involved in attempts at regional cooperation across Central Asia and, along with China, Russia, Kazakhstan and Tajikistan, is a member of the 'Shanghai Five Group', created in 1996, which aims to resolve border issues between the respective neighbours. Kyrgyzstan has been the first and only newly independent state (NIS) country to comply with the World Trade Organization (*The Washington Times*, 1999).

One of the key national development policy objectives introduced by President Akayev focuses on the stimulation of foreign investment to promote economic reform. Between 1997 and 2000 Kyrgyzstan managed to attract foreign investment worth approximately US$368 million; 43% of this was divided between banking, farming, trade, tourism and services (Interfax, 2000). The greatest level of foreign investment has been attracted in 'industry', including gold mining, a noteworthy contributor to the overall gross domestic product of Kyrgyzstan. A joint venture between the Kyrgyz government and Kumtor, a Canadian gold mining operation, was established in 1996. However, concerns have been expressed by environmental groups and politicians over Kumtor's involvement in three incidents of environmental pollution between 1998 and 2000, including a major cyanide spill believed to have contaminated the waters within the Ysyk-Köl oblast.

Kyrgyzstan's acceptance of Western principles appears questionable, tainted by financial dependency rather than shared ideology. Despite membership of international and supranational organizations committed to protecting the human rights of citizens, there remains indecision over whether to maintain or abolish the death penalty in Kyrgyzstan (Interfax, 2000). Further, with respect to commitment to ethical tourism development, hunting (an activity often viewed as

incompatible with other aspects of ecotourism such as wildlife viewing) is approved as an activity for both locals and tourists. One
tourism promotional article published in *The Times of Central Asia*
(2000: 14) informs tourists that, 'the country has many large animals
whose skins and horns will grace the collections of any real hunter'.

Tourism Policy

Jogorku Kenesh has demonstrated a commitment to tourism as a priority policy area for Kyrgyzstan via the establishment of the KSATS,
enabled to perform operations relating to the licensing of every tourist
facility, the revision of tourism legislation, the privatization of tourism
facilities, and ecotourism, mountaineering and trekking investment
projects. Tourism policy has been designed to reflect national development policy with a heavy emphasis on the attraction of foreign investment for development, re-development and reconstruction of
infrastructure and superstructure as well as training of personnel, and
the establishment of partnerships with all countries of the world (The
State Committee for Tourism and Sport of the Kyrgyz Republic, 1994).

Specifically, the latest official tourism policy, outlined in the
'Development of the tourism sector of the Kyrgyz Republic until 2010'
(KSATS, 2000), builds upon the 'Conception of development of
tourism in the Kyrgyz Republic until 2000' (see Lozovskaya, 1996)
and focuses on five key outcomes of tourism development:

- protection of unique natural and cultural heritage resources;
- employment growth;
- increased income levels;
- stimulation of other economic activities;
- increased foreign investment.

However, factors potentially hindering the success of tourism policy
implementation must be acknowledged. These reflect the wider political and economic context of Kyrgyzstan and, in particular, surround:

- privatization, investment and taxation;
- the formulation of separate acts inappropriate to laws;
- the inability of the state to fulfil favourable conditions for tourism
  development;
- visa regulations;
- the weak legal protection of tourists.

Although national regulation of tourism is apparent (officially tourism
operations are only permitted under licence from the government), in
reality it is alleged that corruptive practice frustrates legitimate control and it is perhaps international and supranational organizations

that have the strongest influence on tourism development in Kyrgyzstan. Finance, expertise and training provided by developed (mainly Western) countries as a means of supporting tourism development often comes with conditions attached which may inhibit or encourage adherence to ecotourism principles. For example, large, multinational hotel corporations have constructed luxury hotels in the capital Bishkek, securing favourable rates of taxation and using imported materials and labour with little benefit to the local economy.

With respect to ecotourism, international support may be viewed as appropriate to the geopolitical status of Kyrgyzstan. Although it is difficult to directly compare the region with any other destination presented in the ecotourism literature to date, the views of the International Resources Group with respect to the development of ecotourism in Russia may be relevant:

> ... it was suggested that high local capital investment for ecotourism should be avoided. The reasoning behind this approach is based on the lack of ecotourism infrastructure availability, as well as knowledge of ecotourism and as such it was suggested that any investment funding must come from international organizations or conservation community groups.
> (International Resources Group, 1995: 4)

An example of international investment funding in the development of ecotourism in Kyrgyzstan may be seen in the establishment in 1996 of the first regional tourism office in the city of Karakol, situated in the Ysyk-Köl oblast (*The Kyrgyzstan Chronicle*, 1996). Significantly, it was developed with funding from the United States. A private, non-profit organization, the Karakol Tourism Office (KTO) cooperates with Jogorku Kenesh but remains independent and performs solely as an information provider, dedicated to ethical tourism practices. *The Kyrgyzstan Chronicle* (1996: 20) reports that the members of KTO 'have lived all their lives in the Karakol area and are the experts called upon by Almaty and Bishkek agents to organize treks and expeditions in the Tien Shan mountains and around Lake Issyk-Kul [Ysyk-Köl]'. The most obvious examples of the translation of ecotourism rhetoric into practice may be recognized within the Ysyk-Köl oblast.

The Issyk-Kul Biosphere Reserve Project

In 1997 the Issyk-Kul Biosphere Reserve Project was initiated, financially supported by the German government via the German Agency for Technical Cooperation (Deutsche Gesellschaft für Technische Zusammenarbeit GmbH, GTZ). The project, covering an area of 45,000 km² (approximately the size of Switzerland), encompasses the administrative borders of the Ysyk-Köl oblast and incorporates 60% uninhabited land.

Under the UNESCO 'Man and the Biosphere' programme, biosphere reserves aim 'to contribute to the preservation of natural resources and to cultivate the biosphere carefully and in a way which is oriented to the principle of sustainability' (cited in Biosphere Reserve Issyk-Kul, 1997: 25). The concept of a biosphere reserve may be differentiated from other protected areas on the basis that conservation exists as a goal alongside functionality. Tourism represents only one aspect of the sustainable development aims on which the biosphere reserve concept is founded.

A biosphere reserve may comprise four different zones, corresponding to the impact of human activity (Biosphere Reserve Issyk-Kul, 1997: 25–26):

- The *core area* is characterized by a strict protection regime, which forbids any form of land use. The aim is to conserve biological diversity as well as important natural resources of regional, national and international importance.
- The *buffer zone* can be used extensively as well as seasonally. The aim is to conserve a cultural landscape with its typical and traditional use forms, which have developed over centuries of human use of the area.
- Further development and optimization of sustainable land use is the task of the *transition area.*
- Heavily damaged areas, which urgently need regeneration, are part of the *rehabilitation zone.* Once these areas have recovered they become part of the three other zones listed above.

The overall objective of the Issyk-Kul Biosphere Reserve Project has been stated in accordance with a national policy objective surrounding the development of Kyrgyzstan 'in an ecological and regional sustainable manner'. A specific consequence of the project is intended to be 'the formulation of guidelines for ecological and sustainable land use practices and development of the Ysyk-Köl oblast' (Biosphere Reserve Issyk-Kul, 1997: 19).

There have been reported differences between concept and reality for at least half of the world's 352 biosphere reserves (located in 87 different countries). As a result, UNESCO enforces regulations through the submission of regular reports detailing the status of the reserves. To date, the following results have been reported by the Biosphere Reserve Issyk-Kul (1997: 31–35):

- The objectives of a lasting sustainable development of the future biosphere reserve have been defined, formulated and agreed with the local people, local administration, relevant ministries, NGOs and international organizations.
- The framework plan is being prepared in conjunction with all relevant interest groups in the project area.

- In three selected typical model areas a plan for sustainable land use was drafted on a large scale.
- Environmentally sound land use activities are supported in the model areas through a small-scale project fund with a decision making board.
- Division of the biosphere territory into the four zones and drafting of the future administration structure has taken place.
- Proposals for the strengthening of environmental laws were prepared.
- Public relations work is being undertaken.

One of the three model areas supporting environmentally sound land use is Chon-Kyzyl-Suu, situated to the south-east of Lake Ysyk-Köl near Karakol. Characterized by high levels of precipitation, the area is dominated by a glacial valley surrounded by spruce forests. The construction of a traditional yurt camp for overseas tourists has been initiated as part of the Biosphere Reserve Project. The project proposes the development of:

- ecotourism as an alternative source of income for local people;
- scientific and wildlife watching tourism as an alternative to trophy hunting;
- a decentralized accommodation system (currently lacking);
- environmental awareness of local people through tourists and vice versa.

The implementation of the Issyk-Kul Biosphere Reserve Project is ongoing. Although the full results of the project will only emerge in the long-term, if the self-reported results of the project are to be believed, it would appear that, thus far, the Issyk-Kul Biosphere Reserve Project has demonstrated success in the face of economic hardship and political instability. However, a belief in its success is not shared by all stakeholders. Attempts to regulate the development of the ecotourism aspect in particular have generated criticism from foreign operators and investors. In an article published in *The Bishkek Observer*, (Dudashvili, 2000: 14), the owner of a foreign tour operating company expressed frustration with the implementation of the project:

> When [the] tourist season starts, checking structures such as the ministry of national security, ministry of internal affairs, tax inspection, sanitation epidemic station, environmental protection etc. come alive. Similar actions are continued from year to year and many times within one season. As a matter of fact, it brings only damage. Endless checking of passports, permits, drawing up protocols and others compromise tourism in Kyrgyzstan. This season I succeeded to again visit Sary-Chelek Lake. I saw nothing good: dirt and rude attitude of officials of biosphere reserve. Entrance to the reserve costs 20 dollars per person for one day. We left the reserve with pain in our hearts.

With respect to ecotourism, it is debatable whether Kyrgyzstan's ability to successfully *formulate* policy is frustrated by a failure to successfully *implement* policy. At the very least, Table 9.3 indicates that planning for the development of ecotourism has stimulated inter-departmental cooperation within Jogorku Kenesh.

On an international scale, Kyrgyzstan remains at a relatively primitive stage of development, and the impediment that unresolved political

Table 9.3. Measures to be implemented for the growth of ecotourism. (Source: Kyrgyz State Agency for Tourism and Sport, statistics and documentation provided to the authors.)

Measure	Responsibility	Target date	Funding
Define key regions for growth of ecotourism, taking into account types of tourism and relatively vulnerable territories	Ministry for the Protection of the Environment Kyrgyz State Agency for Tourism and Sport	2001–2002	Within the limits of the Jogorku Kenesh Budget
Plan and create tourist infrastructure in protected territories (paths, reclaimed roads, communication, information centres, observation centres and so on)	Ministry for the Protection of the Environment Kyrgyz State Agency for Tourism and Sport	2000–2001	Self-financing
Set up special ecological tours and excursions to protected territories (national parks, nature reserves, hunting reserves)	Ministry for the Protection of the Environment State Agency for Biospheric Reservations	2000	—
Restore and create woods and park zones, with public participation, in the buffer zones of state nature reserves	Ministry for the Protection of the Environment	2001–2003	Within the limits of the budget, attracting means?
Set up systems of ecological standards and strategies for ecological tourism, main ideas to improve the standard of living in populated places, survey of means of implementing measures for the protection of nature, protecting the environment, culture and traditions of Kyrgyz people	Ministry for the Protection of the Environment Kyrgyz State Agency for Tourism and Sport	2000–2001	—

and economic difficulties present to tourism development must be acknowledged.

Future of Tourism Development in Kyrgyzstan

In January 2000 the Kyrgyz government adopted a resolution to simplify the entry and exit of foreign citizens. From 1 January 2001 no visa should be necessary for citizens of WTO member countries and the USA. There have also been moves, initiated by the WTO Silk Road project, to introduce a multiple visa for neighbouring Central Asian countries, but agreement has not been reached.

It is difficult to imagine that Kyrgyzstan will become a popular tourist destination in the foreseeable future, despite the determined efforts of the public and private sectors. Distance from the main tourist markets, weaknesses in destination marketing and lack of facilities and infrastructure are the main constraints. Nevertheless, the ecological and cultural assets which Kyrgyzstan possesses represent the basis of a strong, specialist ecotourism market. The mountains, in particular, are unique in that the majority have not been climbed and large areas remain unexplored, indeed it is not known how many mountains exist in Kyrgyzstan's territory. Such areas hold an irresistible appeal to mountaineers, and several international mountaineering clubs have organized high altitude expeditions, using local tour operators and guides.

The year 2002 has been designated International Year of the Mountains by the United Nations, following a proposal by the government of Kyrgyzstan. It is hoped that the hosting of international events in the fields of adventure tourism and extreme sports will focus attention on the potential of the region (Chernyak, 2000) and repair the damage caused by two recent incidents in the south of Kyrgyzstan. In August 1999, a group of 17 people were taken hostage by Islamic Uzbek extremists, among them four Japanese geologists. In August 2000 four American climbers were taken hostage by Islamic rebels from Tajikistan, who had crossed the border into Kyrgyzstan. While no foreign visitors were seriously harmed in either of these incidents, they resulted in negative publicity in the West and the issue of a travel warning by the United States State Department.

A cultural tourism market should be sustained by the ongoing efforts of the WTO Silk Road project. In addition, a list of the outstanding monuments of Kyrgyzstan was presented for consideration to the UNESCO World Heritage Centre in January 2001. However, Kyrgyzstan's cultural tourism industry is likely to be most profitable when marketed in conjunction with other, neighbouring Silk Road countries.

Consultancy reports on tourism development in Kyrgyzstan have stressed the requirement for communication between the public and private sectors and local communities and for cooperation within the private sector (Touche Ross; 1995: CBO Consulting, 1996). This has so far been facilitated by common objectives. MacLellan *et al.* (2000) blame heavy-handed control measures (rather than open communication) for alienation of and non-compliance by the private sector and locals in the case of Nepal. Given that environmental and cultural heritage hold such value for native Kyrgyz citizens, tensions are most likely to occur with foreign operators and investors, as illustrated above.

The issue of land ownership is an important one. Private ownership of land has only been permitted since 1997; state ownership has been a significant factor in environmental protection. Natural protected areas still remain under the jurisdiction of the government, but these do not include all of the high altitude sites (see Fig. 9.1). The implementation of protective measures outlined in the government's tourism development plan will rely on sustained communication and training.

The Kyrgyz Minister for Tourism sees the future of tourism for the country as a combination of nature and business tourists. 'Exotic locations attract people. Kyrgyzstan is unexplored and exotic' (Palmquist, 1999). The challenges for Kyrgyzstan are to sustain a profitable nature and adventure tourism industry (for example through premium pricing to reflect the rarity of the landscape) in a region of political instability, while at the same time upholding ecotourism principles and integrating its cultural tourism product with that of its neighbours.

References

Anderson, J. (1999) *Kyrgyzstan: Central Asia's Island of Democracy?* OPA, Amsterdam.

Biosphere Reserve Issyk-Kul (1997) *Guidelines for an Environmentally Sound Development Planning in the Area of the Future Biosphere Reserve Issyk-Kul.* Biosphere Reserve Issyk-Kul, Bishkek, Kyrgyzstan.

CBO Consulting (1996) *Constraints to Tourism Development in Kyrgyzstan.* World Bank/Netherlands TA Trust for Central Asia, Amsterdam.

Chernyak, P. (2000) Extreme adventures in Kyrgyzstan continue in the year 2000. *The Times of Central Asia*, 6 January, p. 15.

Consortium for International Earth Science Information Network (1993) *Kyrgyzstan – Trends in Developing Economies.* www.ciesin.org/datasets/wbank/tde/kyrgyzstan.html

Dudashvili, S. (2000) Tourism needs more than mountains. *The Bishkek Observer*, 1 October, p. 14.

Eckford, P.K. (1997) *International Tourism Potential in Issyk-Kul Oblast the Kyrgyz Republic: Report and Analysis*. WTO, Madrid.

Goodwin, H. (1996) In pursuit of ecotourism. *Biodiversity and Conservation* 5 (3), 277–291.

Guttman, C. (1999) Kyrgyzstan: breaking out of the old shell. *UNESCO Courier*, November, p. 21.

Interfax (2000) Kyrgyzstan attract more direct investors in 2001. *The Times of Central Asia*, 14 December, p. 5.

International Resources Group (1995) *Ecotourism in the Russian Far East: a Feasibility Study*. IRG, Washington, DC.

Khanna, M.K. (1996) *Potential for Attracting More Indian Tourists to Kyrgyzstan*. Government of India, New Delhi.

Kusler, J.A. (1991) Ecotourism and resource conservation: introduction to issues. In: Kusler, J.A. (ed.) *Ecotourism and Resource Conservation: A Collection of Papers*, Vol. 1. Omnipress, Madison, Wisconsin.

Kyrgyz State Agency for Tourism and Sport (2000) *Development of the Tourism Sector of the Kyrgyz Republic until 2010*. KSATS, Bishkek, Kyrgyzstan.

The Kyrgyzstan Chronicle (1996) Karakol Tourism Office and Association are the first of their kind in Kyrgyzstan. *The Kyrgyzstan Chronicle*, 23 February–5 March, p. 20.

Lozovskaya, L. (1996) Tourism: 'poor but prospective'. *Kyrgyzstan Chronicle*, 23 February–5 March, p. 3.

MacLellan, L.R., Dieke, P.U.C. and Thapa, B.H. (2000) Mountain tourism and public policy in Nepal. In: Godde, P.M., Price, M.F. and Zimmermann, F.M. (eds) *Tourism and Development in Mountain Regions*. CAB International, Wallingford, UK, pp. 173–198.

Osorov, Z. (2000) Kyrgyzstan: Clue to regional security. *The Times of Central Asia*, 14 December, p. 2.

Palmquist, K. (1999) Heavenly escapes in Kyrgyzstan's celestial mountains. *The Washington Times*, 6 December, www.washtimes.com/internatlads/kyrgyzstan/12.html

Smith, V.L. (1989) *Hosts and Guests: the Anthropology of Tourism*, 2nd edn. University of Pennsylvania Press, Philadelphia.

The State Committee for Tourism and Sport of the Kyrgyz Republic (1994) *Travels to the Land of the Sky Mountains*. The State Committee for Tourism and Sport of the Kyrgyz Republic, Bishkek, Kyrgyzstan.

Tayanova, T. (2000) Kyrgyzstan looks into lobbying. *The Times of Central Asia*, 7 December, p. 2.

TCA (2001) President Akayev appoints his new team. *The Times of Central Asia*, 4 January, p. 1.

The Times of Central Asia (2000) Hunting in Kyrgyzstan. *The Times of Central Asia*, 7 December, p. 14.

Touche Ross & Co/International School of Mountaineering (1995) *Kyrgyzstan Tourism Development Project: Final Report*. Touche Ross & Co, London/International School of Mountaineering, Gwynedd, UK.

The Washington Times (1999) Kyrgyzstan: an investment climate overview, *The Washington Times*, 6 December, www.washtimes.com/internatlads/kyrgyzstan/2.html

Weaver, D.B. (1998) *Ecotourism in the Less Developed World.* CAB International, Wallingford, UK.

WTO (1997) *WTO Reopens the Ancient Silk Road.* World Tourism Organization press release, Istanbul.

WTO (1998) *Tourism Market Trends, 1998 Edition – South Asia.* World Tourism Organization, Madrid.

Ecotourism Development in Fiji: Policy, Practice and Political Instability

10

Kelly S. Bricker

Division of Forestry, West Virginia University, PO Box 6125, Morgantown, WV 26506-6125, USA

Some of the world's richest and most diverse ecosystems exist within the less developed countries (LDCs). The diversity of their environments forms a natural platform from which to launch ecotourism activities. LDCs, however, are also often characterized by uneven distribution of wealth, cheap unskilled labour, isolated rural communities, dependence on external markets, high economic leakages and political instability. This juxtaposition between the environmental attributes the LDCs have to offer and the potential impediments to the fulfilment of ecotourism policy can result in significant challenges. Fiji typifies this predicament. This chapter proposes to discuss the progression of ecotourism policy adoption *vis-à-vis* the realities of ecotourism development in LDCs, utilizing the recent political events in Fiji as a case study. Additionally, impediments and support to the realization of the ecotourism policy since the coup are discussed.

Fiji: The Country and the Tourism Product

Situated in the south-west Pacific, Fiji comprises an archipelago of over 300 islands, of which only about one-third are inhabited. The 1996 census (the most recent to have been done) indicated that Fiji has a total population of 772,655. The majority of the population resides on the main island of Viti Levu (75% of the population); 18% live on Vanua Levu (north of Viti Levu) and the remaining 7% are spread across approximately 100 outer islands. Just fewer than 40%

are urban dwellers, concentrated mainly in the Suva-Nasouri area, Labasa, Lautoka, Nadi and Ba (Ministry of Information, 1999).

In terms of its tourism product, Fiji offers all of the attributes associated with a tropical island destination: sun, sea, sand and surf, along with a reputation for friendly service and hospitable locals. Most of the tourism development in Fiji is concentrated in the west of Viti Levu, the outer islands (particularly in the west) and the Coral Coast (the southern coast of Viti Levu). Traditionally, Fiji has traded on the idyllic South Pacific paradise tourist image. Increased competition in main source markets (New Zealand and Australia in particular) from destinations with similar attributes (e.g. Queensland, Australia, beach resorts in South East Asia and other South Pacific destinations), however, has meant that Fiji has had to re-assess its tourist product. This has been reflected in recent efforts to expand the tourist image by placing more emphasis on Fiji's cultural richness, environmental diversity and unique attributes, differentiating it from its competitors.

Tourism and sugar are the mainstays of the Fijian economy, although other industries such as textiles, agriculture, mineral water, forestry, fisheries and mining also make significant contributions. Since 1989, tourism has surpassed sugar as the primary source of foreign exchange. In 1998, tourism was directly and indirectly responsible for the employment of over 45,000 people. It also contributed, directly and indirectly, to 20% of GDP. In 1998, Fiji's foreign exchange earnings from tourism were US$568.2 million. In 1999 they were just over US$600 million, after a record number of visitors (409,955) was achieved. It was hoped that 2000 would be a year in which even higher numbers of visitors would be realized. The Fiji Visitors Bureau (FVB) in its Year 2000 Marketing Strategy (FVB, 1999) had set a visitor arrival target of 412,300, a number that it appeared would be achieved, and perhaps even surpassed. However, on 19 May 2000, Fiji experienced its third coup in 13 years, at which time gross earnings from tourism declined nearly 24%, with unemployment reaching 40,000. The People's Coalition government, which had supported these ecotourism initiatives, was deposed, and an interim government was installed.

Ecotourism Policy Development in Fiji

Many potential sites for ecotourism development exist within Fiji. As part of a national strategy to assist in rural development, the Ministry of Tourism and Transport focused on undeveloped rural areas, which play host to nature-based activities such as hiking and trekking, scuba diving, white-water rafting and kayaking, rural village stays and mountain biking trails, to name just a few. With the increasing importance of this type of nature-based tourism to Fiji (Harrison and Brandt,

2002) and the increasing importance of human resource development in rural areas, ecotourism potentially plays an important role in providing benefits to a wider range of stakeholders.

Since 1965, reports from consultants extolled the value of Fiji's physical environment in attracting tourists (Harrison and Brandt, 2002). Yet the underlying focus of tourism development continued to emphasize ecotourism as secondary to mainstream resort tourism. Hall (1994) suggested that 'the selection of particular tourism development objectives represents the selection of a set of overt and covert values' (p. 110). Fiji clearly placed emphasis on resort development, as depicted in the 1998–2005 tourism development plan:

> the appeal of untrammeled nature was seen as additional to the more standard tourist attractions of sun, sea and sand. Put simply, it was an add-on, an aspect of 'secondary tourist attractions,' an extra category of 'special interest' tourism, which was there to be exploited if tourists were looking for something extra to do.
>
> (Coopers and Lybrand Associates, 1989, Vol. II: 15, in Harrison and Brandt, 2002)

As a signatory to the biodiversity strategy introduced at the Rio Summit (1992), Fiji was now compelled to develop an ecotourism policy. In February 1999, the government of Fiji adopted its national ecotourism policy (NEP), titled *Ecotourism and Village-based Tourism: a Policy and Strategy for Fiji.* This was followed by the re-formation of the Fiji Ecotourism Association (FETA), the National Ecotourism Advisory Group and the National Tourism Council, on which ecotourism was represented. In addition, the National Planning Committee (under the auspices of the Ministry of National Planning, Local Government, Housing and Environment) undertook significant consultation with stakeholders and identified ecotourism as a priority area for development over the next 5 years. This was to be presented to the National Economic Summit in June 2000 for inclusion in the National Economic Plan.

The recognition of ecotourism in Fiji

Notwithstanding the importance of tourism to Fiji, several challenges emerged in relation to tourism development in the country and government recognized that these needed to be addressed. Included in these were issues such as the need to increase economic retention from the Fiji tourism product and to spread the benefits of tourism throughout the country, particularly into the rural sector. It was also recognized that there was a need to integrate traditional arts and crafts into the tourism industry. In an industry that was dominated by international interests, there was a need to further the development of Fijian people in line with their own wishes and to minimize the negative impacts of tourism development and promote conservation and environmental awareness in the tourism industry. Finally, it was acknowledged that Fiji needed a way to tap into the growing niche market of international ecotourists (Harrison, 1999).

The NEP (1998) was created to respond to the needs voiced by rural communities, non-governmental organizations and tourism associations. Up until this time, ecotourism had been somewhat loosely defined in the tourism industry in Fiji. Regrettably, almost any operator that took a tourist into the outdoors or to a village was considered an 'ecotourism' operator. Some of this confusion stemmed from a previous lack of coordination of nature and culture-based tourism operators, as well as general misunderstanding at a national level. Indeed, before the NEP was created, the Fiji National Tourism Development Plan (1998–2005) stated that 'to avoid misunderstandings in the Fiji context, we here refer to community-based tourism activities as "ecotourism" and to "nature tourism" under National Parks' (Ministry of Tourism and Transport *et al.*, 1998: 42). To clarify the criteria that comprise genuine ecotourism, the NEP defines it as follows:

> A form of nature-based tourism which involves responsible travel to relatively undeveloped areas to foster an appreciation of nature and local cultures, while conserving the physical and social environment, respecting the aspirations and traditions of those who are visited, and improving the welfare of the local people.
>
> (Harrison, 1999)

In addition to defining ecotourism in Fiji, the NEP recognized the importance of including several stakeholders such as government and statutory bodies, international bodies, landowners, the tourism industry at large, tourists and the general public. The principles outlined in the NEP addressed issues such as ecotourism complementing and not competing directly with mass tourism; restriction of tourism development if it caused irreversible damage to features of natural or cultural importance; bringing together village-based tourism stakeholders; maintenance and development of ecotourism databases; and the support and cooperation of institutions at government and non-government levels (Harrison, 1999).

The NEP also addressed issues concerning conservation of the natural and cultural environment and prioritized an increased awareness of its importance. Since most tourism development in Fiji occurs in island groups and major urban areas, the policy added concern for rural development and enhancement of the quality of life for villages in more remote regions. The NEP document also recognized the importance of developing a central register for all ecotourism endeavours. It suggested that to move towards a nationwide system of best practice or certification, those adopting true ecotourism principles should be rewarded and assisted with marketing. The policy went on to suggest that implementation needed some type of national committee to support principles of ecotourism at a grass-roots level, as well as act in an advisory capacity to the Ministry of Tourism. In essence, the NEP launched a foundation on which a collective understanding of ecotourism within government, non-government organizations, communities and the private sector could be built (see Fig. 10.1).

The Fiji Ecotourism Association

The development and implementation of the NEP was only one of several ecotourism initiatives launched in Fiji. In April 1999, the Ministry of Tourism and Transport called a meeting of tourism industry stakeholders and encouraged the rejuvenation of The Fiji Ecotourism Association (FETA). FETA had originally formed with the primary objective of consolidating all those with an interest in ecological, cultural, historical, nature-based and adventure tourism. However, the mix of actors in promoting ecotourism ranged from those actively involved in nature-based products to large corporate entities, with relatively little direct interest. According to Harrison and Brandt (2002):

> Chaired by an inbound tour operator specializing in trekking expeditions, other members included representatives from Fiji Pine Ltd, the Sheraton Hotel, several airlines, the NLTB and several beach resort hotels. Since its formation, one bulletin seems to have been issued and the Committee has met intermittently. By the end of 1998, it had, in effect, become extinct.

What caused the near 'extinction' of the association? Based on a review of former meetings, there was very little village-based participation and nature-based operator participation in the original membership. Additionally, some of the previous members felt that after the founder and president of FETA left the country many people were simply uninterested in pursuing its development.

National marketing continued to push sun, sand and sea, with little emphasis on nature-based tourism activities. As Harrison and Brandt (2002) established, ecotourism still was not seen as primary to tourism development in Fiji, though efforts continued to move forward. With the announcement of the NEP, a renewed enthusiasm for environmental protection, nature-based tourism activities and increasing interest in village-based tourism development inspired the rejuvenation of the Fiji Ecotourism Association. The first annual general meeting was held in July 1999, during which the president outlined the goals for FETA for 2000 (Bricker, 2002):

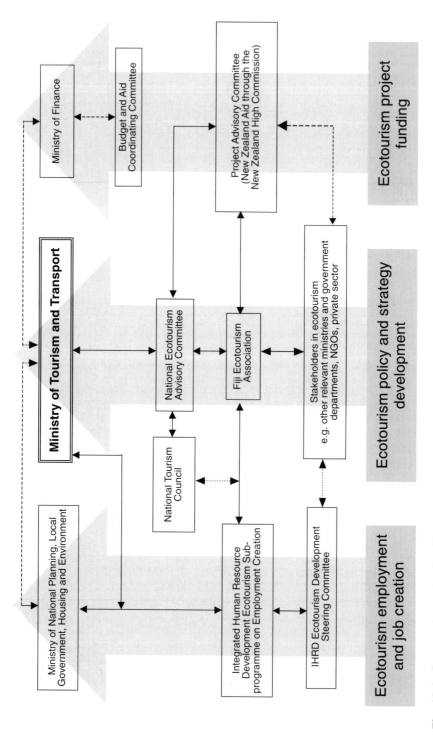

Fig. 10.1. Proposed organizational framework for ecotourism within the Ministry of Tourism, Fiji Islands. (Bricker (2002) in consultation with the Ministry of Tourism and Transport.)

1. *Training and education*: the organization workshops and an annual conference to address training and education needs within the tourism industry regarding implementation of ecotourism and sustainable tourism.

2. *Advocacy*: lobbying of appropriate government ministries on behalf of ecotourism and those engaged in tourism businesses.

3. *Centre for information and resources*: providing a centre for expertise, resources and the identification of experts in a variety of specialized fields related to ecotourism development in Fiji.

4. *Develop a best practice or ecotourism certification programme*: the initiation of an ecotourism certification programme to improve the quality of ecotourism products in Fiji.

During late 1999 and early 2000, the association also developed a 'Code of Practice' for their membership (Box 10.1). The Code of Practice was adopted by the membership at the second annual general meeting in August 2000. At the time of writing, membership had grown to 65 full members (Bricker, 2002).

The Ministry of Tourism and Ecotourism Development

By July 1999, the newly elected People's Coalition party was in government. Fortunately, the new Minister for Tourism was in full support of ecotourism development, as it fitted with the aims and objectives of the incoming administration. Minister for Tourism Adi Koila Mara Nailatikau addressed the first meeting of the recently formed FETA, during which she expressed the People's Coalition's vision of their role in the future of ecotourism developments as one of partnership with stakeholders. This was to be achieved through a range of initiatives including the establishment of an administrative framework for sustainable development and ecotourism activities (including natural and cultural heritage enhancement), and the development of appropriate infrastructure to facilitate remote village-based and nature-based interests (Bricker, 2002).

To help facilitate the achievement of their ecotourism objectives, the Ministry of Tourism and Transport established a National Ecotourism Advisory Committee in December 1999. Membership comprised both public and private stakeholder groups[1]. According to the minister, the goals of this advisory group were to build a strategic

[1] These included the Fiji Visitors Bureau, University of the South Pacific tourism faculty, National Trust of Fiji, Native Land and Trust Board, New Zealand High Commission, Telecom Communications, Ministry of Commerce, Business Development, and Investment, Fijian Affairs Board, Fiji Museum, Ministry of Women and Culture, Fiji Ecotourism Association, Fiji Hotel Association, Fiji National Training Council, Department of Environment, and the 'Keep Fiji Beautiful' Association.

Box 10.1. The Fiji Ecotourism Association Code of Practice. (Adopted August 2000, modelled and adapted from the Australian Ecotourism Association.)

Members of the Fiji Ecotourism Association agree that they will adopt the following Code of Practice:

1. To strengthen the conservation effort for, and enhance the natural integrity of, all places that are visited.
2. Respect the sensitivities of all cultures.
3. Practice and promote the efficient use of natural resources (e.g. water and energy).
4. Prohibit use and/or importation into Fiji of harmful/hazardous materials/chemicals.
5. Ensure waste disposal (human, organic, inorganic, etc.) has minimal environmental and aesthetic impacts and is properly managed and contained.
6. Agree not to use or import goods packaged in non-recyclable or harmful materials, and develop, support and promote recycling programmes.
7. Support and promote tourism operators, plant and suppliers (i.e. hotels, carriers) who believe in and practise conservation of natural and cultural resources.
8. Raise environmental and socio-cultural awareness through the distribution of ecotourism awareness guidelines to visitors, local communities.
9. Support ecotourism education/training for tour operators (marine and land-based), accommodation, government and other tourism industry affiliates/stakeholders.
10. Educate and train staff to be well versed in and respectful of local cultures and environments.
11. Give clients appropriate verbal and written educational material (interpretation) and guidance with respect to the natural and cultural history of the areas visited.
12. Maximize use of locally produced goods that benefit the local community and economy; and support sustainable agricultural and fishing practices.
13. Do not purchase or use goods made from threatened or endangered species.
14. Avoid intentionally disturbing or encouraging the disturbance of wildlife or wildlife habitats.
15. Keep vehicles to designated roads and trails.
16. Commit to the principles of sustainable practices and guidelines in all tourism activities and programmes.
17. Comply with internationally recognized safety standards for all tourism industry activities.
18. Ensure truth in all marketing materials, publications and presentations.

management plan for ecotourism, coordinate appropriate agencies to ensure that the necessary infrastructure requirements were in place and, most importantly, work with FETA to address issues identified by all stakeholders in ecotourism and sustainable development. More specifically, the National Ecotourism Advisory Committee sought the following:

1. To promote ecotourism as a tool for sustainable development and conservation of Fiji's natural resources and cultural heritage.
2. To address the importance of including the values of villages and communities that choose to engage in ecotourism.
3. To improve the quality and diversity of current tourism offerings and ensure quality in all future developments.
4. To promote the importance of maintaining or enhancing the health of all natural environments at all levels through increasing awareness and appreciation of the diverse natural heritage areas in Fiji.
5. To develop training and educational programmes for all tour operators in order to meet the demands of the discerning international tourist (i.e. to acknowledge the fact that tourists' expectations for increased knowledge of local culture and natural environments are increasing and we must meet that demand with well-trained, knowledgeable ecotourism employees).
6. To review the National Ecotourism Policy.

The Ministry of Tourism also believed that, in order to move ecotourism forward in Fiji, it was important for all stakeholders to develop an increased awareness of ecotourism as a viable and valid part of the overall tourism industry in Fiji (Bricker, 2002). The ministry placed emphasis on the need to preserve cultural and natural resources and promote these concerns within their village, community and workplace thereby '[acknowledging] that Fiji is distinctive from other sand, sea, and surf, destinations ... for where else in the South Pacific are waterfalls so prolific, rivers so extraordinary, and seas so pristine?' (Mara-Nailatikau, 1999).

In February 2000, the Ministry of Tourism and Transport also called to order the National Tourism Council (NTC), an advisory committee chaired by the minister, comprising government, non-government and private sectors linked to tourism development in Fiji. While purely an advisory body, the NTC did, however, play an active role in creating the Tourism Sector Plan, a document to be included in the National Development Plan, 2000–2005.

For three months, the Ministry of National Planning held tourism sector meetings to address the major issues arising from the industry. Several issues documented in the Draft Plan were specific to ecotourism development and acknowledged the need for private sector and government to work in close partnership. The results of these meetings were to be presented for discussion by industry and government representatives at the National Tourism Summit, June 2000, followed by presentation at the National Economic Summit, to be held later in that month.

The National Tourism Council convened on 18 May 2000 to make final adjustments to the Tourism Sector Draft Plan and complete the details for presentation to the Prime Minister in June. This, however, was not to unfold as anticipated.

> Political serenity, not scenic or cultural attractions constitute the first and central requirement of tourism.
>
> (Richter and Waugh, 1986: 231)

On 19 May 2000, 12 months after the election of the first Indo-Fijian Prime Minister, Mahendra Chaudhry, Fiji suffered a coup. Shortly before 11 a.m., a group of seven gunmen, led by failed businessman George Speight, entered parliament. Members of the group were later identified as soldiers of the elite First Meridian Squadron, the military's Counter Revolutionary Warfare Unit, which had been established after the 1987 coup. Two gunshots were fired in parliament to warn the speaker and opposition members to leave. Prime Minister Mahendra Chaudhry and his Ministers were then segregated on ethnic grounds, and guarded by rebel supporters toting Uzi and M16 rifles. It was not until 13 July, 56 days later, that the remaining hostages were finally freed. Despite the hopes of the nation, the release of the hostages did not signal the end of political instability and related events in Fiji.

Post Coup Ecotourism: Picking Up the Pieces

The aftermath of the events of 19 May 2000 were immediate and devastating to the economy and to the tourism industry in particular. Many tourists cut short their holidays in Fiji and cancellations for upcoming bookings began to pour in. Fiji's peak tourism months are June–August. The coup resulted in a decline in visitor arrivals in that period of almost 70%. Fiji also missed out on potential trans-pacific stopover traffic during the Sydney 2000 Olympic Games. It was estimated that the loss of tourism revenue by the beginning of August was F$84.6 million (excluding Air Pacific, the national carrier). The hotel sector experienced a 44% reduction in employment as a result. Major tourism operators such as Shotover Jet and the Sheraton Royal closed operations due to low numbers.

While the nation anxiously awaited the release of the hostages, hoping that some semblance of law and order would prevail afterwards, this was not to eventuate. It was not until after 56 days of internment that the hostages were finally freed. This, however, did not mark the end of political unrest and, at the time of writing, Fiji was still under a state of civil emergency and was being led by an appointed (not elected) interim government.

Land ownership also emerged as a major issue during the 2000 crisis, even after the hostages had been freed. On 6 July, Monasavu landowners seized control of the Wailowa power station and plunged Fiji into darkness. It was to be nearly 8 weeks before power was fully restored to the country. Although the power shortage caused only a minimal direct impact on the tourism industry (most operators had their own generators), the tourism industry was affected directly by other land ownership issues. The rhetoric of the coup leaders about indigenous landrights spilt over into disputes over resort-based tourism, which saw several resorts being taken over by hostile landowners. The takeover of the exclusive Turtle Island Resort in the Yasawa Islands (a multiple ecotourism award-winning operation) received extensive international media coverage (Berno and King, 2001).

A cloud of uncertainty hung over the future of tourism in Fiji, including ecotourism. It was unclear what would happen to the significant progress that had been realized in ecotourism policy development in the 2 years prior to the coup, particularly under the auspices of a caretaker government.

Ministry of Tourism support for ecotourism?

Political stability has proved to be an essential circumstance for tourism development (Richter and Waugh, 1986; Richter, 1992; Hall, 1994). This was evident certainly during the aftermath of the events of 19 May and the months following, when tourism and other industries in Fiji were devastated. However, despite the instability and uncertainty of future politics, the Ministry of Tourism continued to support ecotourism actively. During the FETA annual general meeting in August 2000, the newly appointed first interim Minister of Tourism reiterated the Ministry's support for ecotourism development. In his opening address, Minister Jone Koroitamano highlighted the government's support of the tourism sector, increasing the marketing budget by F$4 million for 2000, with ecotourism an important new area. However, the temporary nature of the interim government was open to continuous change.

In January 2001, Koroitamano stepped down as interim Minister for Tourism and a new appointee assumed the position. The fluid nature of the interim administration seemed to cause further setbacks to the rejuvenation of both the National Tourism Council and National Ecotourism Advisory Committee. The National Tourism Council had not reconvened as of January 2001; nor had the National Ecotourism Advisory Committee met since January 2000.

One area in tourism development that continued to receive particular attention from the interim government, however, was the increased involvement of indigenous landowners. The government's support was demonstrated through the organization of the Land Resource Owners Conference held in December 2000, during which indigenous Fijian resource owners made several resolutions and appealed to government to support ecotourism development. Among the nearly 30 resolutions, indigenous landowners requested a range of ecotourism-related support including: provisions for training in cultural, environmental and natural heritage conservation; assistance with infrastructure development in rural areas (where ecotourism was likely to be developed); promotion of sustainability measures in all tourism developments; active marketing of Fiji's ecotourism products by the FVB; establishment of an ecotourism unit by the Fiji Development Bank; and government funding for the upgrading of trails and tracks in the hinterland (to be used for ecotourism activities).

After the events of 19 May 2000, the draft document for the National Strategic Development Plan was in abeyance. As of 30 December 2000, however, the Ministry of National Planning stated that the efforts made on the initial Draft document for the Development Plan were to go forward under the interim administration. The National Strategic Development Plan, Tourism Sector, originally to have been presented at the National Economic Summit in May 2000, was rescheduled until after elections to August 2001.

Fiji Ecotourism Association

Prior to the coup, the Fiji Ecotourism Association had scheduled a conference for June 2000. Due to the political instability, however, the conference was rescheduled for December 2000. With financial support from the New Zealand Official Development Assistance Programme (NZODA), Air Pacific Airlines and the Fiji Visitors Bureau, the conference went ahead as rescheduled, with 100 participants in attendance. A broad range of ecotourism interests were represented at the conference, with participants from government, the private tourism sector, regional non-profit organizations, academic institutions and business (Bricker, 2002).

One of the primary objectives of the FETA conference was to engage participants in discussions about appropriate best practice developments for ecotourism operations in Fiji. This was directly linked to the NEP's call for the development of an industry-based association to establish and implement standards for ecotourism operations. Representatives from the National Ecotourism Accreditation

Program in Australia attended to guide workshops in criteria for developing best practice or certification programmes. From the workshops, FETA was empowered to begin development of a best practices programme in 2001 (Bricker, 2002).

The Fiji Ecotourism Association has been assured of further support for the development of a best practices programme by the Fiji Visitors Bureau, marketing and advertisement enhancements for the coming year, and a dedicated liaison to communicate FETA's programme to the rest of the tourism industry.

Policy, Practice and Politics: What it all Means for Ecotourism in LDCs

Since the late 1990s, Fiji has been very progressive in its approach to the development and implementation of ecotourism policy. With ecotourism defined and accompanying principles in place, the Ministry of Tourism implemented ecotourism as part of its national tourism development strategy.

The NEP provided a comprehensive framework for: outlining the importance of conservation and protecting physical and social environments; diversifying tourism into rural areas; fostering the development of culture; reducing conflicts over land tenure and access to natural resources; the extension of incentives for ecotourism projects; and creating systems for cataloguing, evaluation, licensing and monitoring of ecotourism products. Ecotourism was now well positioned for growth in Fiji.

Through the creation of national tourism committees, advisory units and the revitalization of the industry-based Fiji Ecotourism Association, Fiji was poised for ecotourism policy implementation. This not only included significant recognition in the development of the National Strategic Plan 2000–2005, it also attracted support funding through non-government organizations such as NZODA and the International Labor Organization. There was a conscious movement towards the inclusion of rural communities in shaping Fiji's tourism product.

The NEP provided a framework and reason to move forward. All the essential pieces were in place and the future for ecotourism in Fiji was very optimistic – until 19 May 2000. Fiji, like so many LDCs with immense potential for ecotourism, shares the characteristic of political instability. No matter how salient a country's ecotourism attributes are, without a foundation of political stability, their potential will not be realized. According to Richter (1992), developing countries are particularly vulnerable to the implications of political instability, because they have fewer resources to draw upon and often are more dependent on tourism, which in Fiji's case is the leading industry. Additionally,

Fiji relies most heavily on its international markets, with very little domestic base to 'cushion seasonality or absorb some tourism capacity when international tourism fails' (Richter, 1992: 37).

Although a framework and master plan continue to provide a platform for ecotourism development, it is difficult to ignore the enormous impact of political instability on its actual progress. Through reviewing the development of ecotourism policy in Fiji pre- and post-coup, it is clear that political stability plays a critical role in tourism, with tremendous implications for ecotourism policy implementation.

> It is a value choice, implicit and explicit, which orders the priorities of government and determines the commitment of resources within the public jurisdiction.
>
> (Simmons *et al.*, 1974, in Hall, 1994: 3)

The actual implementation of the NEP remains to be seen; as Harrison and Brandt (2002) have suggested, 'it is yet too early to say how far the Fiji government is prepared to go in setting up and supporting the formal organizations necessary if such policies are to be implemented and monitored over the next few years'. At the ministerial level, several changes occurred between the time the NEP was adopted and after the ensuing political crisis. Tourism ministerial leadership changed three times since adoption of the NEP. Consequently, efforts made by the Minister of Tourism before 19 May 2000 came to a virtual standstill. For example, initiatives slated for the National Strategic Development Plan (inclusive of ecotourism) were deferred and national advisory committees overseeing ecotourism policy implementation became inactive. As Hall (1994) has suggested, 'it is not just the range of objectives that needs to be considered but the relative priority attached to the objectives as they are implemented' (p. 114). The National Strategic Development Plan had the potential to raise ecotourism into a priority area. However, the deferment of initiation of the national strategic plan resulted in loss of momentum to re-focus ecotourism as part of a national development strategy, potentially minimizing further support for the implementation of the NEP.

Since the coup, however, some efforts were made to re-focus tourism alternatives towards empowering indigenous landowners. The conference held for tourism resource owners has helped to progress some aspects of the NEP. However, actual resources to implement these ideas are questionable. Like most LDCs, Fiji relies on funding from overseas aid organizations to assist with development projects. Ecotourism is no exception. Grass-roots ecotourism initiatives in particular have benefited from overseas aid (Harrison and Brandt, 2002). The ongoing political uncertainty in Fiji has resulted in a redeployment of some aid monies. Ross MacFarlane, the Development Program Manager, NZODA, stated that 'in light of reporting on the state of the projects, and the need to release funds for humanitarian and rights support as a result of the coup, ecotourism activity has been further scaled back within the $300,000 allocation' (R. MacFarlane, New Zealand, 2001, personal communication). Yet, at the time of writing, NZODA's commitment to ecotourism support remains steadfast. While support for some of the core projects continued, some assistance was resumed for previously established projects during the spring of 2002.

Conclusion

> Overall conditions of Third World nations make political conditions generally more erratic, and resilience less likely and more difficult to predict.
>
> (Richter, 1992: 35)

Fiji is a country with a great deal of untapped ecotourism potential. From the highland interior of Viti Levu to the marine environments of the coastal regions and outer islands, Fiji possesses a natural platform from which to launch ecotourism activities. With the increasing importance of nature-based tourism to Fiji, and escalating attention given to human resource development in rural areas, ecotourism and community-based tourism are the most viable means of spreading the tourist dollar beyond the industry's major areas of concentration and of increasing the retention of the tourist dollar.

Following the events of 19 May 2000, signs of recovery have begun. Elections were held in August 2001, and demonstrated to some extent that the country is moving forward. According to the Fiji Visitors Bureau, visitor arrivals have exceeded predictions set in 2000 for 2001, although they were still below 1999 by just over 18%. Additionally, a review of the recovery programme suggests an effective approach to events of 19 May:

> A coordinated and collective approach to the recovery process (i.e. 'one voice, one message') was essential. Coordinating the recovery effort through TAG, under the auspices of FVB, meant that a singular message was disseminated and that consistent responses were made towards situations. This was particularly important in dealing with the media, and for downgrading the travel advisories. Although there was some internal debate about the size of the financial contributions demanded by the TAG and the promotion of a single product branding campaign, the strategy has ultimately been successful in attracting visitors back to Fiji.
>
> (Berno and King, 2001: 91)

The national ecotourism policy document (NEP) set Fiji on a course for implementing ecotourism and elevating it to a prominent level within tourism development. Despite this promising start, the coup in 2000 and the ensuing political crisis have had an effect not only on the implementation of the policy document, but also on its beneficiaries, the tourism industry and operators involved in this sector. However, because ecotourism gained prominence in the political arena (i.e. supported indigenous tourism resource owners, community development), supporting efforts to assist indigenous people in rural development, it was not forgotten through months of instability. Instead, it remained an agenda item and was highlighted in conferences, meetings and community programmes (Bricker, 2002). Yet, despite its encouragement, it cannot be disputed that the implementation of the NEP requires a growing 'tourism industry' and real financial commitment to support and further its mission.

This juxtaposition between the environmental attributes that Fiji has to offer and the impediments to the implementation of ecotourism policy typifies the realities of ecotourism development in LDCs. Without the critical foundation of political stability the further advancement of all forms of tourism, especially ecotourism in Fiji (or any other politically unstable LDCs) is fraught with difficulties (Richter and Waugh, 1986; Lea and Small, 1988; Richter, 1992; Hall, 1994).

Fiji is only beginning to experience the longer-term effects of the coup in 2000. The implications of this for ecotourism policy and strategy for the country are yet to be realized. As Richter (1992) notes, 'the issue is not whether the tourism industry will be killed, but whether it will survive in particular places in the face of civil strife and with what costs to the societies that depend on it' (p. 36).

Acknowledgements

The author would like to thank Dr Tracy Berno of the University of the South Pacific for her support and energy, the Fiji Visitors Bureau Ecotourism staff, Severo Tagicakiverata, Manoa Malani from the Ministry of Tourism, and the Fiji Ecotourism Association for assistance and insight into the complexities of this topic.

References

Berno, T. and King, B. (2001) Tourism in Fiji after the coups. *Travel and Tourism Analyst*, 2, 75–92.
Bricker, K. (2002) Planning for ecotourism amidst political unrest: Fiji navigates a way forward. In: Honey, M. (ed.) *Setting Standards: the Greening of the Tourism Industry*. Island Press, Washington, DC.

FVB (Fiji Visitors Bureau) (1999) *Year 2000 Marketing Plan Summary*. Fiji Visitors Bureau, Suva, Fiji.

Hall, C.M. (1994) *Tourism and Politics: Policy, Power, and Place*. John Wiley & Sons, Chichester, UK.

Harrison, D. (ed.) (1999) *Ecotourism and Village-based Tourism: a Policy and Strategy for Fiji*. Ministry of Tourism and Transport, Suva, Fiji.

Harrison, D. and Brandt, J. (2002) Ecotourism in Fiji. In: Harrison, D. (ed.) *Pacific Island Tourism*, Cognizant, New York.

Lea, J. and Small, J. (1988) Cyclones, riots, and coups: tourist industry responses in the South Pacific, paper presented at Frontiers in Australian tourism conference, Australian National University, Canberra, 30 June – 1 July. In: Hall, C.M. (1994) *Tourism and Politics: Policy, power, and place*. John Wiley & Sons, Chichester, UK.

Mara-Nailatikau, K. (1999) Address to the Fiji Ecotourism Association Annual General Meeting, 23 July 1999, unpublished manuscript. Fiji Ecotourism Association, Suva, Fiji.

Ministry of Information (1999) *Fiji Today 2000*. Ministry of Information, Suva, Fiji.

Ministry of Tourism and Transport, Tourism Council of the South Pacific and Deloitte and Touche (1998) *Fiji Tourism Development Plan*. Ministry of Tourism and Transport, Suva, Fiji.

Richter, L.K. (1992) Political instability and tourism in the third world. In: Harrison, D. (ed.) *Tourism and the Less Developed Countries*. John Wiley & Sons, Chichester, UK, pp. 35–46.

Richter, L.K. and Waugh, W.L. Jr (1986) Terrorism and tourism as logical companions. *Tourism Management* 7(4) December, 230–238.

Simmons, R., Davis, B.W., Chapman, R.J.K. and Sagar, D.D. (1974) Policy flow analysis: a conceptual model for comparative public policy research. *Western Political Quarterly*, 27(3), 457–468. In: Hall, C.M. (1994) *Tourism and Politics: Policy, Power, and Place*. John Wiley & Sons, Chichester, UK.

Ecotourism and Protected Areas in Australia

11

John Jenkins[1] and Stephen Wearing[2]

[1]*Department of Leisure and Tourism Studies, University of Newcastle, Callaghan, NSW, 2308, Australia;* [2]*School of Leisure, Sport and Tourism, University of Technology, Sydney, Kuiring-gai Campus, PO Box 222, Lindfield, NSW 2070, Australia*

Introduction

A large proportion of tourism in Australia constitutes nature-based tourism and, in particular, ecotourism in protected areas (e.g. see McKercher, 1998; Nepal, 2000) (see Table 11.1 for a list of high profile ecotourism destinations and businesses in Australia). In theory, ecotourism offers economic opportunities to local populations (indigenous and non-indigenous), maintains or enhances the environment, is a source of funds for protecting endangered species and cultural heritage, and enhances environmental consciousness and understanding. However, as ecotourism destinations experience increasing demand pressures and use becomes more intense, the potential for conflict between maintaining environmental quality and economic development is exacerbated. By no means do we have a detailed understanding of the impacts of ecotourism, or any other form of tourism, on the environment (or vice versa). However, many tourism businesses appear to be becoming increasingly aware of public pressures to have due regard for the environment. Many such businesses participate, more or less, in accreditation programmes, self-regulation programmes and corporate policies that are sympathetic to sustainable environment and ecotourism aims.

In broad terms, the tourism industry could be seen to be working hard to influence public opinion. A large number of agencies attempting

© CAB *International* 2003. *Ecotourism Policy and Planning*
(eds D.A. Fennell and R.K. Dowling)

Table 11.1. Examples of popular ecotourism destinations in Australia (based on Weaver and Oppermann, 2000; Charters *et al.*, 1996).

State or Territory	Location/business
New South Wales	Alpine region Barrington Tops Blue Mountains National Park Tropical and temperate rainforests
Queensland	Fraser Island – King Fisher Bay Resort Gold Coast Hinterland – Binna Burra Lodge and O'Rielly's Guest House Northern Queensland – Undara Lava Lodge Cairns – Skyrail rainforest cableway The Great Barrier Reef Tropical and temperate rainforests
Tasmania	Lemonthyme Lodge Tropical and temperate rainforests
Northern Territory	Kakadu National Park
Western Australia	Eco Beach Resort, near Broome Kimberleys Rottnest Island
South Australia	Kangaroo Island Flinders Ranges
Victoria	Alpine region
Australian Capital Territory	Namadgi National Park

to cash in on the ecotourism market are in fact public sector agencies with a command over extensive areas of land and water in Australia. Outside of the 'tourist gaze' (Urry, 1990), public and private sector executives are actively working to develop strategies and images for their corporate business. The adoption of ecotourism principles can be viewed in many ways, but, to be sure, it is no commercial accident. Ecotourism is both a means of sustaining tourism specifically and the capitalist system generally, and a popular marketing label.

This chapter examines important issues concerning ecotourism in protected areas in Australia. The chapter begins by defining protected areas and explaining their importance to tourism development in that country. The benefits and costs of ecotourism in protected areas are examined, with attention directed to recent public and private sector initiatives in protected areas. The chapter concludes with an examination of critical issues, including: public sector reforms influencing the management of tourism in protected areas; the imposition of user fees and other charges; and the role of the private sector in protected areas in a country whose public policy is heavily oriented to neo-liberal economic philosophies. In the absence of detailed studies of the impacts of ecotourism in protected areas, we stress the need for: (i) scientific research in ecotourism planning and development; and (ii) a fundamental paradigm shift in industry and community attitudes to the environment and economic growth, and to tourism in protected areas.

Ecotourism: Definitions and Initiatives

There is little consensus about an appropriate definition of ecotourism, though basic principles underpin most definitions. According to the Ecotourism Association of Australia (EAA; www.ecotourism.org.au/About_Eaa.htm):

> Ecotourism can be defined as *nature-based tourism that involves interpretation of the natural and cultural environment and ecologically sustainable management of natural areas.* Ecotourism is seen as ecologically and socially responsible, and as fostering environmental appreciation and awareness. It is based on the enjoyment of nature with minimal environmental impact. The educational element of ecotourism, which enhances understanding of natural environments and ecological processes, distinguishes it from adventure travel and sightseeing....

Australia was the first country in the world to develop and implement a National Ecotourism Strategy. In establishing a National Ecotourism Strategy in 1994, the federal Labor government's intention was to formulate an overall policy framework for the planning, development and management of ecotourism, and to contribute to the achievement of sustainable tourism in natural areas. The aims of the Strategy were to:

- identify the major issues that affect, or are likely to affect, the planning, development and management of ecotourism in Australia;
- develop a national framework to guide ecotourism operators, natural resource managers, planners, developers and all levels of government towards achieving a sustainable ecotourism industry;
- formulate policies and programmes to assist interested parties to achieve a sustainable and viable ecotourism industry (Charters, 1996).

The Ecotourism Strategy identified key issues concerning the planning, development and management of ecotourism in Australia, including:

- the development of ecologically sustainable approaches to tourism planning, development and management;
- planning and regulation;

- natural resource management;
- infrastructure development;
- impact monitoring;
- marketing;
- industry standards;
- industry accreditation;
- ecotourism education;
- development of opportunities for Aboriginal and Torres Strait Islanders;
- equity considerations in the allocation and management of natural resources.

Approximately Aus$10 million was committed by the Commonwealth government for nature-based tourism programmes over a 4 year period, commencing in 1993–1994. The programmes were funded under two main areas: regional planning and destination management (e.g. infrastructure projects, baseline studies and monitoring), and business and product development (e.g. accreditation, ecotourism education, market research, business skills, and energy and waste management) (see Charters, 1996; Hall, 1998).

The National Ecotourism Strategy highlighted the importance of government's role in establishing the necessary guidelines to develop ecotourism according to sustainability principles. However, following the victory of the Coalition (Liberal and National) parties at the 1996 federal election, support to implement this strategy was withdrawn (Grant and Allcock, 1998). The impetus for any further actions was left with industry and state governments, whose commitment to such strategies has been varied. None the less, federal funding has contributed to the development of national accreditation programmes, market research, energy and waste minimization publications and programmes, infrastructure, education, visitor management strategies and actions (e.g. interpretation facilities), regional planning and conferences. Commonwealth and, to a lesser extent, state governments invest in industry research (e.g. the Bureau of Tourism Research; the CRC for Sustainable Tourism), and in marketing and promoting the tourism industry (e.g. the Australian Tourist Commission; state, territory and regional agencies). They also provide competitive grants for direct industry and related infrastructure development.

There are approximately 600 ecotourism operators in Australia, most of which (about 85 per cent) employ less than 20 staff (Cotterill, 1996). It is estimated that another 2000, mainly adventure, tourism-oriented operators also offer ecotourism-related opportunities (McKercher, 1998). EAA is the ecotourism industry's peak representative body. EAA's vision is 'to be leaders in assisting ecotourism and other committed tourism operators to become environmentally sustainable, economically viable, and socially and culturally responsible'. National membership of EAA comprises diverse individuals and agencies (ecotourism accommodation, tour and attraction operators; tourism planners from local, state and Commonwealth governments; protected area managers; academics and students; tourism, environmental, interpretation and training consultants; ecotour guides) (Weaver and Oppermann, 2000).

One of EAA's largest and most significant projects is the National Ecotourism Accreditation Program (NEAP). This programme is run in conjunction with its joint venture partner, the Australian Tourism Operators Network (ATON). NEAP is an industry-driven accreditation scheme, which enables industry, protected area managers and consumers to identify 'genuine' ecotourism products. The product accreditation programme is divided into three categories: Nature Tourism, Ecotourism and Advanced Ecotourism Accreditation. EAA has also recently branched out into guide certification.

Protected Areas in Australia

Protected areas (e.g. national parks, nature reserves and marine parks) are defined by the World Conservation Union (IUCN, 1994) as 'areas of land and/or sea especially dedicated to the protection and maintenance of biological diversity, and of natural and associated cultural resources, and managed through legal or other effective means'. In Australia, most protected areas are managed by the states and territories, reflecting land management responsibilities under the Australian Constitution and the federal system of government. In other words, the Commonwealth, state and territory governments and their respective agencies have each developed national parks and associated administrations and policy within their own peculiar histories before and under the present Australian Constitution. There are approximately 60 categories of protected areas (Pittock, 1996) encompassing an estimated 5100 terrestrial and marine sites in Australia (Environment Australia; www.environment.gov.au/bg/protecte/intro. htm). The first substantive laws to protect Australia's scenic areas were passed in Tasmania in 1863, and then in 1879 with the National Park (renamed Royal National Park in 1954), the world's second national park, which was established south of Sydney.

The Commonwealth, through Environment Australia, manages terrestrial parks and reserves established in those parts of Australia which come under its direct responsibility. The Kakadu National Park in the Northern Territory was established in three stages: in 1979, 1984 and 1987. Subsequently, in 1992, the Park obtained World Heritage Status for its natural and cultural values. 'The phased introduction of the Kakadu as a National Park was in part due to the controversy which accompanied its inauguration as the interests of conservation, mining, Aboriginal land rights and tourist potential were reconciled' (Ryan, 1998: 122). The Commonwealth also manages the External Territories and Australian waters beyond the state limit of 3 nautical miles. The Great Barrier Reef Marine Park, for instance, is managed by a Commonwealth agency, the Great Barrier Reef Marine Park Authority (GBRMPA).

The first act to simultaneously establish national parks and a National Parks and Wildlife Service in Australia was brought down in New South Wales (NSW) in 1967. Other agencies, such as the NSW Department of Land and Water Conservation, have substantive responsibilities for public lands (see below), many of which are environmentally sensitive and important recreational resources (e.g. Crown Reserves; Travelling Stock Routes) (Jenkins, 1998, 1999). With respect to tourist access to nature-based resources, no one piece of legislation nor overriding common law right exists within Australia, let alone in individual states and territories. Access provisions are scattered among many pieces of legislation. Even single acts prescribing legislation are often the responsibility of more than one government department. This legal and administrative complexity is reflected in the range of mechanisms by which protected areas are declared and managed, and by which rights of access for nature-based tourism are legitimated (Jenkins and McIntyre, 2001). The situation has been well described by Pitts:

> The management of outdoor recreation resources and facilities in Australia is characterised by a myriad of government departments, authorities and agencies operating in apparent isolation of each other. Outdoor recreation has rarely been recognized, on its own, as a legitimate function of government in this country. Rather, it has been allowed to develop as a secondary function associated with more traditional government activities such as forestry, conservation, water supply and town planning. This approach has inevitably led to problems of coordination, conflicts in connection with overlapping responsibilities and doubts about the effectiveness of the whole delivery system in meeting community [and tourist] needs.
>
> (Pitts, 1983: 7)

Twenty years later, outdoor recreation and tourism receive much greater public attention. However, the care, control and management of protected areas remain the responsibility of a wide variety of public and private agencies, among whom coordination and consultation is lacking (e.g. Jenkins, 1998). Public sector agencies have been very careful to guard their responsibilities and roles, to protect their empires, and to be more than cautious about the innovative ideas of other agencies, leading to the death of many 'good ideas' (e.g. see Pigram and Jenkins, 1994; Jenkins, 1998). Notwithstanding the above situation, some progress has been made in the integration of conservation of biodiversity and nature-based tourism beyond the formal system of parks and protected areas. The Voluntary Conservation Agreements in place in New South Wales are an example, as are the Regional Parks of Perth, Western Australia (Moir, 1995, in Pigram and Jenkins, 1999), some of which remain in private ownership. In Sydney, eight areas have been designated as Metropolitan regional parks. These parks were conceived by the New South Wales government as a way of providing the people of Sydney with green lungs, a justification similar to that used in establishing the National Park in 1879! Regional parks are areas of open space for recreation and for conserving fragile ecosystems. They vary in size (from 4000 ha to less than 50 ha) and in the activities for which they cater, and represent a promising public sector initiative (www.npws. nsw.gov.au/parks/regprks.htm). Very promising private sector initiatives outside the public protected estate include the work of Earth Sanctuaries Ltd, an organization which exploits tourists for the benefit of wildlife (e.g. see Wamsley, 1996). Earth Sanctuaries Ltd has acquired more than 90,000 ha of land, created 10,000 ha of feral-free habitat and successfully re-introduced 16 species of rare, threatened and endangered wildlife into their original habitat. There are currently ten sanctuary projects being developed around Australia. Four sanctuaries are currently open to the public; Warrawong, Yookamurra, Scotia and Hanson Bay (Earth Sanctuaries Ltd; http://esl.com.au/index.htm).

Apart from these initiatives, a generally negative attitude prevails at official levels to the introduction of European-style national parks or US National Reserves in Australia. This reaction, coupled with growing resistance by rural landholders to any further acquisition of park lands of any kind, means that progress towards establishment of integrated public/private 'regional parks' is likely to be slow. In Australia there is an urgent need to adopt an alternative approach to allocating land for parks as the changing nature of agriculture and rural life acts as a disincentive for landholders to maintain the character and quality of the countryside in their keeping. If it can be shown, by successful pilot projects, that the economic and amenity functions of the countryside can be compatible, then a range of park types can be created as and where appropriate. Given time and enlightened management, such parks have the potential to demonstrate the benefits of sharing the countryside as a living communal resource, both for ongoing productive purposes and for outdoor recreation (see Pigram and Jenkins, 1999).

Ecotourism and Visitation to Protected Areas

According to Ceballos-Lascurain (1999), Australia is an international leader in the development of ecotourism, with diverse products offering a number of strengths, including:

1. The relative safety of travelling to and within the country.

2. Immense scale of natural areas (such as the Great Barrier Reef and vast deserts surrounding Uluru).

3. Icon-like animals with strong international recognition (i.e. kangaroo, koala, platypus and wombat).

4. Australia is one of the five mega-biodiversity countries of the world (Mittermeier, 1997, in Ceballos-Lascurain, 1999) and contains two of the world's top priority global biodiversity hotspots (south-western Australia and south-eastern Australia/Tasmania).

5. High levels of endemism (Australia is by far the highest ranked in the world).

Protected areas, and especially national parks and World Heritage Areas, are important destinations for national and international eco-tourists (Dowling, 1991; Driml and Common, 1995; Ceballos-Lascurain, 1999; Weaver and Oppermann, 2000). However, evidence or data concerning the magnitude of ecotourism in Australia is largely circumstantial; the focus of much data collection is on broader nature-based tourism. Recent evidence (e.g. BTR 1997, in Wearing and Neil, 1999) suggests that nature-based resources or natural attractions are the third most popular reason (after VFR (visiting friends and relatives) and business) for international visitors travelling to Australia. In fact, about one-third of all international visitors to Australia are motivated mainly by the opportunity to experience wildlife or nature-based outdoor recreation opportunities. Every year, more than 4 million people visit Australia's national parks (e.g. see Wearing and Neil, 1999; Weaver and Oppermann, 2000). Ecotourism has paralleled this growth in nature-based tourism (Buckley, 2000).

Further indication of the significance of Australia's natural environment to international visitors can be obtained from the Bureau of Tourism Research (BTR). In 1995, international tourists who participated in nature-based tourism activities spent approximately Aus\$6.6 billion (Blamey and Hatch, 1998). In 1998, 47% of all inbound visitors to Australia aged 15 and over (1.7 million people) reported that they had visited national parks. The IVS Ecotourism Supplementary Survey, conducted in the March quarter of 1996, looked at nature-based visitors by country of residence and defined nature-based visitors as those who visited a national park, participated in snorkelling and scuba-diving, whale watching, horse-riding, rock climbing, bushwalking, four-wheel drive tours or outback safaris. The study revealed that Asian visitors were the most numerous of inbound participants in nature-based activities (51%), but were among the least likely to participate in such activities, with only 35% doing so. Conversely, Scandinavian visitors accounted for less than 3% of all nature-based visitors, but had the greatest propensity to participate: 72% of all Scandinavian visitors participated in at least one nature-based activity during their stay in Australia. More than 40% of visitors from Germany and other areas of Europe, the UK and the US participated in nature-based activities (Blamey, 1995; Blamey and Hatch, 1998; also see Hall, 1998: 292).

The main motivations for visitors to participate in nature-based activities were to see the natural beauty of the sites or to experience something new. Seeing wildlife in detail and being close to nature were also prime motivations. Sixty-nine per cent of visitors also felt that an educational experience was an important motivation. Observing animals, plants and landscapes was the most important learning experience, followed by learning about the biology or ecology of a species or region.

A study of the potential domestic ecotourism market by Blamey and Braithwaite estimated that 66% of Australia's adult population would like to spend some of their holidays in the 12 months after the survey to increase their understanding and appreciation of nature (see Blamey, 1995: 40). Despite these findings, Hall (1998: 292) argues that ecotourism may also be seen as 'a relatively small market niche – most people who are described as ecotourists only want a brief sample or a look at the environment rather than a 7-day walk slogging through the wilderness'. From behavioural and impact perspectives, defining an ecotourist lacks consensus, much less scientific precision!

The Impacts of Ecotourism in Protected Areas

Any form of tourism brings about change to an environment. Protecting, much less using, protected places is a challenge to public and private sector resource managers. Rapid growth in tourism in the last 20 years has paralleled increased urbanization, less day-to-day access to natural environments, and a dramatic awaking of large sections of the community to the effects of pollution and the need to preserve the natural environment (McKercher, 1998; Pigram and Jenkins, 1999; Buckley, 2000). Many battles have been fought between the environment lobby, committed to environmental protection, and business groups more interested in short-term profit than considerations of long-term environmental damage and sustainability. Caught up in this battle, the tourism industry has long recognized, though not widely practised, the need to preserve those elements of an area that the tourist finds attractive, simultaneously endeavouring to develop a product or service that yields a sustainable profit (e.g. see McKercher, 1991, 1998; McArthur, 1997). There is considerable debate over the benefits and costs of ecotourism in protected areas, and in particular, the impacts of ecotourism.

A range of economic, environmental, and social and cultural benefits associated with ecotourism have been identified (see Table 11.2).

Table 11.2. Examples of impacts and characteristics of ecotourism in Australia. (Adapted from Commonwealth Department of Tourism, 1994: 19–22; Buckley and Pannell, 1990; Pigram and Jenkins, 1999; Buckley, 2000.)

Type of impact and characteristics	Positive impacts/benefits	Negative impacts/costs
Environmental	An incentive for conserving natural areas	Clearance and damage to vegetation from trampling
	Provide resources for environmental conservation and management	Indirect damage to vegetation from grazing and manure from riding horses
	Provide incentives to maintain or enhance the physical environment	Altered habitats
		Altered fire regimes
	Engender an environmental ethic	Hunting and fishing
		Disturbance of wildlife and road kills
		Soil erosion and compaction, leading to modifications in land cover, modifications of plant cover
		Erosion
		Marine impacts
		Hydrological changes to rivers, estuaries, groundwater and oceans
		Pollution – air, noise and waste
		Introduction of exotic species, including exotic weeds
		Changes to ground and surface water quality and hydrology
Economic	Foreign exchange earnings	Failure of total revenue to match costs of ecotourism impacts
	Economic development and diversification	Increased burden on underfunded resource management agencies
	Distribution of income to local economies/communities	
	Tendency for ecotourists to spend more and stay longer	
	Generation of income for conservation and reserve/ park management	
	Increased employment opportunities	
	Local infrastructure development	

Table 11.2. *Continued*

Social and cultural	Employment opportunities	Overcrowding, especially at peak periods
	Diversification of the economic base, and hence greater diversity of facilities and services	Seasonality
		Diversion of resources (opportunity costs) away from other activities/issues
	Assist in the long-term conservation of cultural heritage	Conflicts over access and appropriate use
	Revitalization of local arts and traditions	Inappropriate commodification of local cultures
	Historical perspectives concerning indigenous peoples and flora and fauna	Improper/inappropriate tourist behaviour
	Conservation of traditional cultural activities	
	Encourage local communities to value, and benefit from, natural and cultural assets	

Davis *et al.* (1997), for example, reported on the careful design and operation of the ecotourism industry developing for whale shark observation in Ningaloo Marine Park in Western Australia (WA). They noted that observation of the whale sharks is closely controlled and monitored by WA's Department of Land and Water Conservation. Research has been used to develop guidelines governing the degree and type of contact with sharks that ecotourists are allowed. Use fees on both the ecotour operators and the ecotourists cover management costs. The visibility of these fees to the operators and the tourists makes an important connection between use and the need for management of environmental quality. These benefits clearly lend support to Pigram's (1990) claim that 'whereas tourism can lead to environmental degradation and therefore be self-destructive, it can also contribute to substantial enhancement of the environment'. Unfortunately, in many instances, ecotourism in Australia has not lived up to these high expectations.

The sensitive natural and cultural environments of protected areas can be easily damaged by unqualified and poorly managed tourist access (see Table 11.2). Unsurprisingly, increasing use levels in Australian parks have prompted calls for reconsideration of existing access and broader management policies (Buckley and Pannell, 1990; Figgis, 1994; Eagles, 1996; Pigram and Jenkins, 1999; Buckley, 2000). Buckley and Pannell (1990) argue that parks and reserves are only suitable for low-impact recreation, such as wilderness travel or natural history tours. Ecotourism has impacts on the environment, regardless of how sensitive or carefully planned it might be. Few self-labelled 'ecotourism' operators understand the full dimensions of their impacts, nor, might one suggest, do some care too much. Claims that ecotourism enhances the environment are not based on comprehensive information, and may be considered spurious in the absence of such information. Measures of compliance with respect to self-regulation, the impacts of four-wheel drive excursions into the bush and on places like Fraser Island, and the development of large-scale eco-resorts and cable cars are all understudied dimensions of ecotourism.

Examples abound where ecotourism has led to an opening of areas to more intensive forms of tourism. Fraser Island, and a host of other national and marine parks and wilderness areas explored by few only a generation ago are now under intense visitor pressure. Fraser Island is the largest (168,000 ha) sand island in the world. It was proclaimed a national park in 1971 and given international recognition as a World Heritage Area in 1992. Yet, Fraser Island's wildlife populations are declining and its vegetation patterns are changing. According to Sinclair (2000):

> This is because the priority and preoccupation of Fraser Island managers has been to maximize its tourist potential – it attracts over 300,000 visitors per annum – and to concentrate almost exclusively on vehicular based recreation. Catering for four-wheel drives is taking over protecting the environment. In 1998–99 the Queensland government spent almost 16% of its total budget for Fraser Island on road and waste management, while less than one per cent ($32,000) was dedicated to natural resource management. Ultimately, the island ecology is being adversely altered by negligence.

Economic returns to ecotourism operators may be limited or modest, or may fluctuate dramatically even in the short term. Low economies of scale may lead operators to thwart or fail to implement and maintain sustainable tourism practices. Those advocating private sector leases and licences in protected areas rarely adequately consider the long-term implications for such areas in the likelihood of business failures, transfers of business ownership, and inabilities to maintain environmental initiatives in periods of low returns. The agricultural and pastoral industries serve as valuable lessons. In times of drought and low incomes, farmers still engage in clearing practices, overstocking and other management practices leading to environmental degradation (e.g. soil erosion and salinity). Evidence suggests that the tourism industry is little different, despite repeated claims that the industry is environmentally responsible. The need for private commercial accommodation in parks has considerable support in the tourism industry (Charters *et al.*, 1996). The prevailing philosophy of such proponents is deeply anthropocentric; a human demand, tourism growth, exists, therefore it must be met. However, arguments for commercialization of parks are couched mainly in terms of 'protecting' the parks through better management of the 'inevitable' demands. Advocacy of so-called eco-resorts is a case in point. According to Dowling (2000: 165):

> An eco-resort is a self-contained, upmarket, nature-based accommodation facility. It is characterised by environmentally sensitive design, development and management which minimizes its adverse impact on the environment, particularly in the areas of energy and waste management, water conservation and purchasing. An eco-resort acts as a window to the natural world and as a vehicle for environmental learning and understanding.

The concept of an 'eco-resort' is open to widespread interpretation and debate, and may well lead to the misrepresentation of development in protected areas in the same way that ecotourism has become a development and marketing cliché. The concept of 'eco-resort' begs the question: 'can the development of a large-scale resort in or very near to a protected area be ecologically sensitive?' Claims of best practice and development devoid of impacts, for instance, warrant considerable scrutiny. Surely 'knocking down trees in a national park to clear 10 m × 10 m tower sites' for a cable car in Australia's wet tropics and claiming 'private enterprise in protected areas can work' is *not* 'leading the world in the introduction of appropriate infrastructure to allow people to experience the rainforest without impacts'. Indeed,

> High impact recreation such as sporting and social activities, the use of motorised vehicles and large-scale engineering and building construction should be discouraged in parks and reserves. It is thus inappropriate for tourist developments in park and reserve areas to include facilities such as large hotels, conspicuous cable cars and ski-lifts, tennis courts and golf courses, or marinas and water-ski areas. It should be necessary to argue these points in relation to every tourist development proposal in park and reserve areas. Generic guidelines for natural areas tourist development, adopted and adhered to by Commonwealth and State governments, and promulgated to all local government authorities, could overcome problems associated with the current piecemeal and ad hoc approach to natural areas of tourist development in Australia.

(Buckley and Pannell, 1990: 29–30; also see Good and Grenier, 1994)

The debate over whether ecotourism in any location is ecologically, economically, socially and culturally sustainable is in fact clouded by a number of problems. In particular:

- there is a lack of appropriate indicators and resources for comprehensively measuring impacts at inherently diverse sites;
- many measures are inherently subjective and contestable as are levels of acceptable change in any environment;
- it is inherently difficult to distinguish tourism impacts from the impacts of other 'causes';

- there are often long periods between cause and effect;
- forecasting is inherently risky (Buckley and Pannell, 1990).

Tourism industry leaders and natural resource managers face significant challenges in the sustainable development of tourism in protected areas and in managing impacts on flora and fauna (HaySmith and Hunt, 1995). Nature-based tourism can only survive when the resources on which it depends are protected. According to Whelan (1991: 4), 'eco-tourism, done well, can be sustainable and a relatively simple alternative. It promises employment and income to local communities and needed foreign exchange to national governments, while allowing the continued existence of the natural resource base'. This last point gives implicit recognition to the need for adequate and appropriate management regimes (see also Valentine, 1991, 1992), which foster environmental and cultural understanding, appreciation and conservation (e.g. see Ecotourism Association of Australia; www.ecotourism.org.au/ About_Eaa.htm). As McKercher (1998: 191) points out:

> A review of the principles of ESD [Ecologically Sustainable Development] offers valuable insights into how the tourism industry must act in relatively undisturbed areas. Underlying the entire ESD philosophy is a commitment to operate within the social and biophysical limits of the natural environment. To abide by this tenet, tour operators may have to trade off economic gain for ecological sustainability and, indeed, will have to accept that there are some places where tourism should be excluded.

Important elements in the development of any management regime or programme, therefore, are appropriate research and adequate monitoring and evaluation. Yet,

> how many tourism businesses and associations still refuse to acknowledge that tourism produces impacts, and shy away from any suggestion of systematic monitoring? In this respect the industry is still remarkably immature, contrasting sharply with mining and manufacturing where impacts are acknowledged, measured, minimized where possible, and managed as a routine part of day-to-day operations.
>
> (Buckley, 2000: 256)

Much tourist activity in natural areas is permitted without a great deal of understanding of tourism's impacts on the ecosystem (e.g. see Liddle, 1998; Buckley, 2000). And, the tourism industry contributes very little to research in and management of the protected estate, which it is so intent on commercializing.

> Regrettably, current research funding mechanisms do not support detailed research on the impacts of tourism. Science funding agencies considered it too applied; tourism funding agencies think it too expensive and don't recognize its significance; national parks management agencies

recognize its significance but have inadequate budgets to support it; and the tourism industry itself has little interest in quantifying its own impacts.

(Buckley, 2000: 35)

With respect to both flora and fauna and the landscape itself, these are very salient points. For instance, the impacts of tourism on wildlife are well documented at some sites, but the findings of such studies and related management strategies can rarely be applied universally. As HaySmith and Hunt (1995: 206) point out:

> Impacts on wildlife from nature tourism are varied, and are often difficult to observe and interpret. Reactions of animals to visitors are complex. Initially, some species or individuals of a species retreat from visual or auditory stimuli caused by humans but become habituated over time. Other species or individuals that are more sensitive may alter their behaviour and activities to completely avoid contact with visitors, with potentially long term effects. Other animals cannot escape the disturbance and may be negatively affected, directly injured ... or killed.

Nature tourism can be blatantly invasive towards wildlife when hundreds of observers congregate to view one rare animal or group of animals, when artificial feeding is used to draw animals for tourist viewing and entertainment, and when relationships between species are disturbed (for a more detailed discussion, including case studies, of the relationships between recreation, tourism and wildlife, see HaySmith and Hunt, 1995; Lindberg and McKercher, 1997; Liddle, 1998; Buckley, 2000).

Nature-based and related forms of tourism will only be successful if comprehensive planning strategies include appropriate and extensive research programmes. Any arguments that nature-based tourist activity, or any form of tourist activity, has a particular beneficial or negative or otherwise relationship with the environment cannot be sustained without related research. Those who choose to argue one way or another could be easily challenged by questions about the precise nature of the tourism–environment relationship (Pigram and Jenkins, 1999).

Land rights are an undeniable aspect of the policy issue relating to protected areas and ecotourism. The recent (1992) High Court (Mabo) decision on 'native title' has brought much focus on to land 'ownership'. 'Mabo' is the name given to the decision on 3 June 1992 by the High Court of Australia recognizing native title to land. 'The High Court rejected the doctrine that Australia was terra nullius (land belonging to no one) at the time of European settlement and said that native title can continue to exist:

● Where Aboriginal and Torres Strait Islander people have maintained their connection with the land through the years of European settlement; and

- Where their title has not been extinguished by valid acts of Imperial, Colonial, State, Territory or Commonwealth Governments'.

(*Mabo: the High Court Decision on Native Title*,
Discussion Paper, June 1993; 1).

The High Court also ruled that the traditional laws and customs of the Aboriginal and Torres Strait Islander people involved were to be taken into account in the conferring of native title rights (*Mabo: the High Court Decision on Native Title*, Discussion Paper, June 1993; 1).

'Mabo is about people not real estate' (Wootten, 1993: 19) and therefore a fundamental shift in the power relationships of involved parties must be achieved. As such, planning for the future of Aboriginal peoples and ecotourism in Australia's national parks must explicitly take into account relations of power and knowledge when designing management regimes. This shift in power relations has occurred with the Wik and Mabo decisions that now provide Aboriginal culture with a legal basis for change. That said, it is apparent that this shift in power has been received expediently by many at a political and managerial level who make appropriate changes according to the legal determinations but then go no further than is required. Progress is limited as park management continue to pursue the original government response to the Wik decision and fulfil the minimum legal requirement of the legislation rather than the intent, thus failing to deal with the ethical or moral issue of indigenous rights (e.g. see Wearing and Gartrell 2000; Wearing and Huskins, 2001).

Tourism within national parks can impact upon indigenous communities in many ways. Aboriginal rights to subsistence living within national parks can be directly and indirectly impinged upon by tourism, and direct competition for resources can occur, for example, with tourists' recreational fishing, as aquafauna is not protected in some parks. Indirect pressure can come from increasing tourist numbers pushing people further into the parks, and thus restricting the areas where hunting and gathering can safely and comfortably take place. These pressures create ' ... an inverse relationship between tourism growth and Aboriginal access to subsistence' (Altman and Allen, 1993: 124). Further, ' ... evidence from Kakadu and elsewhere indicates that *tourism is of marginal economic benefit to indigenous people*. It may even leave them worse off' (Cordell, 1993: 7).

Where the benefits from this industry are to go is a serious ethical consideration. Given that many if not most Aboriginal heritage sites being visited are in national parks or other protected areas, and having recognized Aboriginal land and therefore cultural rights, should non-Aboriginal enterprises and park management authorities be profiting from this cultural tourism?

Aboriginal cultural themes, together with the significance of protected areas to Aboriginal people are increasingly being promoted by government and non-government conservation and tourist agencies as values and attractions of the land as a tourist destination. Aboriginal people are still largely in an advisory role with respect to this appropriation of Aboriginal culture.

(Woenne-Green *et al.*, 1992: 375)

It is timely and appropriate for the inclusion of mechanisms and processes, which will help to end this appropriation, and benefit indigenous peoples in the process. The most direct approach is for Aboriginal communities to be in control of the resource and the tourism. If protected area management arrangements were to incorporate equitable co-management arrangements with Aboriginal communities within national parks, it would be possible for this to happen.

Unfortunately, however, indigenous culture is a much neglected but critical dimension of Australia's ecotourism product:

Australia can market that it has the world's oldest continuous culture and cultural tourism to experience this culture. Significantly for ethical ecotourism markets, Australia's Aboriginal tourism is fast moving away from the performance and object exploitation to more sensitive and sophisticated cultural tourism where Aboriginal people set their own terms for ecotourism opportunities on their own land. Although Aboriginal people are still grappling with tourism as a threat to land rights, sacred sites and social values, this threat is being progressively reduced through small, low-impact, limited duration tours.

(Ceballos-Lascurain, 1999: 7)

Ceballos-Lascurain's view of Aboriginal tourism 'fast moving away from performance and object exploitation to a more sensitive and sophisticated cultural tourism' is an oversimplification. Aboriginal tourism cuts across the heart of historical, cultural and political insensitivity to, and lack of awareness of, indigenous culture. Racism, exploitation of indigenous culture and Aboriginal rights to land are divisive and important political and cultural issues which have yet to be resolved. More specifically, however, several challenges face the growth and consolidation of indigenous tourism in Australia:

- the need for a stronger business ethic;
- complex Aboriginal politics;
- loss of Aboriginal people to government positions;
- maintaining authenticity;
- presenting an evolving culture;
- avoiding too much duplication (Freeman, 1999: 22).

Freeman (1999: 24) summarized the situation thus:

> The challenge of the next century for Indigenous tourism product will be finding a balance... This means finding a compromise that allows customers to fulfill their romantic expectations of Aboriginal culture, and at the same time, modernizing the presentation... Beyond a dollar value is an experience in which the client has direct and personal contact, and thus feels that they are supporting an ethical venture and giving something back. However, this experience is incredibly volatile, because it involves a fragile and primitive culture and a degree of frankness that can be unappreciated and even misplaced. The result of this reaction can be damaging to the Indigenous people, and of course, the business.

Our understanding of protected areas is on the whole very limited. Hence, the precautionary principle ought to be invoked in protected area management. If the private sector wishes to use protected areas for tourism but cannot afford to subsidize research, then it should forfeit its right to pursue such development in public lands. Unfortunately, however, public sector agencies also lack funds to manage extensive parks systems, and are being driven by political corporate directives to develop nature-based tourism opportunities when the survival and inherent qualities of the areas they are obligated to protect are at stake. This raises questions about the appropriate roles and responsibilities of public and private agencies, which have been influenced by important global and local shifts in political ideology, and public administration and management.

The Roles of the State and Government

The major economic revolution of the 1980s transformed Australia from one of the most regulated, interventionist countries in the world, to a de-regulated, free-market economy. The effects within Australian society have been enormous, and have radically influenced the delivery of community services from healthcare to income support. In addition, the 'hands off' or 'step back' attitude of government and the penetration of the 'market' into every aspect of Australian life has affected and is continuing to affect the management of conservation resources and access to outdoor recreational/tourism opportunities. Specifically, public sector management of protected areas and nature-based tourism resources in Australia is faced with mounting resource use claims (including recreational demand), and yet is receiving commensurately less government funding and assistance.

The economic recession of the 1990s, and the substantial restructuring of many national and local economies, especially since the early 1980s, gave considerable light to tourism as an economically significant industry. Tourism is widely considered as a means of generating international trade links, foreign investment, industrial diversification, income and employment (e.g. see Office of National Tourism, 1998). Many economies are moving, or have moved, away from a dependence on primary industries such as agriculture, mining, gas or oil, towards a greater reliance on the industrial and service sectors, significantly influenced by the globalization of economies and the spread of multinational corporations who look to tax incentives, subsidies and lower wages to offset the costs of producing goods, and to increasing their market penetration and shares. Australia has been no exception.

The period following the Second World War has been one of generally rising per capita incomes and diminishing rates of population increase. Reductions in the size of the public sector and concomitant reductions in expenditures with macroeconomic restructuring have resulted from a public sector focus on market-led recovery and economic efficiency as a precursor to social welfare. Governments have been prompted to encourage private sector investment in tourism projects such as casinos, waterfront development, coastal resorts and ecotourism, while much greater emphasis and resources have been given to international marketing and promotion to increase international visitor numbers and receipts.

Parks and Public Sector Reforms

Parks management is currently grappling with issues associated with the realignment of their organizations in a more demanding public sector environment (Wearing and Bowden, 1999). Governments are using a number of mechanisms to increase accountability for resource use while also applying pressure on public sector organizations to adopt more cost-effective ways of delivering community services. While much of the reform pressure to date has focused on public sector activities with a demonstrably commercial potential, national parks and other protected area organizations have not been exempt from this pressure. The economic benefit of encouraging tourism has only relatively recently begun to be recognized within national parks organizations, and parks management must deal with the issue of increasing the economic benefit of tourism while simultaneously (and perhaps conflictingly) meeting traditional conservation and preservation objectives.

It is suggested here that one of the primary functions of national parks services is to control behaviour in designated areas in order to preserve their natural state. National parks services are now less able to rely on traditional, reactive ways of responding to increased demand as funding is reduced and a wider audience increases expectations for the use of parks (Wearing and Bowden, 1999). Prasser

(1996) notes the discrepancy in the publicly demanded increase in total areas of national parks and a stagnation (and subsequent short-fall) in resources allocated by governments to manage them, as rising numbers of visitors demand more access, facilities and activities. In the case of escalating use from areas such as tourism, there is a grow-ing realization that specific site carrying capacities are going to be exceeded. As a result, site-hardening becomes essential along with ranger patrolling, and for most organizations this is believed to be a less than effective use of limited park resources particularly when the return from the tourist is often marginal, and when the achievement of conservation objectives is jeopardized.

To date, there are no wholly privatized national parks in Australia, yet private enterprise currently plays a variety of roles in different national parks. In some parks there is no private sector involvement at all, while in others private companies hold head leases or concessions over certain areas and are responsible both for the running of tourist operations and for many park management functions.

The ideological commitment to small government in Australia and the lack of adequate resources for management is also fuelling a substantial push for a greater role for the private sector in managing and operating in national parks (Christoff, 1998). The privatization debate ranges over a wide field, from the privatization of services within national parks, such as the delivery of parks maintenance, accommodation, food transport and tour services, to a far more radical approach where the government's role would retreat to setting stan-dards and monitoring outcomes while the private sector provided all services.

Another issue arises from the increasing dependence of parks authorities on tourism charges to meet budget shortfalls (also see below). There remains widespread concern in the environmental movement that this creates an inexorable shift towards tourism-cen-tred management (Figgis, 1994, 2000). Some recent examples include the following:

- The controversial decision of the Commonwealth government in 1996 to raise the Great Barrier Reef Environment Management Charge ('reef tax') from Aus$1 to Aus$6 immediately brought calls for greater representation of tourism interests in management.
- In mid-1996, work commenced in the Otway National Park in Victoria on an extension to the Great Ocean Walk that will involve cutting of tracks through pristine coastal environments. No prior consultation was conducted with conservation groups.
- In Western Australia the strict nature reserve status of Two Peoples Bay is being altered to allow for a tourism centre despite

serious scientific concerns over impacts on the rare species found in the reserve.

A significant portion of Australia's tourism industry is based on protected areas, creating what Figgis (2000) described as a 'double-edged sword'. On the one hand, it creates a powerful economic argument for the dedication and proper management of protected areas for their value as a 'tourism product'. On the other hand, it creates substantial pressures for tourism-centred management of protected areas, where the interests of tourism prevail over nature conservation (Figgis, 1994, 2000). Figgis suggests that the current trend for tourism developments in protected areas is one of the major threats identified by environmental groups around Australia. That such developments are aggressively promoted by their supporters as 'models' of good ecotourism and in the interests of good management, ignores the fact that commercial development and the assumption that demand must be met, will inevitably compromise ecological integrity. It completely distorts the idea of ecotourism as tourism that supports nature conservation, and inevitably supports the growth of commercial tourism.

> While ecotourism creates support for the proper management of protected areas, it can also cause protected areas to be regarded primarily as economic resources. Park managers increasingly see tourism development rights, licences, entry fees and levies as the answer to government budget cuts.

> (Figgis, 2000: 24)

User Fees

There is considerable evidence that Australia's national parks and other protected area agencies face funding crises as dramatic expansion in the number and area of parks since the mid-1970s has not been matched by funds for management (e.g. Figgis, 1994; Wescott, 1995; Dickie, 1995; McArthur, 1999). According to Dickie (1995), there was little infrastructure in established parks, much infrastructure was in poor repair, and some facilities had to be closed because of imminent safety risks and insufficient resources for enforcement. Meanwhile, visitor numbers to well-known protected areas (namely, national parks and World Heritage sites) continue to swell, budgets to manage protected areas are becoming scarcer across the protected estate, more is needed for roads, toilets and interpretation programmes, and less is available for conservation-related research and management.

User fees and other charges have been gaining increased consideration from protected area managers as governments fail to provide adequate funds and as protected areas have become more popular for recreational use. Fees and charges include:

- User fees: charges on 'users' of an area or facility such as park entrance or tramping fees.
- Concessions: a fee for the permission to operate within a location for groups or individuals that provide certain services to visitors such as food, accommodation and retail stores.
- Sales and royalties: fees levied on a percentage of earnings that have been derived from activities or products at a site such as photographs or postcards.
- Taxation: an additional cost imposed on goods and services that are used by ecotourists, such as airport taxes.
- Donations: tourists are often encouraged to contribute to maintaining a facility (cf. Hedstrom, 1992; Marriott, 1993; Lindberg and McKercher, 1997).

Fees can provide an important source of revenue for resource managers (Swanson, 1992). One rationale supporting user fees is that most foreign visitors travel to remote protected areas to experience their very isolation and unspoilt natural features. The visitors should be willing to contribute to the costs of maintaining such conditions (Bunting, 1991). Ecotourists travelling in tour groups pay a fee, which is usually incorporated into the price of the tour. Fees may be seen as inequitable, but tiered schedules of fees can help to overcome this problem.

Clearly, the imposition of fees provides an important source of revenue for protected area agencies. But the political situation is complex and one well described by Buckley (2000). In summary, keeping funds in local parks where they are paid, has led to pressures to operate parks on a commercial basis and to grant concessions for large-scale tourism development. Shifting revenues to a consolidated purse might remove some of this pressure, but might also mean that heavily used sites do not get the finance needed to support a tourism base. Buckley (2000: 37) suggests a hybrid funding model,

> where funds for conservation management and the provision of basic recreation facilities are provided centrally through the government budget process, and funds required for marginal expenditure associated with increasing levels of tourism may be raised from tourists and tour operators, and retained locally for immediate management expenditure.

Frameworks: Carrying Capacity, ROS, LAC

Park managers have moved to place greater emphasis on managing park visitors through the development of visitor management frameworks. Examples of decision-making frameworks that have emerged include Carrying Capacity (CC), Recreation Opportunity Spectrum

(ROS), Limits of Acceptable Change (LAC), Visitor Impact Management (VIM), Visitor Activity Management Process (VAMP), and Visitor Experience and Resource Protection Plan (VERP). Whereas some frameworks focus more on the resource (e.g. CC, ROS and LAC) others have shifted the emphasis back to the resource users (e.g. VIM, VAMP and VERP). When implemented these frameworks help to protect the natural and cultural heritage, enhance public appreciation of the resource, and assist in managing the competing values between a range of interests in protected areas. Simultaneously, park managers have also embraced visitor education and interpretation programmes as important techniques in visitor use planning in recent years. The benefits of such programmes relate to an increasing awareness and appreciation of the natural environment by park visitors.

Despite the inherent value of these visitor management frameworks, their practical application has had problems (for a broader discussion of these frameworks and models, see Pigram and Jenkins, 1999). Some commentators (cf. Dearden and Rollins, 1993; Pigram and Jenkins, 1999; Wearing and Bowden, 1999; McArthur, 2000) have noted that park managers continue to rely on strategies that regulate and control visitor behaviour, and rely on site hardening and positioning of visitor amenities. Reactive in nature, these strategies typically force park managers to wait until visitors are on-site before taking (any) action to manage their impacts and behaviour. Jenkins and McArthur (1996) argue that the focus of park managers represents a preoccupation with the 'stock control' aspects of resource management and ignores any potential to manage natural resource conservation through strategies that proactively manage demand. That is, in the main, park managers have remained passive to market forces that have an increasing impact on demand for use of protected areas, thereby placing considerable pressure on visitor impact minimization strategies.

Marketing

Marketing as it applies to protected area agencies remains surrounded by much confusion. However, this need not be the case. Since the 1980s especially, ecological and social marketing approaches have been increasingly acknowledged by marketing theorists as important elements of a more holistic marketing perspective.

Protected area agencies may need to discourage and reduce demand for a setting or service if excess demand is evident. Many national parks and protected areas in Australia and other countries are facing crowding and carrying capacity problems across a range of visitor experiences and types of recreation. With park visitation being based on limited

supply, park agencies may effectively use the marketing mix for discouraging participation. This discouraging of demand has been coined
'demarketing' (Kotler, 1971) to emphasize that marketing may be used
to decrease as well as increase demand for access to particular settings.
Demarketing is not a negative concept as 'a decrease in visitor numbers
can lead to an increase in clientele satisfaction, through preserving a
higher quality experience' (Crompton and Howard, 1980: 333). A
demarketing plan may be appropriate in a number of situations. For
example, Groff (1998) identified three different circumstances where a
protected area agency may utilize demarketing strategies:

- *Temporary shortages*: due to either lack of supply or underestima-
  tion by management of demand for particular settings or programmes.
- *Chronic overpopularity*: can seriously threaten the quality of the
  visitor experience and also damage the natural resource that
  attracts the visitors.
- *Conflicting use*: encompasses issues of visitor safety, compatabil-
  ity of use with the available resources, and the different uses and
  programmes demanded by the public.

Sometimes a park agency may be engaged in marketing and demarketing activities at the same time (Crompton and Lamb, 1986). Methods
of demarketing can include:

- increasing prices in a manner so they increase disproportionately
  as time spent in the park management destination increases;
- creating a queuing system to increase the time and opportunity
  costs of the experience;
- limiting the main promotional strategy to select and specialized
  media channels;
- promoting the importance of the area by educating the public and
  emphasizing the need to conserve the area through minimal
  impact and sustainable development;
- promoting a range of alternative opportunities in surrounding
  areas which may satisfy needs and wants;
- highlighting the environmental degradation that could occur if too
  many people frequent the area;
- highlighting any restrictions or difficulties associated with travel
  to the area.

Marketing and demarketing offer protected area agencies a visitor
management approach incorporating existing management techniques that are supply oriented and existing marketing tools that are
demand oriented. The convergence of these ideologically different
approaches has led to very limited use of marketing as a management
tool by protected area agencies and the loss of opportunities to

develop new models for park management. Marketing and demarketing strategies can directly address the causes of visitor use problems in a proactive and positive manner and provide off-site management control not open to protected area agencies through other management techniques.

Conclusion

Like Peter Garrett (1999: 11), speaking as President of the Australian Conservation Foundation, 'we are alarmed at the discernible trend towards commercialisation of national parks'. This chapter endorses calls for a more biocentric view of the environment. Powerful tourism interests, not to mention many researchers and academics, have much to gain by advocating sustainable ecotourism development in protected areas. To paraphrase Mowforth and Munt (1998), sustainability (and ecotourism as a form of sustainable tourism) is such a vague, contested concept, that it is easily manipulated to support and enhance the power of industry interests and those who stand to gain. Therefore, we stoically advocate adoption of the precautionary principle and argue for a stronger conservation/preservation bias in protected area management. Neo-liberal economics have served to influence protected area policies in ways that afford little room for intangible values. Yet, a biocentric view of the environment is just as valid as an anthropocentric one, perhaps even more so when knowledge of the environment is so very limited. Moreover, protected areas should not be required to yield financial returns (from ecotourism or any other industry) in order to be created and/or maintained. In other words, sustainable tourism development in protected areas may mean 'no development' or even the removal of structures. Any tourism in protected areas should be carefully evaluated and, where permitted, carefully regulated and monitored. We are a long way from a nature-based tourism industry that can claim to be ecologically sustainable, much less containing a well established and extensive core of mature businesses that can claim to be ecotourism operators.

References

Altman, J. and Allen, L. (1993) Living off the land in National Parks: issues for Aboriginal Australians. In: Birckhead, J., De Lacy, T. and Smith, L. (eds) *Aboriginal Involvement in Parks and Protected Areas*. Aboriginal Studies Press, pp. 117–136.

Blamey, R. (1995) *The Nature of Ecotourism*, Occasional Paper No. 21, Bureau of Tourism Research, Canberra.

Blamey, R.K. and Hatch, D. (*c.*1998) *Profiles and Motivations of Nature-Based Tourists Visiting Australia.* Bureau of Tourism Research, Canberra.

Buckley, R. (2000) Tourism in the most fragile environments. *Tourism Recreation Research* 25(1), 31–40.

Buckley, R. and Pannell, J. (1990) Environmental impacts of tourism and recreation in national parks and conservation reserves. *The Journal of Tourism Studies* 1(1), 24–32.

Bunting, B. (1991) Nepal's Annapurna conservation area. In: *Proceedings of the PATA 91 40th Annual Conference, Bali, Indonesia, 12 April.*

Ceballos-Lascurain, H. (1999) The future of ecotourism into the millennium: an international perspective. In: *Australia – The World's Natural Theme Park, Proceedings of the Ecotourism Association of Australia National Conference.* Ecotourism Association of Australia and the Bureau of Tourism Research, Canberra, pp. 1–9.

Charters, T. (1996) The state of ecotourism in Australia. In: Riching, H., Richardson, J. and Crabtree, A. (eds) *Ecotourism and Nature-Based Tourism: Taking the Next Steps, Proceedings of The Ecotourism Association of Australia National Conference, Alice Springs, Northern Territory.* The Ecotourism Association of Australia and The Commonwealth Department of Tourism, Canberra.

Charters, T., Gabriel, M. and Prasser, S. (eds) (1996) *National Parks: Private Sector's Role.* University of Southern Queensland (USQ) Press, Toowoomba.

Christoff, P. (1998) Degreening government in the Garden State: Environment Policy under the Kennett Government, 1992–1997. *Environment Planning and Law Journal* 15(1), 10–32.

Commonwealth Department of Tourism (1994) *National Ecotourism Strategy.* Commonwealth Department of Tourism, Canberra.

Cordell, J. (1993) Indigenous Involvement in Australian Protected Areas, Unpublished paper.

Cotterill, D. (1996) Developing a sustainable ecotourism business. In: Richins, H., Crabbe, A. and Richardson, J. (eds) *Proceedings of the 1995 Ecotourism Association of Australia Conference.* Bureau of Tourism Research, Canberra and EAA, Sydney.

Crompton, J.L. and Howard, D.R. (1980) Financing, managing and marketing. In: Cronin, L. (ed.) *A Strategy for Tourism and Sustainable Developments.* Government of Canada, Ottawa, pp. 12–16.

Crompton, J. and Lamb, C. (1986) *Marketing Government and Social Services.* John Wiley & Sons, New York.

Davis, D., Banks, S., Birtles, A., Valentine, P. and Cuthill, M. (1997) Whale sharks in Ningaloo Marine Park: managing tourism in an Australian marine protected area. *Tourism Management.* 18(5), 259–271.

Dearden, P. and Rollins, R. (eds) (1993) *Parks and Protected Areas in Canada: Planning and Management.* Oxford University Press, Toronto.

Dickie, P. (1995) Money squeeze on parks. *Brisbane Sunday Mail,* 7 May.

Dowling, R. (1991) Tourism and the natural environment: Shark Bay, Western Australia, *Tourism Recreation Research,* 16(2), 44–48.

Dowling, R. (2000) Ecoresort. In: Jafari, J. (ed.) *Encyclopedia of Tourism.* Routledge, London.

Driml, S. and Common, M. (1995) Economic and financial benefits of tourism in major protected areas. *Australian Journal of Environmental Management.* 2(2), 19–39.

Eagles, P.F.J. (1996) Issues in tourism management in Parks: the experience in Australia. *Australian Leisure,* June, p. 29.

Figgis, P. (1994) Eco-Tourism. *Habitat Australia.* 21(1), 8–11.

Figgis, P. (2000) The double-edged sword: tourism and national parks. *Habitat Australia.* October, 28(5), 24.

Freeman, J. (1999) An operator's perspective on how we can further grow indigenous tourism in Australia. In: *Australia – The World's Natural Theme Park, Proceedings of the Ecotourism Association of Australia National Conference,* Ecotourism Association of Australia and the Bureau of Tourism Research, Canberra, pp. 1–9.

Garrett, P. (1999) Perspectives on tourism and the environment in Australia. In: *Australia – The World's Natural Theme Park, Proceedings of the Ecotourism Association of Australia National Conference.* Ecotourism Association of Australia and the Bureau of Tourism Research, Canberra, pp. 10–13.

Good, R. and Grenier, P. (1994) Some environmental impacts of recreation in the Australian Alps. *Australian Parks and Recreation,* Summer, pp. 20–24.

Grant, J. and Allcock, A. (1998) National planning limitations, objectives and lessons. In: Lindberg, K., Epler Wood, M. and Engledrum, D. (eds) *Ecotourism: A Guide for Planners and Managers,* Vol. 2. The Ecotourism Society, Bennington, Vermont, 119–132.

Groff, C. (1998) Demarketing in park and recreation management. *Managing Leisure* 3, 128–135.

Hall, C.M. (1998) *Introduction to Tourism in Australia – Development, Dimensions and Issues,* 3rd edn. Longman Cheshire, Melbourne.

HaySmith, L. and Hunt, J.D. (1995) Nature tourism: impacts and management. In: Knight, R.L. and Gutzwiller, K.J. (eds) *Wildlife and Recreationists: Coexistence through Management and Research.* Island Press, Washington, DC, 203–219.

Hedstrom, E. (1992) Preservation or profit? *National Parks* 66 (1–2), 18–20.

International Union for the Conservation of Nature (1994) *United Nations List of National Parks and Protected Areas.* IUCN Publications Unit, Cambridge.

Jenkins, J.M. (1998) *Crown Lands Policy Making in NSW, 1856–1991: The Life and Death of an Organisation, its Culture and a Project.* Centre for Public Sector Management, Canberra.

Jenkins, J.M. (1999) Politics and management of recreational lands: a case study of Crown land policy-making in NSW, 1856–1980, *Annals of Leisure Research* 2(1), 1–27.

Jenkins, J. and McIntyre, N. (2001) Access to the countryside: a case study of global and local forces in New South Wales, Australia. Unpublished paper presented to Council of Australian Tourism and Hospitality Educators (CAUTHE), Canberra, February.

Jenkins, O. and McArthur, S. (1996) Marketing protected areas. *Australian Parks and Recreation* 32(4), 10–15.

Kotler, P. (1971) *Marketing Decision Making: a Model Building Approach.* Holt, Rinehart and Winston, New York.

Liddle, M. (1998) *Recreation Ecology.* Chapman and Hall, London.

Lindberg, K. and McKercher, B. (1997) Ecotourism: a critical overview. *Pacific Tourism Review* 1(1), 65–80.

Marriott, K. (1993) Pricing policy for user pays. *Australian Parks and Recreation* 29 (3), 42–45.

McArthur, S. (1997) *Ecotourism Resource Booklet: Review of Australian Ecotourism Plans and Strategies.* Ecotourism Association of Australia, Brisbane.

McArthur, S. (1999) Embracing the future of ecotourism, sustainable tourism and the EAA in the new Millennium. In: McArthur, S. and Weir, B. (eds) *Developing Ecotourism into the Millennium, Proceedings of the Ecotourism Association of Australia National Conference 1998.* Ecotourism Association of Australia, Brisbane, and the Bureau of Tourism Research, Canberra.

McArthur, S. (2000) Visitor Management in Action: an Analysis of the Development and Implementation of Visitor Management Models at Jenolan Caves and Kangaroo Island. Unpublished PhD Thesis, University of Canberra, Belconnen, ACT.

McIntyre, N., Jenkins, J.M. and Booth, K. (2001) Recreational access in New Zealand. *Journal of Sustainable Tourism*, 9(5), 434–450.

McKercher, B. (1991) The unrecognized threat to tourism: can tourism survive sustainability? In: Weiler, B. (ed.) *Ecotourism: Incorporating the Global Classroom.* Bureau of Tourism Research, Canberra, pp. 4–10.

McKercher, B. (1998) *The Business of Nature-Based Tourism.* Hospitality Press, Melbourne.

Moore, S. and Carter, B. (1993) Ecotourism in the 21st century. *Tourism Management*, April, 123–130.

Mowforth, M. and Munt, I. (1998) *Tourism and Sustainability: New Tourism in the Third World.* Routledge, London.

Nepal, S.K. (2000) Tourism in protected areas: the Nepalese Himalaya. *Annals of Tourism Research* 27(3), 661–681.

Office of National Tourism (1998) *Tourism – a Ticket to the 21st Century: National Action Plan for a Competitive Australia.* Commonwealth of Australia, Department of Industry, Science and Tourism, Canberra.

Pigram, J. (1990) Sustainable tourism – policy considerations. *The Journal of Tourism Studies.* 1(2), 2–9.

Pigram, J.J. and Jenkins, J.M. (1994) The role of the public sector in the supply of rural recreation opportunities. In: Mercer, D. (ed.) *New Viewpoints in Outdoor Recreation Research in Australia.* Hepper Marriott and Associates, Williamstown, Australia.

Pigram, J.J. and Jenkins, J.M. (1999) *Outdoor Recreation Management.* Routledge, London.

Pittock, J. (1996) The state of the Australian protected areas system, paper presented at CNPPA workshop, Sydney, June.

Pitts, D. (1983) Opportunity shift: Development and application of recreation opportunity spectrum concepts in park management, unpublished PhD thesis, Griffith University, Nathan Campus, Queensland.

Prasser, S. (1996) Foreword. In: Charters, T., Gabriel, M. and Prasser, S. (eds) *National Parks: Private Sector's Role*. USQ Press, Toowoomba, Queensland, p.iii.

Ryan, C. (1998) Kakadu National Park: a site of natural and heritage significance. In: Shackley, M. (ed.) *Visitor Management: Case Studies from World Heritage Sites*. Butterworth Heinemann, Oxford, 121–138.

Sinclair, J. (2000) The fatal shore. *Habitat Australia*, February, 28(1), 26.

Swanson, M.A. (1992) Ecotourism: embracing the new environmental paradigm. Paper presented at the *International Union for Conservation of Nature and Natural Resources (IUCN) IVth World Congress on National Parks and Protected Areas, Caracas, Venezuela, 10–12 February*.

Urry, J. (1990) *The Tourist Gaze: Leisure and Travel in Contemporary Societies*. Sage, London.

Valentine, P.S. (1991) Ecotourism and nature conservation: a definition with some recent developments in Micronesia. In: Weiler, B. (ed.) *Ecotourism: Incorporating the Global Classroom*. Bureau of Tourism Research, Canberra, pp. 4–10.

Valentine, P. (1992) Nature-based tourism. In: Weiler, B. and Hall, C.M. (eds) *Special Interest Tourist*. Belhaven Press, London, pp. 105–127.

Wamsley, J. (1996) Wildlife management: the work of Earth Sanctuaries Ltd. In: Charters, T., Gabriel, M. and Prasser, S. (eds) *National Parks: Private Sector's Role*. University of Southern Queensland Press, Toowoomba, pp. 156–163.

Wearing, S.L. and Bowden, I. (1999) Tourism and a changing public sector culture for parks. *Parks and Leisure* 1(3), 6–8.

Wearing, S.L. and Gartrell, N. (2000) Exploring ecotourism and community based research applications for joint management of national parks. *World Leisure and Recreation Journal* 43, 25–32.

Wearing, S.L. and Huskins, M. (2001) Moving on from joint management regimes in Australian national parks. *Current Issues in Tourism* 2–4, 182–209.

Wearing, S. and Neil, J. (1999) *Ecotourism: Impacts, Potential and Possibilities*. Reed International, London.

Weaver, D. and Oppermann, M. (2000) *Tourism Management*. John Wiley & Sons, Brisbane.

Wescott, G. (1995) Victoria's national park system: Can the transition from quantity of parks to quality of management be successful? *Australian Journal of Environmental Management* 4(2), 210–223.

Whelan, T. (1991) Ecotourism and its role in sustainable development. In: Whelan, T. (ed.) *Nature Tourism*. Island Press, Washington, DC.

Woenne-Green, S., Johnston, R., Sultan, R. and Wallis, A. (1992) *Competing Interests, Aboriginal Participation in National Parks and Conservation Reserves in Australia: A Review*. Australian Conservation Foundation, Melbourne.

Wootten, H. (1993) Green and black after Mabo. *Habitat Australia* 21(4).

The Scope and Scale of Ecotourism in New Zealand: a Review and Consideration of Current Policy Initiatives

12

James Higham and Anna Carr

Department of Tourism, University of Otago, PO Box 56, Dunedin, New Zealand

Introduction

The New Zealand tourism industry has experienced growth in international visitor arrivals of up to 14.0% per annum since the mid-1980s (Tourism New Zealand, 2001). During this time inbound arrivals to New Zealand, currently 1.7 million per annum, have trebled (New Zealand Tourism Board, 2000). While growth in New Zealand tourism has been constant and robust, the prominence of different New Zealand inbound tourism markets has fluctuated dramatically in recent years (New Zealand Tourism Board, 2000). Despite this, one constant in an otherwise changeable New Zealand tourism industry has been the importance of nature-based tourism experiences.

A number of issues arise from the rapid course of tourism development in New Zealand (Warren and Taylor, 1994). The growth and proliferation of nature-based tourism and ecotourism operations has created difficulties for the tourism sector that mirror the international context (Fennell, 1999). Most particularly, the inadequacy of current definitions of ecotourism in New Zealand remains unaddressed. Definitions of ecotourism as applied in New Zealand have proved to be too broad to capture the essence of ecotourism experiences while, at the same time, too limited to embrace the scope and scale of the ecotourism sector. The definitional conundrum is exacerbated by the fuzzy boundaries that exist between nature-based tourism operations within the fields of adventure tourism, outdoor recreation, nature-based recreation/tourism and ecotourism. Currently no clear dividing lines exist between ecotourism and other forms of nature-based tourism. Furthermore, and perhaps as a consequence, no administrative council such as the Adventure Tourism Council (formed in 1994) exists in New Zealand to represent and co-ordinate the interests of the ecotourism sector.

This situation is a current focus of the New Zealand Tourism Industry Association (NZTIA). NZTIA, which represents the interests of the tourism industry in New Zealand, established a Nature Tourism Working Party in 2000. The working party has been charged with the task of investigating the definition of ecotourism in New Zealand and with accreditation, including the adoption of Green Globe 21. A second agency at the forefront of the New Zealand ecotourism sector is the Department of Conservation (DOC), a government department which manages, by issuing of concessions, commercial tourism operations in New Zealand's extensive conservation estate (national parks, forest parks, marine parks and DOC reserves). This chapter considers the scope and scale of ecotourism in New Zealand and examines the directions and implications of current NZTIA and DOC policy initiatives for ecotourism operations.

Background to the New Zealand Ecotourism Sector

Nature-based tourism experiences form the centrepiece of the New Zealand tourism industry (New Zealand Tourism Board, 2000). The majority of inbound visitors to New Zealand seek various nature-based travel experiences (Tourism New Zealand, 2001). By its broadest definitions (e.g. Ballantine and Eagles, 1994) most if not all visitors to New Zealand may be viewed as ecotourists rendering the term meaningless. This hypothetical situation mirrors the conceptual viewpoint documented by Orams (1995) in which, at one pole of the ecotourism continuum, he states that all tourism is ecotourism. The reality in New Zealand is quite different. Most visitors to New Zealand experience New Zealand's natural environment in one form or another during their tour itinerary. However, those seeking experiences that meet the principles of ecotourism, such as effective visitor interpretation (Tilden, 1967; Orams, 1997), an educational/learning component (Boo, 1991; Whelan, 1991; Ecotourism Association of Australia, 1992; Allcock et al., 1993; Wight, 1993; Eagles, 1997; Sirakaya et al., 1999) and contributing to conservation (Williams, 1992; Young, 1992; Orams, 1997), form an important and specialized subset of nature-based tourism experiences. Within the parameters of ecotourism, this subset is broad in scope and increasing in scale.

The absence of a definitional foundation for New Zealand ecotourism operations has various implications for the industry. For example the current situation is counter-productive to the effective management of impacts arising from distinct forms of nature-based tourism. Self-proclaimed ecotourism operations that provide transport services, or rely on mechanized forms of recreation such as four-wheel driving and scenic flights in helicopters, demonstrate impact mitigation issues that are quite distinct from operations that comply with the principles of ecotourism. In the absence of a clearly defined ecotourism sector it remains impossible to identify niche visitor markets and foster product development in response to the experiences sought by visitors to ecotourism operations (Sirakaya *et al.*, 1999; Bjork, 2000).

In response to this matter, NZTIA has implemented a range of initiatives that include the Quality Tourism Standard (QTS) and Qualmark (quality rating system). The QTS is to be applied initially to adventure tourism operations such as horse trekking, off-road vehicle pursuits, kayaking, rafting and jetboating in the pursuit of adequate safety standards. The application of QTS to the ecotourism sector provides a more complex challenge. The NZTIA Quality Tourism Standard for ecotourism is currently at the draft stage.

QTS is currently being developed by the NZTIA in association with the possible adoption of Green Globe 21. Green Globe 21 is an international accreditation scheme that harnesses a powerful market-driven demand to assess the environmental performance of commercial operations as part of the purchase decision-making process. It provides companies with an action plan for improving environmental performance. The merits of this scheme have been critically reviewed by the World Wide Fund for Nature (WWF) (2000). In theory accreditation schemes are intended to allow consumers to appraise the environmental performance of commercial operators and it is here that one recurring criticism of Green Globe 21 arises. Green Globe 21 has in the past allowed its logo to be used as soon as a company undertakes to complete the certification programme (although this situation has recently changed). This, according to the WWF (2000: v), devalues the scheme quite simply because 'certification is "process" rather than "performance" based'. Lack of auditing and verification, combined with consumer confusion arising from the international proliferation of certification programmes, undermines the effectiveness of this scheme.

The WWF (2000) review of Green Globe 21 and other accreditation programmes has contributed to further development of accreditation processes. The response from Green Globe 21 has been comprehensive (Green Globe 21, 2001). The Green Globe Path to Sustainable Travel and Tourism introduced in 2001 now involves a three-step process (referred to as A, B, C) involving affiliation, benchmarking and certification. Operators are able to use the licensed Green

Globe logo only following the benchmarking process. The use of the full Green Globe licensed logo (with an additional tick) takes place only when the operation has developed an environmental management system and has been audited on-site by an accredited third party auditor. These recent developments confirm the view that certification schemes, of which Green Globe 21 is the most internationally prominent, are 'one of a suite of tools required to make tourism sustainable … (but these) need to be complemented by education, regulation and comprehensive land use planning' (WWF 2000: v).

The application of Green Globe 21 and QTS to the ecotourism sector is a complex challenge. The scope and scale of ecotourism, while defined by a series of founding principles, is sufficiently diverse to heighten this challenge. While horse trekking, rafting, kayaking and other predominantly adventure-based operations are relatively self-apparent, ecotourism operations, as has been stated repeatedly in the tourism literature, are less readily defined (Butler, 1992; Orams, 1995; Blamey, 1997). Green Globe 21 has been applied particularly within the accommodation, transport and tour operations sectors of the tourism industry. While recycling and energy efficiency within international hotel chains, international aviation carriers and the like are most creditworthy, these initiatives are far less applicable to the ecotourism sector within which most operators consume few products and very little energy. Furthermore, it is verging on the impossible to accredit a highly personalized experience that is reliant on intrinsic factors such as the personality, knowledge and values of the operator for its success. The intrinsic qualities of an ecotourism operation, which are critical to providing high quality and sustainable ecotourism experiences, vary in ways that are immeasurable in terms of accreditation. The ecotourism sector, therefore, demonstrates a distinct and diverse range of human and environmental management issues, such as the mitigation of site and species specific visitor impacts. The extent to which these can be measured and addressed through accreditation programmes such as Green Globe 21 remains an open question.

The focus of current Department of Conservation policy towards visitor experiences in the conservation estate lies with the delivery of effective visitor interpretation. DOC policy is established in its Conservation Management Strategy (2000), which requires commercial operators to meet stated standards of visitor operation. The current focus of the DOC (2000) policy with relevance to ecotourism operations imposes the condition that operators will 'provide interpretation for clients on the natural and historic resources in use' (DOC, 2000: 246). The policy makes 'industry training in conservation interpretation' a requirement of ecotourism operators and states that 'operators shall ensure that customers have the opportunity to learn about cultural, natural and historical aspects of the product through interpretation' (DOC, 2000: 246). The DOC (2000) policy clearly states that concessionaires providing ecotourism experiences in the conservation estate will be monitored not only in terms of mitigating adverse visitor effects, but also in the extent to which learning opportunities and high quality visitor interpretation are provided.

Methodology

Our research project (1999–2001) involved the implementation of a mixed methods approach, in order to achieve more detailed insights into the ecotourism phenomenon in New Zealand than have been produced to date. This research project initially required that a national nature-based tourism operations database be created. This was generated using various sources including requests for information from all regional tourism organizations and Visitor Information Network (VIN) visitor centres in New Zealand, Internet searches and content analysis of national and regional tourism directories, guide books, magazines and regional tourism information brochures. Information was obtained on 410 nature-based visitor operations throughout New Zealand.

From this operator database ecotourism operations were identified employing the selection criteria outlined in Table 12.1. The selection criteria employed in this process were developed with the aim of identifying the wide spectrum of operations that collectively comprise the ecotourism sector in New Zealand. This exercise allowed the researchers to identify 257 ecotourism operations in New Zealand that form a specialized subset of nature-based visitor operations, and are the subject of the current research.

The same selection criteria were then used to identify 12 ecotourism operations at which primary data collection took place. The selection process was designed to include both 'best practice' operators (see Table 12.1, items 7–12) and a balanced representation of the operations in the ecotourism sector. The latter was achieved through consideration of additional factors such as type of operation, core product, focus on conservation, scale of operation and domestic/international visitor focus in the selection of study operations. Twelve study sites organized into three geographical clusters (Fig. 12.1) were drawn from the ecotourism operator database ($n = 257$).

This chapter draws on qualitative data collected in the first phase of the research project (1999–2000). Qualitative methods were developed to allow the researchers to enter the tourists' sphere of experience, thus gaining insights into all aspects of both the visitors' on-site experiences and the visitor operation. Two research techniques were utilized in

Table 12.1. Principles of ecotourism drawn from international literature, applied to the New Zealand context and employed as selection criteria in the current research.[a]

1	Educational component
2	Contribution to conservation
3	Active research
4	Local ownership
5	Department of Conservation concessionaires
6	Limited vacancies on daily excursions
7	Developed/adopted and implement code of ethics
8	National tourism award winner (Ecotourism category)
9	International tourism award winner (Ecotourism category)
10	Finalist national tourism awards (Ecotourism category)
11	Finalist international tourism awards (Ecotourism category)
12	Recommended in publications such as travel guides
13	Visits/manages protected species/habitat
14	Nature Trust/foundation status

[a]Nature-based tourism operations were assessed on the basis of these selection criteria in order to identify those operations that could be viewed to comply with the general principles of ecotourism.

undertaking this phase of the research project. Anonymous participant observations were conducted through an observations guideline that was developed and pilot tested by the researchers. The guideline included seven sections as follows: demographic profiling, tourist behaviour, interpretation, compliance with behavioural guidelines, temporal aspects of site visit, spatial aspects of site visit and visitor impacts. Between 4 and 10 researcher days were spent at each participating operation during which between 3 and 15 tours or excursions were observed at each site (as determined by the duration and regularity of tours). Detailed written notes were recorded on copies of the observation guideline and transcribed. The researchers conducted field work in pairs during all site visits throughout the field season to minimize observer bias. Participant observations provided valuable insights into various aspects of ecotourism operations and visitor experiences.

Semi-structured interviews were then conducted in the latter part of each site visit following the completion of participant observations. Interviews with tourists were designed to provide a more detailed understanding of visitor experiences, and the environmental values held by visitors. An interview schedule was developed at the start of the field season. It was designed to allow visitors to describe in their own words aspects of the on-site experience. Interview questions were structured along the means-end (attribute–benefit/consequence–value) chain (Reynolds and Gutman, 1988) to produce insights into existing values held by participants. Questions were open-ended and worded simply to aid comprehension. In total 76 interviewees were randomly

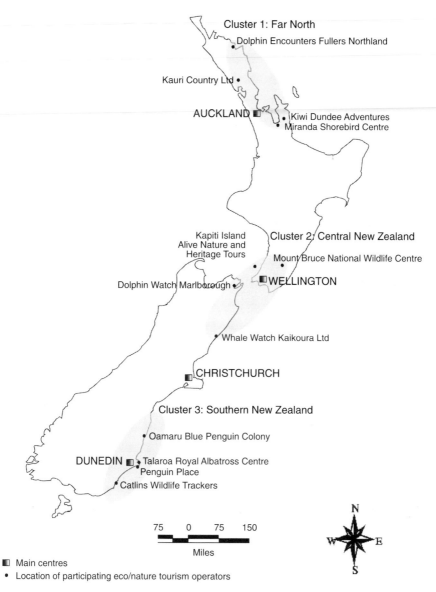

Cluster 1: Far North
Dolphin Encounters Fullers Northland

Kauri Country Ltd •

AUCKLAND ■

Kiwi Dundee Adventures
Miranda Shorebird Centre

Kapiti Island
Alive Nature and
Heritage Tours

Cluster 2: Central New Zealand

Mount Bruce National Wildlife Centre

Dolphin Watch Marlborough •

■WELLINGTON

• Whale Watch Kaikoura Ltd

■CHRISTCHURCH

Cluster 3: Southern New Zealand

• Oamaru Blue Penguin Colony

DUNEDIN ■ • Talaroa Royal Albatross Centre
Penguin Place
• Catlins Wildlife Trackers

75   0   75   150

Miles

N
W   E
S

■ Main centres
• Location of participating eco/nature tourism operators

Fig. 12.1. New Zealand map identifying ecotourism operations that participated in the research project.

selected and interviews ranged in duration from 5 to 70 minutes. All interviews were tape-recorded, transcribed, annotated and analysed using a thematic guideline. These methods enabled the researchers to gain a detailed understanding of ecotourism operations at the 12 study sites. The length of time spent at each study site also allowed the opportunity for detailed and valuable discussions to take place with ecotourism operators. This phase of the research was conducted during the 1999–2000 southern hemisphere summer field season.

The Scope and Scale of Ecotourism in New Zealand

An overview of the operator database generated at the outset of the research project affords an appreciation of the scope and scale of both the nature-based and ecotourism sectors in New Zealand. The nature-based tourism sector included operations ($n = 410$) for which the primary visitor experience included adventure (e.g. bungee jumping and rafting), nature observation, photography, physical challenge (e.g. mountain biking, tramping and hiking), mechanized recreation (e.g. four-wheel driving and skidoo operations), soft adventure (e.g. experiencing nature in a kayak) and wilderness experiences. The extent to which these *nature-based* operations and activities complied with the principles of ecotourism was measured employing the selection criteria outlined above (see Table 12.1). While this selection process reduced the database to 275 *ecotourism* operations, the scope and scale of these operations remained notably broad.

Types of ecotourism operations, for example, focused primarily on guiding, observation, conservation advocacy, science and research, interpretation and visitor education, and, particularly in the case of trusts, increasing memberships and raising public awareness of environmental issues. The resources on which ecotourism operations were based demonstrate similar scope including one or more among marine environments, particularly marine mammals, botanical, geological and ornithological resources. Operations that complied with the study selection criteria varied markedly in terms of scale from operations receiving hundreds of visitors daily, to others catering generally for groups of 2–4 visitors over a period of several days. The relative significance of domestic and international visitors at ecotourism operations varied widely although the international visitor mix was significant at most study operations. An operation occupying one extreme focused its product in terms of marketing entirely on the United States visitor market. As one might expect, the experiences provided by the operations considered to comply with the principles of ecotourism reflected this diversity.

Case Studies

The scope of the ecotourism sector in New Zealand is best illustrated with the use of two case studies that demonstrate many of the contrasts alluded to above. Noteworthy contrasts include the scale of operations, the levels of visitation that they receive, the diversity of the values held by visitors, approaches to interpretation, levels of physical and intellectual involvement of visitors, the behaviours that visitors demonstrate during their on-site experiences and impact management techniques. Despite various stark contrasts, it is argued that both case studies demonstrate exemplary practice in ecotourism. The case studies are intended to provide one illustration of the scope of ecotourism operations in New Zealand, and the diversity of strategies that may be applied to ecotourism operations. The cases also provide the basis for a discussion of definition and accreditation issues in the New Zealand ecotourism sector.

Case study 1: Catlins Wildlife Trackers

Catlins Wildlife Trackers (CWT) is, by most definitions, the archetypical ecotourism operation. It is small scale, locally owned and operated by long-term residents of the Catlins region, energy efficient, low impacting, high in interpretative content, provides a highly personalized visitor experience and challenges visitors to consider a range of environmental issues of local, national and international significance. CWT operates from a small private residence set on the coastal fringes of the Catlins. The Catlins Coastal Rainforest Park is the last vestige of native forest on the more populous east coast of New Zealand's South Island (Fig. 12.1). The lowland forest environment in this area was intensively cleared for farmland until recent years, with some forested areas held in private hands remaining under the threat of milling and/or clearing for farmland. The Catlins coastline is sparsely populated in small and remote rural communities. The regional economy has traditionally relied heavily on primary industry, particularly timber milling and farming.

CWT was established in 1990 as a private venture. The operation has received widespread recognition through national and international tourism and media awards. It is effectively marketed via a carefully planned Internet site and promoted through its self-perpetuating national and international reputation. CWT is unique in New Zealand in the manner and extent to which visitors are able to achieve intimate nature experiences. Visitor groups ranging from one or two to a maximum of eight individuals are accommodated at the private residence of the operators. The duration of tours typically ranges from 2 to 4 days with longer visits customized to meet demand. The temporal aspect of the visitor experience ensures that visitors are able to achieve a degree of learning and depth of understanding of natural processes beyond the scope of most commercial ecotourism operations. It also allows visitors to enjoy nature experiences in a leisurely and relaxed family environment. High physical and intellectual involvement of visitors is a central aspect of the experiences provided by this operation.

The visitor profile encountered at CWT is diverse. For the most part visitors may be described as nature enthusiasts, many with specialized and long established interests in the conservation of nature. The majority of observed visitors were well equipped with waterproof clothing, appropriate footwear, cameras, binoculars and bird identification manuals. However, CWT also receives more generalized visitors from a range of international origins. One interviewee of Irish nationality described himself as 'a professional who has worked in Tokyo for several years looking for the ultimate nature experience during my return journey to Europe' (anonymous interviewee, London, UK, 2000). Another described himself as 'a professional urban-dweller from Houston, Texas' (anonymous interviewee, 2000) and explained that he knew nothing about New Zealand and chose to visit CWT having stepped off the plane in Christchurch as, based on their Internet profile, they seemed 'the most informed and reliable source of advice on places to visit in this country' (anonymous interviewee, Texas USA).

The experiences that visitors to CWT are able to enjoy are notable in terms of both the diversity of activities and the opportunities for in-depth interpretation. Visitors are able to engage in a range of half-day guided activities or to take the advice of the hosts and experience the natural environment independently. Activities include dawn and dusk twilight tours to observe wildlife, particularly the endangered yellow-eyed penguin, beach walks to observe New Zealand sea lions, dunes and/or the forested coastal zone, scenic tours to waterfalls and forested coastal environments, visits to fossilized and regenerating native forests and geological tours. All excursions are guided and richly interpreted. Behavioural guidelines are delivered personally by the field guide. The potential impacts of visitation are explained and codes of behavioural conduct outlined. For example prior to crossing dunes to observe New Zealand sea lions, the guide demonstrated the body language and roars of the species that relate to a hierarchy of visitor disturbance. Codes of conduct to avoid such disturbance (e.g. to avoid walking between marine mammals and the route to where they are going) were then taken up by a receptive audience.

The impacts of the CWT commercial operation are carefully considered by the operators. Environmental impact procedures such as consideration of potential impacts and alternative activities are constantly to the fore, both in routine daily activities and the strategic development of the operation. The impacts associated with field excursions are managed effectively through the constant presence of a guide who holds abundant experience in both tour guiding and the principles of conservation. Both the persona and conservation experience of the field guide was observed to command the immediate respect of visitors. The content of interpretive talks and the manner in which they were delivered ensured that visitors complied with minimum impact codes of conduct. The intrinsic qualities of the experience were observed to be central to its success.

The functions of accommodating and transporting visitors are also undertaken with a critical view to minimizing visitor impacts. All visitors are transported in one vehicle and accommodated in the private residence of the hosts. This minimizes transport and accommodation costs and the consumption of energy. Visitors dine at one sitting in one room. Home produce and composting complement the impact mitigation and conservation goals of the operation.

Six interviews were conducted at CWT, a number limited by small tour group sizes and the lengthy duration of the visitor experience. Visitors were particularly captivated by the tour guide who, according to one interviewee from Texas, USA, 'provided detailed scientific and anecdotal insights and demonstrated an ability to answer all questions fully and in detail' (anonymous interviewee, 2000). The enthusiasm of the field guide was evident through the unprompted discussion of points of interest throughout the field excursions. This was considered to '... add immensely to the educational value of the tour'. Most interviewees described the fact that they '... received informed and appreciative nature experiences, and insights from the guide the likes of which could never be achieved traveling independently' (anonymous interviewee, Cornwall, UK, 2000).

The development of close personal relationships with the hosts was considered a special feature of the ecotourism experience. One interviewee from Alaska, USA, explained that '... only in association with such intimate nature experiences can lifelong friendships be forged in a relatively short space of time' (anonymous interviewee, 2000). Interviewees stated emphatically that they had gained an appreciation of various environmental issues (not limited to issues specific to the Catlins Coastal Rainforest Park) and of the need for conservation of the New Zealand environment (Table 12.2).

Case study 2: Mount Bruce National Wildlife Centre for Tourism

The Mount Bruce National Wildlife Centre is a scientific and captive recovery centre for endangered native species operated by a central government department, the Department of Conservation. It is located in one of the few remaining areas of native lowland forest in New Zealand set within a wider agricultural landscape. The significance of the setting arises from the fact that the vast majority of native lowland

Table 12.2. Environmental issues and conservation advocacy: Catlins Wildlife Trackers and Mount Bruce National Wildlife Centre.

Operation	Contributions to conservation
Catlins Wildlife Trackers	Scientific research and monitoring
	Data collection
	Protection of marine mammals
	Visitor interpretation
	Conservation advocacy
	'Mainland Island' concept
	'Restoring the Dawn Chorus'
	Submissions to Conservation Board
	Designation of national and marine parks
	Education (tertiary student placements)
Mount Bruce National Wildlife Centre	Predator eradication programmes
	Approaches to predator management
	Captive recovery and management of
	critically endangered species
	Damming of New Zealand rivers
	Species science and research
	Visitor interpretation
	Conservation advocacy
	Donations/sale of souvenirs
	Membership, 'Friends of Conservation'
	Petitions/conservation campaigns

forest, along with many species of native lowland birdlife, was systematically destroyed and milled to make way for agriculture in the 19th century (King, 1984).

Mount Bruce National Wildlife Centre is situated alongside national State Highway One (SH1), providing ease of access for visitors travelling by private vehicles. The fact that Mount Bruce is a convenient roadside stop, providing parking, public toilets and refreshments, ensures a high rate of casual visitation, particularly at certain times of the day. This site, therefore, offers the contrast of a large-scale, highly accessible visitor operation. Mount Bruce provides visitors with both natural and developed attractions. A visitor centre is set alongside an extensive car park providing access for visitors of all levels of mobility. This spacious facility includes an attractive reception desk (with national tourism awards prominently featured), a lecture and audio-visual theatre, interactive and static interpretation displays, indoor/outdoor tea room, public convenience facilities and office space for Department of Conservation staff. Access to the visitor centre is free. Conservation funds are generated through a small fee for entering the Mount Bruce forest reserve, the sale of souvenirs from a shop alongside the reception desk and through the placement of a donation box in the foyer of the reception centre. A proportion of all souvenir shop purchases is designated towards conservation at Mount Bruce and visitors are given the opportunity to become a 'Friend of Conservation'.

Observations and interviews confirmed a varied visitor profile. Independent visitors included people travelling alone, couples, friends and family groups. A substantial number of visitors carried camera equipment; fewer but still significant numbers were equipped with binoculars and bird-watching field guides. The majority, however, indicated that they were, in fact, visiting Mount Bruce quite incidentally. Many visitors stated that their visit to Mount Bruce was primarily a social outing with friends, a place to have a cup of tea in pleasant surroundings, or a rest break during a long journey. One domestic respondent stated in an interview that 'Mount Bruce provides the opportunity to get out of my hot car, enjoy the cool of the forest and break my journey from Wellington to Auckland' (anonymous interviewee, 2000).

Organized tours were also prominent; groups of up to 50 people were encountered by the researchers. Examples of groups tours included domestic visitors from a regional senior citizens club and international visitors on a shore excursion from a cruise ship visiting Wellington. High physical and/or intellectual involvement of visitors is optional and, based on observations completed by the researchers, generally the exception rather than the rule.

The Mount Bruce forest reserve provides varied walking opportunities, access to aviaries and complete freedom from time constraints. Information and visitor guidelines for successful bird viewing are presented to Mount Bruce visitors on the site map (brochure), which is given to visitors at reception, and in various interpretive panels. The site map advises visitors to remain on marked tracks, not to feed the birds, educates visitors on 'Essential conservation methods employed at Mt Bruce' (anonymous interviewee, 2000) and invites visitors to help with research by recording the birds they saw or heard. Numerous extensive aviaries are encountered along the tracks. The aviaries were originally developed for the purpose of captive breeding programmes for endangered native bird species. The birds that may be viewed within each aviary, detailed descriptions (including bird song) and bird-watching techniques (particularly relating to the need for silence and patience) are provided on information boards. Prominently featured on the map are the sites and times at which two interpretation programmes involving endangered native species are presented by conservation staff.

Informal observations were undertaken outside aviaries, within the visitor centre and during the two daily interpretation programmes. The interpretation programmes are delivered by scientific staff employing personal, field-based interpretation. Both programmes employ rare wildlife species to focus visitor attention on environmental issues of regional and national significance. The first addresses the native New Zealand river eel, during a daily feeding programme. Interviews conducted at this site identified that this programme challenged widely held perceptions of eels as 'slimy unattractive creatures with a nasty bite' (anonymous interviewee, Auckland, New Zealand, 2000). The act of learning about the native eel, particularly relating to the longevity of the species and breeding migrations across the Pacific Ocean that take place in the last year of life, was considered by interviewees to generate a new respect borne from awareness. Environmental issues relating to the damming of rivers and the implications for migratory river species such as the native eel were drawn to the attention of visitors. The opportunity for visitors to observe and converse informally with the interpreter was considered an important feature of the visitor experience.

The second interpretation programme takes place in a small forest clearing at which kaka (native parrot) feeding stations are located. The secluded setting serves the purpose of minimizing possible outside distractions during the interpretation programme. Bait traps and nesting boxes are displayed for visitors in the kaka feeding area to provide insight into methods used for the kaka breeding programme and predator control at Mount Bruce. Kaka feeding allows interpretation to take place utilizing demonstration techniques much, it seems, to the benefit of the visitor experience. Both observations and interviews revealed the extent to which many visitors had their eyes opened to issues of species protection, breeding, predation and such like. The multiple benefits of kaka feeding in the interests of scientific research (recording the presence of individual birds), breeding success (providing energy rich food sources) and visitor education is a feature of this aspect of the Mount Bruce experience. These interpretation programmes take place in the late morning and mid-afternoon, respectively, when the highest concentrations of visitors are at Mount Bruce. In both cases the compelling nature of the interpretive programmes was generated by the enthusiasm and skill of the interpreter.

Minimal negative impacts by tourists were observed due to active management of the attraction with hardened walking tracks, diverse facilities (e.g. cafeteria, restrooms and souvenir shop) and extensive interpretation (map guide, information panels, display room and audio-visual). Impacts that were observed included raised voices, rapid movements and flash photography in the vicinity of the aviaries. This appeared to be because visitors were unfamiliar with wildlife viewing; those visitors observed undertaking these behaviours were, in most cases, school children and members of large tour groups.

Visitor experiences were addressed during 17 interviews that were conducted at Mount Bruce. Interviews ranged in duration from 10 to 70 minutes. The primary experiences that interviewees identified at Mount Bruce included one or a combination of the following: socializing, photographic opportunities, rest/relax in the forest environment or at the visitor centre/café, view rare/endangered wildlife, view specific critically endangered species (e.g. kiwi and kokako) and educational/interpretive experiences. A number of visitors explained that they had no preconceived expectations of the Mount Bruce visitor experience, but rather, as one female respondent from England stated, '... it seemed a nice place to look around and have a nice leisurely stroll' (anonymous interviewee, 2000). Regardless of visitor motivations, one British visitor observed that 'At Mount Bruce, even if visitors came for a social time in the first place, you educate them to actually think ... and they are likely to learn something from the AV or displays' (anonymous interviewee, 2000).

Interpretive talks by Mt Bruce staff at advertised times were popular; the five sessions that were observed each attracted between 10 and 25 members of the public. The talks had high educational value with opportunities after the main activity for members of the public to converse with the staff, offering a personalized experience for more interested parties seeking to further their knowledge. The personal interpretation programmes conducted by Department of Conservation staff undoubtedly contributed to the education of visitors and heightened awareness of the Department of Conservation's role in conserving critically endangered native species. At the Mount Bruce experience, according to a retired couple from Yorkshire, England, '...I get the impression that they are trying to make people aware of how acute the problem of native bird loss is. It seems to be such an uphill battle. Seeing that there will soon be no kiwi left on the New Zealand mainland sort of brings it home to you' (anonymous interviewee, 2000).

The informative nature of the Mount Bruce experience, which was achieved via audio-visual facilities, displays, interpretation panels and, most particularly, interpretation programmes, was highly valued by the majority of interview participants.

Discussion

This research confirms the scope of the ecotourism sector as measured in terms of size of operations, levels of daily visitation, visitor experiences and the profiles and values of visitors. The scope and scale of ecotourism in New Zealand affords the opportunity for operators to contribute to conservation in various valuable and meaningful ways.

The two case study operations presented in this chapter stand in stark contrast to each other. They utilize diverse strategies and management techniques to achieve high levels of visitor satisfaction and, in addition, important conservation goals. Both operations meet the DOC (2000) policy of engaging visitors in carefully designed and delivered interpretation programmes, albeit using distinct interpretation strategies and techniques. Notwithstanding this point, both provide visitor interpretation programmes that comply with Tilden's (1967) classic principles of interpretation. It is noteworthy that these operations deliver 'ecological interpretation' (Tilden, 1967). This involves investigation of the delicate biological relationships that exist between observed species and their wider ecologies (including human influences on the ecology). The alternative, to focus interpretation solely on the observed species, is resisted and, as a consequence, the conservation value of the interpretation programme is immeasurably enhanced. The case studies confirm that interpretation programmes, if carefully designed and delivered, play a critical role in the ecotourism experience. This statement applies regardless of the size of the audience but, in either case, requires a high level of experience, ability and enthusiasm on the part of the interpreter.

Both of the case study operations presented in this chapter, and various others involved in the research project, deliver to their audiences clear conservation messages, fulfil stated conservation goals, contribute to conservation, science and/or research, and challenge the environmental values of visitors. These goals are achieved in different ways by operations that contrast with each other in terms of scale. These aspects of the study operations set them apart from the wider field of nature-based tourism and form the basis on which the ecotourism phenomenon in New Zealand may be defined. The case studies demonstrate several dimensions of the scope and scale of the ecotourism phenomenon and illustrate the varied strategies implemented by differing operations. These strategies relate to environmental management, visitor impact mitigation, interpretation and clearly stated conservation goals.

Given these points, it is critical that a definition of ecotourism is achieved that recognizes the points of distinction between ecotourism and the wider field of nature-based tourism, while also recognizing the scope and scale of the ecotourism sector. Bjork (2000) acknowledges that defining ecotourism is necessary to provide a conceptual basis from which planning and development can proceed. 'Only by having a strict theoretical definition (an ideal situation) is it possible to go on and adjust the dimensions in accordance with the unique characteristics of a specific tourism area' (Bjork, 2000: 190). This research acknowledges the viewpoints articulated by Pearce (1994), Blamey (1997) and Bjork (2000), that developing a single definition of ecotourism may be futile. This arises from the fact that ecotourism takes place in many varied contexts. It is, therefore, necessary to adopt definitions that reflect national/regional/local tourism contexts. Facilitating this process requires consensus on definition parameters that can be applied with different weighting in differing regional or national contexts.

So, for example, limiting the potential for product development and growth and remaining faithful to a small-scale operation is often associated with ecotourism (Butler, 1990; Place, 1991; Thomlinson and Getz, 1996; Ryan *et al.*, 2000). This is necessary and important in many tourism contexts. However, some of the largest and best developed ecotourism operations in New Zealand (such as Mount Bruce National Wildlife Centre) demonstrate great potential to rehabilitate critically endangered species, generate revenue for conservation and capture the imaginations of diverse audiences, while conducting a minimum impact visitor operation.

The issue of accreditation in the New Zealand ecotourism sector also poses great challenges. Central to this challenge is the vexing issue of precisely what to accredit. Ecotourism operations, while meeting many of the parameters of ecotourism used as selection criteria in the current project (Table 12.1), are sufficiently diverse in scope and scale to make this a critical point of issue. The two cases presented in this chapter suggest that the diversity of the ecotourism phenomenon is sufficient to require the accreditation of ecotourism operations on grounds that vary on a case-by-case basis. Indeed recent developments in the Green Globe 21 accreditation process confirm that this is precisely what will be required when auditing the environmental performance of ecotourism operations (Green Globe 21, 2001). This research confirms the importance of on-site visits to ecotourism operations and the development of case specific auditing guidelines, if accreditation in the ecotourism sector is to be meaningful and worthwhile. The current research also confirms the importance of the knowledge, values and interpretation skills of the operator (or ecotourism operations staff). Intrinsic factors play a central part in the development of the ecotourism product and the success of an ecotourism operation. Accrediting such personalized aspects of the ecotourism experience poses a genuine challenge to the accreditation process.

Conclusions

This chapter argues that important points of distinction exist between ecotourism operations in New Zealand and the wider field of nature-based tourism. It is also noted that much scope exists within the ecotourism sector in terms of the scale of ecotourism operations, levels of visitation and visitor management strategies that exist within these operations. Current New Zealand Tourism Industry Association policy directions place priority on the definition of ecotourism in the New Zealand context and the development of accreditation in this sector. It is imperative that a definition of ecotourism in New Zealand is achieved that recognizes important points of distinction between nature-based and ecotourism operations. It is equally important that the scope and scale of the operations within the ecotourism sector are recognized in defining ecotourism in New Zealand. These processes are fundamental to the advancement of policy and planning, government support, product development, sustainable tourist experiences and the development of an international reputation in the field of ecotourism.

This is also relevant to the development of an accreditation programme that may be applied meaningfully to ecotourism operations. The authors highlight the difficulties of accrediting intrinsic aspects of a visitor experience. In the light of these points it might be argued that the achievement of a definition of ecotourism that is appropriate to the regional or national context within which it is applied is a fundamental precursor to the further development and maturation of this tourism industry sector. Until the ecotourism sector and its visitor markets are more clearly defined this goal will be unattained.

Acknowledgements

The authors acknowledge the research funding provided by the Foundation of Research, Science and Technology, New Zealand, and the logistical support of Liz Mender (Mount Bruce National Wildlife Centre) and Fergus and Mary Sutherland (Catlins Wildlife Trackers). The authors also acknowledge the support of all 12 study operations (north–south): Fullers Dolphin Encounters, Kauri Country Ltd, Miranda Shorebird Centre, Kiwi Dundee Tours, Mount Bruce National Wildlife Centre, Kapiti Tours, Dolphin Watch Marlborough, Whale watch Kaikoura, Oamaru Blue Penguin Colony, Royal Albatross Colony, Penguin Place and Catlins Wildlife Trackers.

References

Allcock, A., Jones, B., Lane, S. and Grant, J. (1993) *Draft National Ecotourism Strategy*. Commonwealth Department of Tourism, Canberra.
Ballantine, J.L. and Eagles, P.F.J. (1994) Defining Canadian ecotourists. *Journal of Sustainable Tourism* 2(4), 210–214.

Bjork, D.P. (2000) Ecotourism from a conceptual perspective: an extended definition of a unique tourism form. *International Journal of Tourism Research* 2(3), 189–202.

Blamey, R.K. (1997) Ecotourism: the search for an operational definition. *Journal of Sustainable Tourism* 5(2), 109–130.

Boo, E. (1991) Planning for ecotourism. *Parks* 2(3), 4–8.

Butler, R.W. (1990) Alternative tourism: Pious hope or Trojan horse? *Journal of Travel Research* 28(3), 40–45.

Butler, R.W. (1992) Ecotourism: its changing face and evolving philosophy. *IVth World Congress on National Parks and Protected Areas, Caracas, Venezuela.*

DOC (Department of Conservation) (2000) *Conservation Management Strategy 2000.* Canterbury Conservancy, Christchurch, New Zealand, www.doc. govt.nz

Eagles, P.F.J. (1997) International ecotourism management: using Australia and Africa as case studies. *IUCN World Commission on Protected Areas, Protected Areas in the 21st Century: From Islands to Networks, Albany, Australia.*

Ecotourism Association of Australia (1992) *Newsletter, Ecotourism Association of Australia.*

Fennell, D.A. (1999) *Ecotourism: an Introduction.* Routledge, London.

Green Globe 21 (2001) *Green Globe 21. Path to Sustainable Travel and Tourism.* www.greenglobe21.com

King, C. (1984) Immigrant Killers. *Introduced Predators and the Conservation of Birds in New Zealand.* Oxford University Press, Melbourne.

New Zealand Tourism Board (2000) Tourism Statistics Online. www. tourisminfo.govt.nz/Default.htm

Orams, M.B. (1995) Towards a more desirable form of ecotourism. *Tourism Management* 16(1), 3–8.

Orams, M.B. (1997) The effectiveness of environmental education: can we turn tourists into 'greenies'? *Progress in Tourism and Hospitality Research* 3, 295–306.

Pearce, D.G. (1994) Alternative tourism: concepts, classifications and questions. In: Smith, V. and Eadington, W. (eds) *Tourism Alternatives. Potential and Problems in the Development of Tourism.* John Wiley & Sons, Chichester, UK, pp. 15–30.

Place, S.E. (1991) Nature Tourism and rural development in Tortugero. *Annals of Tourism Research* 18(2), 186–201.

Reynolds, T.J. and Gutman, J. (1988) Laddering theory, method, analysis and interpretation. *Journal of Advertising Research* February/March, 11–31.

Ryan, C., Hughes, K. and Chirgwin, S. (2000) The gaze, spectacle and ecotourism. *Annals of Tourism Research* 27(1), 148–163.

Sirakaya, E., Sasidharan, V. and Sonmez, S. (1999) Redefining ecotourism: the need for a supply-side view. *Journal of Travel Research* 38 (November), 168–172.

Thomlinson, E. and Getz, D. (1996) The question of scale in ecotourism: case study of two small ecotour operators in the Mundo Maya region of Central America. *Journal of Sustainable Tourism* 4(4), 183–200.

Tilden, F. (1967) *Interpreting our Heritage.* The University of North Carolina Press, Chapel Hill.

Tourism New Zealand (2001) *Tourism New Zealand Market Guides 2001.*
    100% Pure New Zealand. Tourism New Zealand, Wellington, New
    Zealand.
Warren, J.A.N. and Taylor, C.N. (1994) *Developing Eco-tourism in New*
    Zealand. New Zealand Institute for Social Research and Development,
    Wellington.
Whelan, T. (1991) *Nature Tourism: Managing for the Environment.* Island
    Press, Washington, DC.
Wight, P. (1993) Ecotourism: ethics or eco-sell? *Journal of Travel Research*
    Winter, 3–9.
Williams, P.W. (1992) A local framework for ecotourism development.
    Western Wildlands 18(3), 14–19.
World Wide Fund for Nature (2000) *Tourism Certification: an Analysis of*
    Green Globe 21 and other Tourism Certification Programmes. A report by
    Synergy for WWF-UK, August 2000.
Young, M. (1992) *Ecotourism – Profitable Conservation?* Ecotourism Business
    in the Pacific: Promoting a Sustainable Experience. Environmental
    Science, University of Auckland.

Ecotourism Policy and Practice in New Zealand's National Estate

13

Ken Simpson

School of Management and Entrepreneurship, UNITEC Institute of Technology, Private Bag 92-025, Auckland, New Zealand

Introduction

Three defining characteristics have combined to position the small South Pacific nation of New Zealand in the forefront of any discussion related to the costs and benefits of ecotourism. Firstly, the country's isolated island status and generally benign climate has fostered the emergence of a unique natural infrastructure, both floral and faunal. Secondly, a well established, highly efficient and economically vital farming sector has lent emphasis to a dominant image of rurality, widely expressed and enthusiastically promoted by means of a 'clean and green' environmental descriptor. Finally, international tourism is an integral contributor to the country's economic performance, and has been recognized as a key element in future national development. In circumstances such as these, it is perhaps inevitable that the concept of ecotourism should have generated widespread local interest.

Though neighbouring Australia provides approximately one-third of all its international visitors, New Zealand is a comparatively remote destination for many primary visitor groups. In particular, almost half of all foreign exchange receipts can be attributed to visitors from four long-haul origin markets (Tourism New Zealand, www.tourisminfo.govt.nz) and, for inhabitants of these highly industrialized and often densely populated countries, ecotourism in its broadest sense may often seem to be an appropriate label for much of the local tourism industry's activities (see Table 13.1).

© CAB *International* 2003. *Ecotourism Policy and Planning*
(eds D.A. Fennell and R.K. Dowling)

Table 13.1. International visitor expenditure in New Zealand for year ended March 2001 (source: Tourism New Zealand, www.tourisminfo.govt.nz).

Country of origin	Total spending (NZ$ million)	% of total spend
Australia	907	18.4
USA	807	16.4
Japan	702	14.2
UK	692	14.0
Germany	199	4.0
Other	1600	33.0

New Zealand's core tourism product includes no noticeable built heritage to attract visitors, there are few if any significant man-made attractions, and night-time entertainment is commonly regarded as low-key at best. However, substantial pockets of outstanding scenic beauty exist alongside a unique and vibrant bi-cultural heritage, and these natural factors are presented to visitors within a somewhat romantic framework of claimed environmental purity. The local tourism industry has thus been able to turn its national isolation to advantage in proclaiming New Zealand to be the ultimate eco-friendly tourism destination, with its most recent marketing initiatives based on a '100% Pure' brand positioning to emphasize a supposedly pristine national environment (Tourism New Zealand, 1999).

Whilst '100% Pure' is little more than a promotional slogan, and should be clearly recognized as such, it is nevertheless appropriate to acknowledge consistent New Zealand government recognition of the intrinsic value of those publicly owned parks, reserves and water-based resources that together comprise a 'national estate' of environments available for recreation purposes. The undoubted jewel in the crown for these 'lands of high conservation value' (Department of Conservation, 1996: 4) is an impressive network of 14 formally designated national parks, two of which enjoy World Heritage Area status. Consequently, there is a significant role for government in the maintenance of high quality natural environments in, and the management of appropriate visitor access to, these critically important areas (Kearsley, 1997).

At the same time, government has adopted a bulk funding approach to the promotion of an international tourism industry which relies heavily on these environments, and supplies the vast majority of resourcing for a marketing effort based on a clean, green and healthy outdoors lifestyle. The potential for either cooperation or conflict between these two philosophies is clearly evident.

The Policy Background

New Zealand is a small island country in the south-west Pacific and comprises three main islands with a combined land area of 270,000 km^2 (about the same size as the UK or Japan). It has been an independent nation since 1948 and one which, for the first 35 years of its sovereign existence, enjoyed high levels of prosperity through the activities of a highly effective and efficient farming sector. However, in more recent years, the diminishing worldwide importance of primary industry has been recognized in a relatively new-found interest in the services sector, here primarily represented by a flourishing and nationally significant tourism industry.

Given its physical isolation from the Northern Hemisphere locus of worldwide tourism activity, it is somewhat surprising to note that New Zealand was in 1901 the first country in the world to establish a national tourist office (Collier, 1994). From this auspicious beginning, growth in tourism was initially hampered by the country's distance from potential generating markets, and it was not until the introduction of large passenger jet aircraft that consistent and significant annual increases could be achieved. Thus, though it took 90 years to achieve an annual arrival figure in excess of 1 million visitors (TIANZ, www.tianz.org.nz), achievement of the second million has been forecast to take place in the 2001/2002 year, just 10 years later.

New Zealand's modern-day visitor industry has to date managed to avoid the worst excesses of charter based sun, sand and sea tourism. Though limitations in data collection processes conspire against accurate evaluation, there is a well performed domestic tourism sector with strong links to outdoor recreation, and much of the international visitor market is founded on the attractions of scenic beauty, indigenous culture and geothermal activity. In addition, these nature-based elements of New Zealand's tourism product are reflected in the composition of a local visitor industry which is solidly based on small business enterprise: of the 16,500 tourism organizations in the country, 13,500 employ less than five people, and only ten are substantial enough to merit stock exchange listing (TIANZ, www.tianz.org.nz).

Though this approach offers considerable potential for enhanced sustainability through community participation, there are a number of accompanying drawbacks associated with industrial immaturity. There are few barriers to entry to an industry which is superficially attractive to many and, in the absence of any formal regulation of operators, new entrants are free to pursue their business operations to a frequently variable quality standard. As such, there is a dominant ethos of competition rather than cooperation among participants, and a consequent tendency towards unplanned and uncoordinated industry development. Until the release of the Tourism Industry Association of New Zealand's National Tourism Strategy document in May 2001 (TIANZ, 2001), there had never been any form of national tourism plan in place, and the past two decades of emphasis on international product marketing have created an environment in which strategies for post-arrival visitor care are ill-defined and inconsistent.

Though the international tourism industry remains tiny by world standards, it is growing rapidly and enjoys a position of critical importance within the context of a small-scale national economy. In the year to March 2001, approximately 1.8 million international visitors contributed around NZ$4.9 billion in foreign exchange earnings (Tourism New Zealand, www.tourisminfo.govt.nz), and these figures are claimed to represent 15.8% of total export earnings (TIANZ, www.tianz.org.nz) and to directly support around 118,000 full-time jobs throughout the country (OTSp, 1999). The magnitude of these statistics is sufficient to classify tourism as New Zealand's leading export industry (Statistics New Zealand, 2000).

At a national level, environmental protection issues are the preserve of central government's Department of Conservation (DoC), while the marketing and promotion of New Zealand as a tourist destination is carried out by the New Zealand Tourism Board, trading as 'Tourism New Zealand'. Though both of these agencies owe their existence to statutory provision – the Conservation Act, 1987 and the New Zealand Tourism Board Act, 1991 respectively – their structure and philosophies are radically different. In particular, the Minister of Conservation is supported by a fully fledged, traditional and centrally located government department and a network of regionally based field offices to pursue the operational implementation of departmental policies. In contrast, the Minister of Tourism has to contend with a much more loosely structured bureaucracy.

In 1991, the government of the day had legislated for the creation of the New Zealand Tourism Board, applying the finishing touches to a 7 year privatization programme which had progressively divested all operational involvement in the national tourism industry from the public to the private sector (Pearce, 1992). The then Minister of Tourism, the Hon. John Banks, set the scene for what was to follow, by predicting that the Tourism Board would henceforth be responsible for what would quickly become a key element in New Zealand's economic portfolio, and by drawing attention to the Board's founding mission 'to ensure that New Zealand is developed and marketed as a competitive tourism destination to maximise the long term benefits to New Zealand' (New Zealand Government, 1991a, s.6).

The years from 1991 until the present day have seen a deliberately engineered decline in the status and influence of central tourism bureaucracy. The 1991 Ministry of Tourism was restructured in 1995 as a new Tourism Policy Group, with a total staff of seven people and seriously lacking the resources to fulfil the objectives with which the 1991 ministry was originally charged (Page and Thorn, 1998). In 1998, yet another reconfiguration was effected through the establishment of an Office for Tourism and Sport (OTSp), to replace the Tourism Policy Group and to adopt the additional responsibility of advising the government on sport, leisure and recreation policy for New Zealanders. Though the new office has a slightly larger staff of nine, it is perhaps indicative of current central government attitudes that nine public servants are responsible for counselling the future directions of both the country's largest export industry and what is arguably its consuming national obsession.

The cumulative effects of these years of reform have been substantial. During that time, a well-established institutional structure for tourism policy determination has been systematically dismantled, and the broadly based concerns of a centralized public sector establishment replaced by increasing adherence to a demand-driven and industry-led marketing approach. Central government has chosen to establish a clear distinction between issues of policy and issues of management in terms of tourism development; the Minister and his OTSp advisers are charged with the former responsibility and a private sector Tourism Board bulk funded by central government delivers the management function.

At a sub-national level of administration, primary governance responsibility is assigned to a network of 12 regional councils, while the day-to-day activities of local politics are carried out by a total of 74 district councils. This twin-tiered approach has emerged from more than 50 years of disagreement over the most appropriate structure for lower level politics, and is essentially a slightly diluted version of original attempts to establish a regional tier of government concerned solely with regulatory activity, and a cohort of local authorities responsible for the provision of services (McKinlay, 1994). However, a considerable degree of local level resistance to these proposals has resulted in an amended version of the original philosophy, with an eventual allocation of both planning and regulatory functions to regional councils, and a corresponding requirement for local authorities to combine a degree of planning and regulation with their predominant service provision responsibilities.

There are no specific requirements for either regional or local government to become involved with tourism, and the regional sector in particular has consequently chosen to dissociate itself with this issue. At the time of writing, only one of 12 regional councils had prepared a specific tourism strategy for its community, while another three had made some form of financial contribution to local tourism development (Simpson, 2001). Even at the territorial (local) government level, councils displayed considerable ambivalence in their attitudes to the local benefits of tourism: nearly 40% of the 74 local councils chose to devolve their tourism industry involvement to a network of 26 regional tourism organizations (RTOs), a local government sponsored initiative to maximize tourism's economic benefits within a local sphere of political influence (Simpson, 2001).

At present, therefore, although development and management of the fragmented and highly diversified tourism industry has necessarily involved other government agencies, such as the Ministries of Transport and of Agriculture and Fisheries, it is possible to visualize a simplified model of New Zealand ecotourism policy as a three-level hierarchical pyramid: a national level policy making function overseen by both the Minister of Conservation and the Minister of Tourism, along with their respective departmental advisers; a national level policy interpretation function headlined by DoC and Tourism New Zealand; and a sub-national implementation and monitoring function supplied by the regional and district council network. It is through this organizational network that the future of ecotourism in New Zealand will be determined.

Ecotourism Policy in Theory

The operations of the Department of Conservation are guided by the Conservation Act of 1987, though its activities in National Parks lands are also influenced by the National Parks Act, 1980. Under the former legislation, the Department's overriding duties relate to the protection of Crown Lands against unacceptable levels of environmental degradation, and its single largest operational function is related to the eradication of introduced flora and fauna. In the face of extreme floral and faunal threats to environmental purity, it is perhaps inevitable that DoC's role in the provision of public recreation opportunities appears to be very much a secondary consideration.

The Conservation Act requires DoC to adopt a regional approach to conservation issues, with primary management responsibilities discharged by a consortium of 13 regional 'conservancies'. Each conservancy must prepare a long-term Conservation Management Strategy (CMS) for its local area, supported by a series of shorter-term Conservation Management Plans (CMPs). Both CMSs and CMPs are required to regard environmental conservation as the paramount objective, with considerations of visitor access relegated to a secondary though still important status. Under the provisions of a 1996 amendment to the Act, prospective tourism operators are required to apply to the Department for a tourism concession – official authorization to operate a tourism business for a specified period of time – at which point the proposed activity will be evaluated for its conformity with both the CMS and the CMP.

The concessions system catches the full range of all commercial activities and, for the tourism industry, includes attractions, activities and supporting services. The applications process is normally initiated through a written approach to DoC by the operator, indicating the nature of the proposed activity, its predicted environmental effects, and a statement of action to be taken to eliminate or minimize negative outcomes. Each case is judged on its merits and, if the concession is approved, will be formalized through a written contract which identifies the nature of the authorized operation, the term of the concession and the fees payable. Fees are negotiated between DoC and the applicant, and may be expressed as a flat annual fee, a percentage of turnover, or a levy based on activity volume.

In addition to its conformity with the concessions process, the likely effects of any proposed tourism business must be assessed according to the provisions of a landmark piece of New Zealand legislation, the Resource Management Act (RMA) of 1991. The RMA is a complex and ground-breaking piece of legislation, which rejects the traditional zoning approach to land use planning in favour of a focus on the environmental impacts of any proposed development. No longer do applicants seek a 'planning' permission to use land for a stated purpose; rather they seek a 'resource consent' permission to undertake an activity that will necessarily produce some form of impact on the natural resources involved in the application. Thus, though the RMA is a New Zealand statute designed for New Zealand conditions, it is clearly possible to discern the influence of internationally relevant sustainability principles as espoused by the Brundtland Report (WCED, 1987).

The DoC Visitor Strategy 1996

The Department of Conservation's approach to strategic planning for visitor management, as represented by its 1996 Visitor Strategy document, has been a thoroughly professional one, to the extent that the development process adopted offers valuable lessons to other would-be tourism planners. The final version is based on a conventional business planning model as commonly advocated by the generic management literature, and opens with a vision statement that is an important criterion in any review of the strategy's success or otherwise: 'by the year 2000, New Zealand's natural ecosystems, species, landscapes and historic and cultural places have been protected; people enjoy them and are involved in their conservation'.

In support of this vision, the document establishes five broad goals, related to environmental protection, fostering of recreational visits, management of tourism activity, provision of visitor information and education, and maintenance of visitor safety. In this context, it is important to establish two important issues which underpin the strategy. First there is an implicit and explicit dominance of the environmental conservation goal over all others, and second the distinction is made between free access recreation and revenue generating tourism. In terms of the Department's consequent approach to visitor management, the latter distinction is critical.

The Strategy document recognizes that, if its stated goals are to be achieved, it will be necessary to achieve an acceptable compromise between the conflicting demands of environmental conservation and visitor access. In pursuit of this objective, the degree of environmental risk present in each regional conservancy is to be assessed against its current and potential level of visitation, and the Recreational Opportunity Spectrum (Clarke and Stankey, 1979) used to determine an appropriate visitor policy and management regime for the individual site.

In this manner, the nature of each site is classified along a continuum which ranges from high volume and easy access areas, with high levels of mainstream visitor appeal, to inaccessible wilderness regions thought to appeal solely to the dedicated and experienced outdoors person. Most of the available infrastructure and other resourcing is to be provided at those sites where higher numbers of less experienced visitors seek a relatively superficial contact with nature, while management of the more remote infrastructural facilities is progressively being relinquished to non-profit incorporated societies such as the Federated Mountain Clubs and the Royal Forest and Bird Protection Society.

The basis of DoC's visitor segmentation is therefore clear, and it is again possible to detect the operation of a tourist/recreationist distinction. Thus, the more mainstream visitors are seen as frequently well-meaning but potentially destructive ecotourists who require substantial infrastructure to be provided at a cost by commercial tourism concessionaires, while the solitude seeking recreationists are an essentially beneficial influence who can be entrusted to look after themselves appropriately while within the wilderness environment which originally attracted them.

Ecotourism Policy in Practice

Hall (1994) has cautioned against regarding ecotourism as a human behaviour which takes place solely on publicly owned lands specifically set aside for that purpose. In Hall's view, this is an overly simplistic Western approach which suggests the concept of physical resources that matter and physical resources that do not. 'In most Western societies ... ecotourism exists in designated areas in which natural heritage values predominate and in which evidence of modern human settlement is often removed or not interpreted to visitors' (Hall, 1994: 140). In New Zealand however, though the local version of what ecotourism means may be philosophically sympathetic to Hall's approach, practical considerations of site appeal and activity potential have ensured that a substantial majority of outdoors tourism takes place on public lands under DoC control.

The Conservation Act provides an eloquent justification for environmental conservation through its identification of 'three reasons for the preservation and protection of natural and historic resources – maintaining their intrinsic values, providing for their appreciation and recreational enjoyment by the public, and safeguarding the options of future generations' (Department of Conservation, 1996: 7). In order to examine the degree to which the DoC Visitor Strategy has been successful in achieving its goals, it is useful to present these three reasons as temporally constructed dimensions of an overall approach to the relationship between the natural environment and those who choose to visit.

From this perspective, 'safeguarding the options of future generations' can readily be visualized as the underpinning and long-term strategic objective of the DoC Strategy, while 'providing for their appreciation and recreational enjoyment' would appear to qualify as a medium-term tactic which contributes to inter-generational equity through the establishment of an effective working relationship with the visitor industry. However, neither of these goals is achievable in the absence of a robust and resilient natural environment, and the 'maintaining their intrinsic values' element is an operational construct which appears to command much of DoC's short-term attention.

Nevertheless, the Department's involvement with visitor recreation (the distinction between free of charge recreation and revenue generating tourism is again emphasized) is substantial. By DoC's own calculations (www.doc.govt.nz), the management of recreation consumes NZ$53 million annually. There are reckoned to be 3627 individual conservation sites nationally, and a total of 15,400 'structures' (bridges, boardwalks, stairs and platforms) and, given this substantial infrastructural investment, it was inevitable that some form of prioritization exercise would be necessary to determine the most appropriate application of available recreation funding.

Each of the individual sites has been assessed under a complex rating formula which evaluates its visitor profile under a weighted four-part criteria set (Table 13.2). As a result of this approach a

Table 13.2. DoC site rating criteria (source: Department of Conservation, 1996).

Criterion	Maximum score
Current visitor numbers	10
Expected visitor number increases	3
Recreational and educational importance	9
Value in enhancing heritage appreciation	6

national priority classification for DoC managed sites has been determined, and this is shown in Table 13.3.

Following the tragic events of the 1995 Cave Creek disaster, near Punakaiki on the West Coast of the South Island, in which 14 young people were killed in the collapse of a poorly engineered and illegally constructed DoC viewing platform, the focus of attention was abruptly shifted to the issue of public safety (Isaac, 1997). This may reflect a simple case of closing stable doors after horses have bolted, or there may be more cynical implications. One recent report (Hunt, 2000) suggests that the Cave Creek judicial inquiry may have deliberately engineered an open verdict aimed at shielding the government of the day from massive compensation claims based on the deliberate underfunding of the Department. In any event, all structures were inventoried in 1996 and prioritized for engineering inspection and, to date, some 3000 inspections have been completed, most of these in the higher traffic count sites.

The position in relation to tourism, as opposed to recreation, is more complex. On one hand, Tourism New Zealand has enthusiastically embraced a clean and green motif in its portrayal of New Zealand as an exotic and isolated island of intrigue; on the other, it is statutorily and operationally committed to the generation of ever-increasing visitor numbers. As a result, as Hall (1994) has stated, the results of tourism promotion are beginning to be a concern. Though it may be a little premature to form a definitive judgement, there have already been a number of occasions where the potential conflict between conservation and visitor access has proved contentious. For example, in fragile areas such as the Waitomo Glow-worm Caves and the world-renowned fjords of Milford Sound, peak season tourist activity can sometimes be excessive, though there has not to date been any perceived necessity for DoC to flex its legislative muscles in support of environmental protection. However, these pressures appear to be building.

As early as the mid-1990s, Warren and Taylor (1994) had noted considerable inconsistency in the nature of relationships between DoC conservancies and their local tourism industries. A full range of attitudes was evident in the four areas surveyed, from mutual respect

Table 13.3. National Priority Classifications (source: Department of Conservation, 1996).

Site classification	Number of sites
'High Priority' sites, scoring between 16 and 28 on the site rating scale, in which the maintenance of facilities is regarded as absolutely essential.	453
'Medium Priority' sites, 6–15 on the rating scale, in which the maintenance of facilities is seen as highly desirable. As much attention is paid to facilities maintenance as budgetary constraints will allow.	1623
'Low Priority' sites, 0–5 on the rating scale, in which facilities maintenance is regarded as desirable but unlikely. At these sites, facilities maintenance is a fiscal luxury, to the extent that deteriorating structures are likely to be removed rather than replaced.	1551

and cooperation to equally mutual suspicion and distrust. In the more negative cases, DoC was seen to be inconsistent in the way concessions were administered, caring about the environment to the exclusion of all other considerations, and inequitable in the methods used to disburse tourism concession fees. Eight years later, annual overseas visitor numbers have increased dramatically, and friction between DoC and the tourism industry has grown in tandem, particularly in the context of 'policing' what tourism concessionaires were actually doing with the authorizations extended to them through concessions.

In one notable case, a local tour operator attempted to take advantage of a DoC imposed visitor limit to an offshore nature reserve, by speculatively pre-booking all weekend passes to the reserve over a 6 week period, and subsequently attempting to resell them to the public as an inclusive guided tour with a higher price tag (Ewan, 2000). In another instance, an approved concession for the formation of a 3.6 m wide tram track through established rainforest had resulted in the construction of a roadway which was 19 m wide in parts. Even then, pressure to remedy a clear abuse of concession conditions did not result from any form of DoC intervention, but was instead attributable to a whistle-blowing initiative by the Royal Forest and Bird Protection Society (Madgwick, 2000).

An element of confrontation is also perceptible in the relationship which exists between DoC and local government agencies. In a recent case (Keen, 2000) involving a major South Island tourism operator, the relevant regional council had granted resource consent to operate day and night lake cruises using a 44 m vessel with an unspecified number of kayaks and amphibian craft. This decision was formally appealed by the Department, and the industry applicant was effectively hamstrung by a dispute between two governmental institutions that, at least theoretically, share a common goal of environmental resource protection in their relevant geographical domains.

In some instances, New Zealand's determination to foster public involvement in decision making has resulted in the emergence of the full spectrum of possible public opinion. In a series of submissions to the proposed Conservation Management Plan for Taranaki (Mount Egmont) National Park, DoC was simultaneously accused of stifling visitor activity and of endangering the natural resource through the encouragement of overuse. Such was the divergence of opinion that a private sector environmental group had vowed to oppose any potential overuse by visitors, while the relevant local government agency was advocating helicopter landing pads and the construction of a gondola to assist disabled access (Warrander, 2000).

Taranaki, a perfect cone-shaped mountain of some 2500 m, enjoys iconic status amongst local *tangata whenua* (original Maori inhabitants of New Zealand) and exemplifies the additional conflicts that can sometimes become apparent when Maori opinion is accorded its rightful degree of influence over future development decisions. In this context, the intended relationship between *tangata whenua* and the first European immigrants was codified through the signing of the Treaty of Waitangi in 1840, a document which has since been accepted almost as a de facto constitution by many New Zealanders. However, the spirit and intent of the Treaty has historically been shrouded in ambiguity, and it often appears to function as a focus of cultural discord rather than as the intended symbol of national unity. Thus, though the *tangata whenua* voice was one of moderation and reason in Warrander's (2000) example, requiring only that visitors respect Maori perspectives, values and traditions when planning and carrying out their activities, similar disputes in other parts of New Zealand threaten to be more contentious.

Warren and Taylor's (1994) report had highlighted such tensions in the Northland region, based on disputes between DoC and local Maori about who should administer the land and, in the same year, Hall (1994) had noted that successful Maori land claims were potentially threatening to the existing public access rights to New Zealand's wilderness lands. In the intervening years, though it is fair to say that Maori have enjoyed mixed success at best in their choice of economic investments, the South island's Ngai Tahu *iwi* (main tribal grouping) has become a major corporate player in the New Zealand economy.

Having cut their tourism teeth as a substantial minority shareholder in the highly successful whale watching operation at Kaikoura, on the east coast of South Island, Ngai Tahu found themselves at the centre of a 1994/1995 constitutional debate over the value of traditional resource rights as originally guaranteed by the Treaty of Waitangi. Although some doubts remain over the extent to which commercial considerations were a factor in this particular case, the *iwi* backed tour operators had sought Court of Appeal support for protection of traditional resource rights, and the overturning of an earlier High Court decision to allow the granting of competitor concessions for whale watching operations.

In an apparent attempt to steer some form of middle course, the Court eventually ruled in favour of a temporary freeze on competitor activity, and the consequent endorsement of a medium-term monopoly position for the existing operator. Attempting to contain the impacts of its judgement solely to the Kaikoura case, the Appeal Court found that conservation values should be regarded as paramount, that *iwi* did not enjoy ownership or harvesting rights over whales and, in any case, the Treaty of Waitangi could not, in 1840, have foreseen that whale watching would be a commercially valuable activity! (Gillespie, 1999).

Buoyed by this success, and the considerable business acumen shown in other investment projects, Ngai Tahu are, at the time of writing, committed to membership of a consortium which plans to build the world's longest tourist gondola, costing NZ$80 million and stretching through 12.6 km of South West New Zealand World Heritage Area between Queenstown and Milford (Hutching, 2000). In this case, institutional decision makers are faced with a dilemma of classic proportions – the apparent necessity to choose between preserving the environment and preserving the rights of indigenous people – and this is a situation clearly fraught with significant political, social and cultural pitfalls.

It is relatively straightforward to make a compelling economic case for tourism projects such as these, and equally easy to produce an environmentally based condemnation. As Newth (2000) has observed, 19th and 20th century national parks were created to protect natural resources from being plundered or destroyed, and their designation consciously favoured environmentally and socially based values at the expense of economic interests. A key 21st century question is whether the widely supported though definitionally ethereal concept of ecotourism can really be effective in reconciling these competing perspectives. Whatever the answer to that question may eventually prove to be, it is clear that the mere adoption of an ecotourism label does not automatically guarantee a cost-free approach to overall tourism development. In fact, the New Zealand version of ecotourism could potentially prove to be more destructive, and less sustainable, than any other tourism style the country has encountered to date.

Environmentalism + Tourism = Ecotourism?

As Lew (1998) has commented, there are multiple conflicting definitions for the concept of ecotourism, and as a result there has been growing academic interest in what is an ill-defined and imperfectly understood construct. However, the thrust of the literature is to suggest that the relationship between tourism and the natural environment can exist anywhere along a continuum with synergistic cooperation and destructive conflict at its poles. Making the former case, some authors (e.g. Priestley *et al.*, 1996; Middleton and Hawkins, 1998) have argued that the tourism industry will inevitably be positive towards the maintenance of environmental purity for purely commercial reasons. Developing tourism in harmony with the physical surroundings can generate a substantial competitive advantage for participating firms, as a pristine environment will tend to satisfy visitor motivations and generate both repeat visitation and word-of-mouth referral.

In contrast, a number of authors (e.g. McKercher, 1993; Bramwell *et al.*, 1996) have cautioned against the intuitively attractive appeal of ecotourism development, noting that the unavoidable impacts of human nature can transform theoretical harmony into practical discord. From a negative perspective, the core component of any tourism activity is the consumption of host community resources for financial reward, a type of exchange that offers little hope for a constructive relationship between the environment and its visitors. In combination with an often cited tourism industry preoccupation with short-term profits, the consequent opportunities for unethical and destructive business practice constitute a very real threat to the long-term viability of ecotourism.

These comments highlight a common misconception in the tourism literature, the idea that ecotourism is inherently more benign than other industry sectors. As Swarbrooke (1999) has argued, in a colourful view of what he calls 'ecotourism locusts', the supposedly sensitive, aware and higher spending ecotourist is easily discouraged by the arrival of volume visitation, and will simply abandon the destination to the callous, careless and cashless package tourist. In these circumstances, the arrival of greater visitor numbers poses a significant threat to the factors which originally attracted them, and the destination runs a very real risk of becoming a victim of its own popularity. Swarbrooke's view seems particularly appropriate for New Zealand.

Whilst operational realities related to environmental protection have been instrumental in delineating the nature of DoC involvement with tourism, Tourism New Zealand (TNZ) has been highly successful in ensuring that the flow of international visitation has shown consistent growth over the past 10 years. Successive versions of a global marketing strategy have served only to reinforce a single-minded devotion to international marketing and, as generic branding is progressively replaced by specific product promotion, the undeniable success of its marketing endeavours has placed TNZ in a position of considerable strength in a marketing sense, but one of limited influence in other, related areas.

In consequence, there is a clearly perceptible threat of polarization, with both DoC and TNZ under some political pressure to 'stick to their knitting' in terms of environmental protection and visitor marketing, respectively. If either or both of these institutions were to surrender to temptation in this respect, the resulting lack of liaison would pose a substantial threat to the goals and objectives outlined in the DoC Visitor Strategy. In these circumstances, ecotourism in New Zealand would demonstrate much more of a culture of confrontation than one of consensus. If advocates of the national estate's intrinsic values cannot, or will not, recognize the claims of recreational visitors and tourists, or conversely if these tourists determinedly resist or ignore the attitudes of the environmental lobby, there can be no ecotourism, no matter what the definition adopted.

In this respect, little appears to have changed since Warren and Taylor (1994) identified the key issues for New Zealand ecotourism as:

- international trends away from mass tourism and towards free independent travel (FIT) experiences;
- the concept of managing national parks for economic return as well as resource conservation;
- growing demands for host community participation in tourism development;
- indiscriminate marketing of the clean and green image;
- misappropriation and misuse of Maori cultural images;
- maintenance of safety and service standards;
- destructive levels of industry competition;
- increasing sophistication of tourist expectations;
- increasing dependence on fragile and finite natural resources.

Warren and Taylor (1994) somewhat disturbingly conclude that New Zealand's clean and green image can essentially be attributed to a happy combination of good luck and low population density, and it is difficult to avoid the conclusion that their analysis may well be accurate. In this context, while previous authors (e.g. Elliot, 1997; Hall *et al.*, 1997) have endorsed the powerful potential of appropriate policy interventions as a positive influence for sustainability in tourism development, the operational implementation of policy can frequently be threatened by real world political considerations. As such, though many governments have enthusiastically endorsed sustainable development theory, in many cases their words speak louder than their actions; a claimed commitment to sustainable development principles is often an easier option than the affirmative action required to support that claim (Butler, 1998).

In New Zealand, at both national and sub-national level, it appears that the initial foundations for effective ecotourism policy are indeed in place, but that a predominance of permissive, rather than compulsive, legislation has contributed to an element of ambiguity in the developmental direction chosen. At one end of the reactive scale, although DoC (like virtually all government departments anywhere in the world) will continue to confront resourcing constraints as a perennial issue, it has appeared to regard its formative legislation as little more than a platform from which to strive towards achievement of a conservation/recreation consensus. On the other hand, operating from a similar legislative base, Tourism New Zealand has often seemed to see its founding statute as a limitation over what it is able to do, rather than a fluid framework within which it can determine the nature and scale of its own operations. This level of contrast, both in terms of policy allocation and interpretation, is a consistent issue throughout much of the New Zealand tourism establishment.

This theme is characterized by the frequent allocation of almost mutually exclusive responsibilities to individual actors in the tourism policy arena, a practice that can be exemplified by a series of critically important questions related to equity issues. For example, is it equitable, fair and sustainable long term that:

- DoC should be responsible for protecting the physical environment from degradation by human visitors, while simultaneously being required to encourage those visitors to visit;
- Tourism New Zealand should be responsible for maximizing the recruitment of international visitors to New Zealand, while simultaneously being statutorily and fiscally discouraged from involvement with what happens to them after arrival;
- local government should be responsible for the long-term sustainability of their assigned territories, while simultaneously being permitted to avoid consideration of (arguably) the most threatening human interface with the environment?

However valid these questions may prove to be, it is relatively easy to mount a compelling defence; after all, the existing legislation does provide DoC with an adequate justification for proactive involvement in the visitor industry, and the 1996 Strategy is a valid and valuable conceptualization of what is required in that respect. Similarly, Tourism New Zealand enjoys continued success in its efforts to attract visitors to New Zealand and, in the main, these visitors return home to enthusiastically endorse the quality of the country's natural attractions. On the surface, it is persuasively easy to assume that everything in the ecotourism garden is relatively rosy.

It is this apparent lack of urgency that may pose the biggest threat to sustainability in New Zealand's ecotourism industry. Whilst the necessity for close cooperation between DoC, TNZ and the industry itself is readily apparent, widely acknowledged and publicly endorsed, the absence of any unifying crisis has led to a relationship where cooperation may be more imagined than real. In this respect, it often appears that the tourism industry, including some so-called ecotourism operators, continue to regard environmentalism as a necessary evil, something that is promotionally useful but operationally restrictive; at the same time, that industry continues to exploit clean, green and unspoiled imagery in its external promotion, threatening to provide a textbook example of killing the goose that laid the golden egg.

So what is likely to lie in the future for New Zealand ecotourism? There is some evidence to suggest that the belated arrival of the country's first national tourism strategy may make a significant contribution to the resolution of the types of uncertainty referred to earlier. As Collier (1994) has commented, there appears to have always been substantial agreement that such a strategy was desirable, but agreement has never actually been translated into action. However, a more cynical interpretation could envisage a situation in which industry stakeholders were able to express unanimous support for the value of a nationally unified approach to tourism development, while simultaneously continuing to conduct their business operations in essentially the same manner as they had done prior to strategy release.

Thus, although considerable progress has been made, during 2000 and 2001, towards a more unified approach to delivery of the national tourism product, and though it appears reasonable to conclude that greater internal cooperation between participants may well result from adoption of a national tourism strategy, the jury is still out on the eventual results of current developments. In the meantime, New Zealand continues to enthusiastically promote its clean and green image to the world, while the long-term sustainability of a world-class ecotourism product remains seriously in doubt.

References

Bramwell, B., Henry, I., Jackson, G., Prat, A.G., Richards, G. and van der Straaten, J. (eds) (1996) *Sustainable Tourism Management: Principles and Practice.* University Press, Tilburg, The Netherlands.

Butler, R.W. (1998) Sustainable tourism – looking backwards in order to progress? In: Hall, C.M. and Lew, A. (eds) *Sustainable Tourism – a Geographical Perspective.* Addison Wesley Longman, Harlow, UK, pp. 25–34.

Clarke, R. and Stankey, G. (1979) *The Recreational Opportunity Spectrum: a Framework for Planning, Management and Research.* General Technical Report PNW-98. US Department of Agriculture and Forest Service, Pacific North West Forest and Range Experiment Station.

Collier, A. (1994) *Principles of Tourism: a New Zealand Perspective*, 3rd edn. Longman Paul, Auckland.

Department of Conservation (1996) *Visitor Strategy*. Department of Conservation, Wellington.

Elliot, J. (1997) *Tourism: Politics and Public Sector Management*. Routledge, London.

Ewan, A. (2000) DoC steps in to Kapiti tourism row. *Evening Post*, 17 August p. 5.

Gillespie, A. (1999) *The bicultural relationship with whales: between progress, success and conflict*. Te Matahauariki, Hamilton (http://lianz.waikato.ac.nz).

Hall, C.M. (1994) Ecotourism in Australia, New Zealand and the South Pacific: appropriate tourism or a new form of ecological imperialism?' In: Cater, E. and Lowman, G. (eds) *Ecotourism: a Sustainable Option?* John Wiley & Sons, Chichester, pp. 137–157.

Hall, C.M., Jenkins, J. and Kearsley, G. (eds) (1997) *Tourism Planning and Policy in Australia and New Zealand: Cases, Issues and Practice*. McGraw-Hill, Sydney.

Hunt, G. (2000) Call to re-open Cave Creek tragedy. *National Business Review*, 28 April.

Hutching, C. (2000) $80 million monorail planned. *National Business Review*, 22 September.

Isaac, A. (1997) The Cave Creek incident: a REASONed explanation. *Australasian Journal of Disaster and Trauma Studies* 3, pages un-numbered.

Kearsley, G. (1997) Tourism planning and policy in New Zealand. In: Hall, C.M., Jenkins, J. and Kearsley, G. (eds) *Tourism Planning and Policy in Australia and New Zealand: Cases, Issues and Practice*. McGraw-Hill, Sydney, pp. 49–60.

Keen, R. (2000) DoC appeals consent for tourism operation. *Southland Times*, Invercargill NZ, 14 June, p. 13.

Lew, A. (1998) The Asia-Pacific ecotourism industry: putting sustainable tourism into practice. In: Hall, C.M. and Lew, A. (eds) *Sustainable Tourism – a Geographical Perspective*. Addison Wesley Longman, Harlow, UK, pp. 92–106.

Madgwick, P. (2000) Tramline damage angers groups. *The Press* (West Coast New Zealand) 28 September, p. 3.

McKercher, B. (1993) Some fundamental truths about tourism: understanding tourism's social and environmental impacts. *Journal of Sustainable Tourism* 1(1), 6–16.

McKinlay, P. (1994) *Local government reform: what was ordered and what has been delivered*. Local Government New Zealand, Wellington.

Middleton, V.T.C. and Hawkins, R. (1998) *Sustainable Tourism: a Marketing Perspective*. Butterworth-Heinemann, Oxford.

Newth, K. (2000) Too many tourists – too much pressure. *Sunday Star-Times*, Auckland, 24 September, p. 4.

New Zealand Government (1980) National Parks Act New Zealand Government, Wellington.

New Zealand Government (1987) Conservation Act New Zealand Government, Wellington.

New Zealand Government (1991a) New Zealand Tourism Board Act New Zealand Government, Wellington.

New Zealand Government (1991b) Resource Management Act New Zealand Government, Wellington.

OTSp (1999) *Briefing to the Incoming Minister of Tourism.* Office of Tourism and Sport, Wellington.

Page, S. and Thorn, K. (1998) Sustainable tourism development and planning in New Zealand – local government responses. In: Hall, C.M. and Lew, A. (eds) *Sustainable Tourism – a Geographical Perspective.* Addison Wesley Longman, Harlow, UK, pp. 173–184.

Pearce, D.G. (1992) *Tourist Organizations.* Longman, Harlow, UK.

Priestley, G.K., Edwards, J.A. and Coccossis, H. (eds) (1996) *Sustainable Tourism: European Experiences.* CAB International, Wallingford, UK.

Simpson, K. (2001) *Community based strategic planning for sustainable regional tourism development in New Zealand.* Unpublished doctoral thesis, Massey University, Auckland, New Zealand.

Statistics New Zealand (2000) *Tourism Related Industry Statistics.* Statistics New Zealand, Wellington.

Swarbrooke, J. (1999) *Sustainable Tourism Management.* CAB International, Wallingford, UK.

TIANZ (2001) *Tourism 2010: A Strategy for New Zealand Tourism.* Tourism Industry Association of New Zealand, Wellington.

Tourism New Zealand (1999) *Tourism New Zealand – 100% pure.* Tourism New Zealand, Wellington.

Warrander, R. (2000) Tourism key issue for many. *Daily News*, New Plymouth, New Zealand, 19 April, p. 3.

Warren, J.A.N. and Taylor, C.N. (1994) *Developing Ecotourism in New Zealand.* NZ Institute for Social Research and Development, Wellington.

WCED (1987) *Our Common Future.* World Commission on Environment and Development, Oxford University Press, Oxford.

Ecotourism Management in Europe: Lessons from the Biosphere Reserves in Central and Eastern Europe

14

Dimitrios Diamantis[1] and Colin Johnson[2]

[1]Les Roches Management School, Bluche, Crans-Montana, Valais, CH-3975, Switzerland; [2]Ecole Hoteliere De Lausanne, Le Chalet-à-Gobet, 100 Lausanne 25, Switzerland

Introduction

Crisis management strategies are a new topic in tourism research. If one considers the enormous size of the tourism industry as well as the variety of tourism products that exists it is noteworthy that these strategies have not come into play until fairly recently. Crises in tourism development may include both negative economic, social and environmental effects and a lack of proactive management strategies to overcome these negative effects. As in any industry, the important issues with crisis management are if and how are crises predicted. Generally speaking, there are four types of crisis that can be predicted (Fink, 1986):

1. The *prodromal crisis stage*: when a crisis is starting to exist in an organization's products and services but the organization is unable to respond.
2. The *acute crisis stage*: when a crisis exists and demands urgent action.
3. The *chronic stage*: when a crisis exists and a company devises certain strategies to overcome the situation.
4. The *crisis resolution stage*: when a crisis exists and a company has developed proactive and/or reactive strategies to deal with it.

In the case of ecotourism, there are a number of crises, ranging from the actual meaning of ecotourism and sustainability up to the effective application of eco-products in destinations. Therefore, ecotourism

© CAB *International* 2003. *Ecotourism Policy and Planning*
(eds D.A. Fennell and R.K. Dowling)

products and destinations seem to be situated in the first three stages of crisis management, those of prodromal, acute and chronic stages. The prodromal crisis stage refers to the inconsistencies that exist in the definitional perspective of sustainability and ecotourism (Diamantis and Ladkin, 1999). Different destinations and organizations define ecotourism in distinct ways, which makes ecotourism an amorphous or slippery term. The acute crisis stage can well be suited to mass tourism destinations or other areas, which due to their existing problems with carrying capacity, use ecotourism as a means to attempt to rejuvenate their destination status. The chronic stage may be seen in destinations that have implemented certain strategies to overcome perceived negative effects of ecotourism development. It would appear, however, that the final crisis resolution stage does not exist to any great degree as the majority of ecotourism areas tend to react to a crisis (at the chronic stage) rather than adopting a proactive stance. In that respect, the question that still remains is how these types of crisis will affect ecotourism destinations in practice, such as in the area of Central and Eastern Europe.

Sustainability in Central and Eastern Europe

The articulation of the concept of sustainability within tourism may be found across the world, a clear indication that the concept has attracted wide attention from both the public and private sectors (Murphy, 1994; WTO, 1995, 1996, 1997; EC, 1995; WTTC/WTO/EC, 1995; Mowforth and Munt, 1998: 105–111). Since the political transition and reforms of the late 1980s, environmental and sustainable practices in Central and Eastern Europe (CEE) have come to light, (Staddon, 1999; Warner, 1999) and these practices have paralleled the increased global awareness of sustainability issues (Hall, 1992; Smeral, 1993; Hall and Kinnaird, 1994; Yarnal, 1995; Fletcher and Cooper, 1996; Hall, 1998a). This wave of interest not only emerged as a result of government commitment to environmental protection during the early years of their political reforms, but also filtered through to local levels with the help of practitioners who re-promoted rural and natural tourism products (Hall and Kinnaird, 1994; Hall, 1998a: 351–353; Ilieva, 1998; Kombol, 1998; Ratz and Puczko, 1998). It is worth noting that rural tourism was a significant feature of Central and Eastern European tourism even before communism, especially in Hungary, accounting for between 35 and 40% of all tourism (Suli-Zakar, 1993). In these days, however, a number of important initiatives have been introduced in the region in an effort to enhance its environmental status.

Sustainability Initiatives

There have been several initiatives in Central and Eastern European countries to promote a more sustainable and 'greener' image. Almost all countries have identified ecotourism as an important market niche. Soon after independence in 1991, Slovenia formulated a comprehensive tourism strategy and strong positioning related to nature and farm tourism (Sirse, 1995; Sirse and Mihalič, 1999). The country recently adopted the slogan of 'A green piece of Europe', emphasizing environmental awareness and responsibility (Hall, 1999).

In Hungary, the Commission on Sustainable Development was established in 1993 as a permanent interministerial level for the coordination of national and international programmes. In Poland, the Committee for Sustainable and Regional Development of the Board of Ministers was introduced in 1997 with an aim to devise long-term strategies for sustainability.

Following the formulation of the Rainbow environmental protection programme in the Czech Republic, environmental issues came to prominence, highlighting the effect of national level influence. As a result new institutions were formulated, environmental legislation was adopted and new economic instruments were introduced. In 1995, the State of the Environment Policy strategy was created and the Environmental Remediation Program was launched to monitor environmental conditions.

Poland, which has strong environmental assets with 20 national parks, has appointed a brand manager for rural tourism and has had a strategic rural tourism master plan since 1996 (TTI, 2000).

Romania's environmental policy came into force in 1996, with the broad aim of maintaining biodiversity and protecting designated areas such as nature reserves and national parks along with monuments and settlements (Turncock, 1999).

An assessment of national decision making structures and of national instruments and programmes, two indicators which may be seen to reflect progress made towards sustainability in selected countries of CEE, show a generally positive picture (see Tables 14.1 and 14.2):

- *National decision making structures*: The examination of national decision-making structures with regard to sustainability revealed that most of the selected countries have 'national sustainable development coordination bodies' or are in the process of creating one. In a similar vein, countries score positively on all of the other indicators, except in the case of the Local Regional Agenda 21 (WTTC, 1999).

- *Strategies and policies*: With respect to the policies and strategies indicator, the picture that emerged was rather more mixed (see Table 14.2). Most of the selected countries have environmental educational systems in schools as well as eco-labelling and recycle/reuse programmes. At the other end of the performance scale, environmental indicator programmes and green accounting practices are not generally applied (WTTC, 1999).

Given this positive performance at the national level, the apparent question is whether sustainability issues have permeated the local levels of operation. In an attempt to answer this from the biosphere reserves perspective, three selected case studies will indicate the role of biosphere reserves as well as the state of their sustainability practices.

Table 14.1. National decision making structure with regard to sustainability (source: WTTC, 1999).

Country	National sustainable development coordination body	National sustainable development policy	National Agenda 21	Local Regional Agenda 21	Environmental impact assessment law
Bulgaria	In progress	In progress	In progress	Yes	Yes
Czech Rep.	Yes	Yes	Yes	Yes	Yes
Federal Rep. of Yugoslavia	No	No	Yes	No	Yes
Hungary	Yes	Yes	Yes	Yes	Yes
Poland	Yes	Yes	Yes	Few	Yes
Romania	Yes	Yes	In progress	No	Yes
Slovenia	Yes	No	Yes	No	Yes

Table 14.2. National instruments and programmes with regards to sustainability (source: WTTC, 1999).

Country	Environmental education in schools	Environmental indicators programme	Eco-label regulation	Recycle/ reuse programme	Green accounting programme
Bulgaria	No	No	Yes	Yes	No
Czech Rep.	Yes	Yes	No	No	No
Federal Rep. of Yugoslavia	No	No	Yes	Yes	No
Hungary	Yes	No	Yes	Yes	No
Poland	Yes	In progress	In progress	Yes	No
Romania	Yes	No	No	Yes	No
Slovenia	No	In progress	Yes	Partial	No

Biosphere Reserves: an Overview

Biosphere reserves are a special type of conservation area, created with the aim of achieving a sustainable balance between conserving biological diversity, promoting economic development and maintaining associated cultural values (UNESCO, 1996; Bridgewater and Cresswell, 1998).

The biosphere reserves concept was launched in 1976 as a key component of UNESCO's Man and the Biosphere (MAB) programme.

Biosphere reserves are nominated by national governments and have to meet certain criteria and adhere to a minimum set of agreements before being admitted to the worldwide network of UNESCO. In particular, each biosphere reserve should perform the following tasks (UNESCO, 1984, 1996; Phillips, 1998; Bridgewater and Cresswell, 1998):

- conserve biological diversity;
- maintain healthy ecosystems;
- learn about traditional forms of land use;
- share knowledge on how to manage natural resources in a sustainable way;
- co-operate in solving natural resources problems.

There is presently a network of 352 biosphere reserves in 87 countries, of which 129 are in Europe, comprising a mix of terrestrial and marine elements. Approximately 90% of the biosphere reserves cover some form of designated protected area while the remaining 10% have no nationally designated area associated with them.

Overall most of the territorial activities in biosphere reserves take place in three areas or zones (see Fig. 14.1) (Bioret *et al.*, 1998):

- *The core areas for nature conservation*: legally protected areas devoted to long-term protection.
- *The buffer or support zone*: areas surrounding the core zone protecting them from any human impact. Different research activities are undertaken in this zone such as training, education and recreation as well as different outdoor recreational activities.
- *The transition or community zone*: the largest of all areas, covering the wider community around the biosphere reserves, where research initiatives are promoted and developed.

One key problem faced by biosphere reserves is that of definition. Biosphere reserves have been viewed as areas designated for the protection of valuable ecosystems and where various forms of ecological research could be conducted (Batisse, 1993). This has led to conflicts with different tourism activities that often exert a major economic influence on biosphere reserves. This situation initially affects the structure within which tourism operates and, subsequently, the type

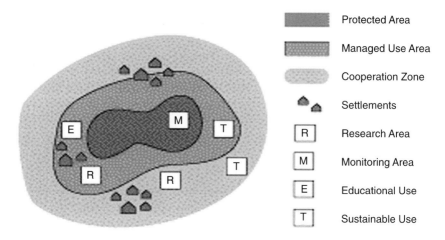

Fig. 14.1. Biosphere reserve zonation structure (source: Unesco, 1996).

of tourism products that are traded and offered to visitors. In addition, the concept of biosphere reserves was created as an ideal working example of sustainably protecting endangered environments and was intended as a benchmark for other destinations. This has not always been the case, however, as numerous negative impacts have become apparent in the reserves. The remainder of this chapter examines the state of environmental practices in biosphere reserves in Hungary, the Czech Republic and Poland and Slovenia where sustainable practices have been proposed.

In each case there is a review of the biosphere reserves structure followed by their current sustainable and ecotourism practices. Each biosphere reserve has different tourism priorities and sustainable practices, but they share a common notion in their intention to channel their environmental consciousness and awareness into the development of tourism.

Aggtelek Biosphere Reserve, Hungary

The Aggtelek was declared a biosphere reserve in 1979, occupying 19,247 ha in the northern part of Hungary (see Fig. 14.2). Approximately two-thirds of the biosphere reserve is covered by forest; the remainder is attributed to other natural interests such as grasslands (Toth, 1998).

There are two main villages (Aggtelek and Josvafo) inside the biosphere reserve and 18 adjoining villages. The cultivated areas include croplands, old orchards and vineyards and tend to form an ecological buffer around the more urbanized section of the villages.

Fig. 14.2. Biosphere reserves in Hungary.

One of the key problems in the Aggtelek reserve was the decline of traditional agricultural land use combined with a decline in the population. In particular, the following weaknesses were observed for this region (Toth, 1998):

- the age structure of the human population has become skewed in favour of the older proportion of the population;
- economic circumstances have become more limited in the rural regions;
- the traditional source of income (i.e. agriculture) has declined;
- a high percentage of cultivated areas have been abandoned;
- the level of health service, education and other public utilities has declined.

One of the main challenges faced by the biosphere reserve was to preserve the traditional land-use patterns by maintaining the balance between natural areas and cultivated lands. In this respect a management programme was formulated with the following objectives (Toth, 1998):

- the preservation of the so-called harmonious landscape;
- the preservation of habitats and species diversity;
- the involvement of the local population at all levels of the management programme;
- the development of a partnership among the communities;
- the preservation of the cultural heritage through promotion of traditional handicrafts;
- the promotion of rural tourism in the region.

Each of these objectives was divided up into three different planning levels, nature conservation, community development and tourism. With respect to tourism development, emphasis was placed on the infrastructure of rural tourism development and identifying the carrying capacity for tourism development. Results so far have indicated that local citizens are involved in the programme development and implementation, through the growth of a new local business predominantly selling traditional products. In addition, a number of guesthouses have also opened, pointing to the positive contribution of the biosphere reserve initiative on the local area.

Trebonsko Biosphere Reserve, the Czech Republic

The Trebon Basin or Trebonsko was declared a biosphere reserve in 1977 (Jelinkova, 1998). It occupies an area of 700 km² in the southern part of the Czech Republic adjacent to the Austrian border with a total population of approximately 25,000 inhabitants (see Fig. 14.3).

At the centre of the biosphere reserve is the historic town of Trebon (9000 inhabitants) where the biosphere reserve administration is located. The medieval core of the town with its unique architecture was declared a national monument in 1976. Týeboå also has a special status as a spa town with a tradition of medical treatment using peat from local deposits.

Since 1979 the core area of the reserve has had the legal status of a protected landscape area according to Czech legislation. This region is

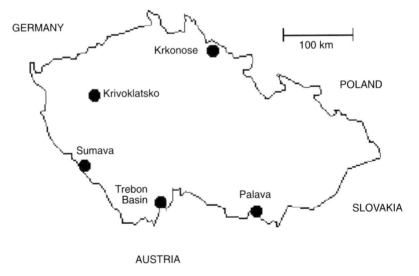

Fig. 14.3. Biosphere reserves in Czech Republic.

also part of Important Bird Areas of Europe (IBA) as well as incorporating 20 small-scale protected territories (nature reserves or monuments of national or regional importance). The western part of the
Trebon region was also declared a Protected Area of Natural
Accumulation of Water in order to preserve the quality of its groundwater, which is accumulated in thick sediment layers of the Trebon
Basin (Jelinkova, 1998).

Although the Trebon region is partly a man-made landscape, its
natural potential and values are very high. The concentration of
numerous animal and plant species living in a relatively small area is
something unique to this type of countryside in Central Europe.
Species native to both the northern tundra and the warm continental
lowlands grow and live in close proximity as well as species associated with both extremely wet and extremely dry biotopes.

The most important ecosystems are protected within the core area
of this biosphere reserve such as wetlands, fishponds, floodplains and
wet meadows as well as specific biotopes of old sand dunes which are
among the most valuable ecosystems in this biosphere reserve
(Jelinkova, 1998).

Unfortunately, the biosphere reserve has suffered during the last
20 years from serious tourism development impacts and other human
activities. Some of its traditional human activities (agriculture,
forestry and fish farming) have been practised with extreme intensity
and with modern technologies that do not respect the natural ecological limits and the carrying capacity of the landscape. This resulted in
negative changes in the fragile ecosystems such as eutrophication and
loss of biodiversity as well as visual degradation of the countryside. In
addition, the improper dumping of waste and intensive hiking by visitors were practised in the core zone, disturbing the legally protected
zone of the reserve.

Ecotourism and sustainable practices in the reserve occur in the
form of tours, mainly for the purpose of observing the flora and fauna
of endemic and natural species. Adventure tours, as well as bio-
gardening (the cultivation and processing of medical herbs and plant
species), are also part of the range of ecotourism products offered to
visitors. Unfortunately, a negative side to these proactive practices is
that there are no guidelines for the development of ecotourism and no
frameworks have been developed for maintaining its attractions.

As a result of the increasing number of initiatives being developed
in the reserve, a number of programmes to estimate the carrying
capacity of the zones were initiated. The results of this study with
respect to tourism attractions indicated the following:

- potential tourists have no information about the tourism attractions or the facilities of the region;

- the biosphere reserve has only limited opportunities to influence the turnover of ventures operating in the area;
- there is a lack of data with regard to tourism arrivals in the reserve;
- there is a lack of programmes to monitor the effects of tourism in the zones of the reserve.

Overall the sustainable and ecotourism practices in the examined area are at the initial planning stages. With the lack of comprehensive management frameworks concerning the practice and the effects of ecotourism, the situation could have disastrous effects, especially if one considers the potential direct and indirect negative impacts in the core zone of the reserve.

Tatra Biosphere Reserve, Poland and Slovenia

Since 1992, five national biodiversity protection projects have been implemented in Central Europe aiming to encourage the creation of transfrontiere biosphere reserves. One of these reserves is that of the Tatra situated on the borders of Poland and Slovakia (see Fig. 14.4).

The Tatra is the highest mountain massif of the alpine folding of the western Carpathians. The total area amounts to about 750 km^2, of

Fig. 14.4. Biosphere reserves in Poland.

which 174 km^2 is within the boundaries of Poland. The idea of establishing a MAB biosphere reserve in this location originated in the Polish Tatra National Park (TPN) and the Slovakian Tatra National Park (TANAP).

A group of experts appointed by the local Council worked on the elaboration of a concept for future biosphere reserves in the area of the two national parks (Kokovkin, 1999). One critical issue concerned the zoning of such transboundary reserves. Zoning became a subject of numerous consultations and negotiations aimed at obtaining a single dense core representing both parts of the reserve. The final result was a biosphere reserve with a designated area of 145,600 ha, of which the core zone shared by both countries constitutes one-third. In this particular reserve ecotourism activities are of a very limited nature, usually taking place in cultural and transitional zone areas. It has been estimated that ecotourism accommodates around 10,000 visitors per year, most of which are independent travellers. There are no nature guides or training for nature guides, interpretation programmes or specific conservation objectives. The majority of the ecotourism clientele are occasional in nature, in that they are likely to be involved in a number of other tourist activities such as cultural and rural tourism. The recent strategy has focused on five key areas (following Kokovkin, 1999):

- international coordination and infrastructure;
- economic activities and population;
- protection and assessment;
- education and promotion;
- implementation support.

The strategy, which does not have a primary focus on tourism, is based on 19 goals and has a detailed planning mechanism for managing the ecosystems between the two countries.

One of the most significant aspects of the strategy is the emphasis on both the *in-situ* and the *ex-situ* protection of the reserve. Across the two countries strict protection covers over 11,500 ha, and includes summits and zones such as the alpine meadows and dwarf pine zones, and also the upper and lower forest belts. This indicates that the issue of *geographical equity* takes into account the resources that are critical to the biosphere reserve as well as the resources that surround the reserve. Another important feature of the strategy is that of the integrated monitoring scheme of the reserve's territory as well as the assurance that the strategy does not come into conflict with other local environmental policies.

In an era where sustainability in tourism is claimed to be extremely tourism centric (Hunter, 1995a,b, 1997) this biosphere reserve seems to approach its management issues from a general sectorial perspective devising strategies that are free from tourism development stances.

An Assessment of Ecotourism and Sustainable Practices in the Biosphere Reserves

The sustainable and ecotourism practices in the examined biosphere reserves have been demonstrated by a variety of actions. As such, these initiatives are different from reserve to reserve reflecting the range of perspectives and paradigms. Ecotourism may be practised in a continuum from an active pole (actions of protecting the environment) to a passive pole (ecotourism development actions which do not create negative impacts) (Orams, 1995).

In addition, ecotourism definitions vary considerably across the tourism literature; authors raise numerous perspectives of ecotourism ranging from the welfare of the local community up to the sustainable management of resources. It seems that ecotourism as a concept contains three common components: a natural-based component, a sustainability component and an educational component (Diamantis, 2000). Further, there are three main types of crisis, suggested in the introduction, that seem to have relevance in the concept of ecotourism:

- *The prodromal crisis stage*: when a misunderstanding exists among destination practitioners over the definitional perspective of ecotourism but the destination is unable to respond;
- *The acute crisis stage*: when a crisis exists in the tourism destination and ecotourism is adopted as strategy to rejuvenate their destination status.
- *The chronic stage*: when a crisis exists in destinations that have implemented certain strategies to overcome perceived negative effects of ecotourism development.

By using the active/passive application of ecotourism, the three components of ecotourism and the three stages of crisis as indicators to assess ecotourism practices in the examined reserves, it may be seen that the biosphere reserves exercised a rather passive stance towards ecotourism and are therefore situated at the prodrominal crisis stage.

With the exception of the Tatra Biosphere Reserve, where a management programme was formulated (chronic crisis stage), the reserves have not implemented any programmes to support the sustainability and educational elements of ecotourism (acute crisis stage) (see Table 14.3).

This suggests that when it comes to implementing certain strategies at a local level all the examined reserves have a rather passive

Table 14.3. Sustainable and ecotourism performance matrix of biosphere reserves.

Indicators	Tarta	Trebonsko	Aggtelek
Active/passive application of ecotourism	Passive	Passive	Passive
Natural-based element of ecotourism	Yes	Yes	Yes
Sustainability element of ecotourism	Yes	No	No
Educational element of ecotourism	Yes	No	Partial
Prodominal crisis	Yes	Yes	Yes
Acute crisis	No	Yes	Yes
Chronic crisis	Yes	No	No
Crisis resolution	No	No	No

role towards an ongoing sustainable management development plan. If one considers that biosphere reserves are areas where ecotourism and sustainable practices could flourish, the current situation is not encouraging.

When it comes to overcoming these kinds of crises and implementing the crisis resolution stage, ecotourism destinations in general and the examined biosphere reserves in particular need to devise strategies which include strategic forecasting and scenario analysis. Strategic forecasting should be part of the ecotourism destination planning and should include methods that predict and assess the impact of major changes on the economic, social and environmental agenda of the area. The use of situation analysis could also assist ecotourism destinations in avoiding different crises, and could be applied in the following manner:

1. Select certain scenarios for sustainability and/or ecotourism purposes (best and worse case).
2. Select certain indicators to fit that scenario (i.e. economic, social, environmental).
3. Conduct research in the destination or on the product to see the applicability of the selected indicators.
4. Consult a number of stakeholders to obtain their views and develop a list of new indicators.
5. Summarize the key concepts of the crisis scenarios.
6. Compare results for each of the critical indicators analysed.
7. Consult a number of stakeholders to obtain their views on that programme.
8. Consult an independent verifier to acknowledge the scenario effectiveness.
9. Develop a feedback process.

Overall, crisis management strategies should involve the wide participation of the local community and include research that demonstrates a proactive rather than a reactive stance towards ecotourism. This could prove problematic in certain circumstances as a result of half a century of state planning and control.

Although it is imperative to have a mix of state intervention and ventures involving both private and public interests to plan for overall sustainable protection, there is something of an 'allergic reaction' to state planning and centralized control after the communist era. This makes it extremely difficult to involve key stakeholders in the consultation process (Hall, 1999).

Conclusion

Implicit in the evolution of sustainability in CEE are the efforts made to transform negative environmental images to policies that embody sustainable principles, through declarations of cooperation and the development of sustainable committees. Although sustainable tourism and ecotourism are emerging as important products, there still remain a number of key challenges at both national and local levels.

Tourism markets in Central and Eastern Europe are likely to become more polarized between the high added-value, post modern, environmentally aware niche markets for (mainly) Western tourists and the mass market demands of tourists from within the region, reminiscent of tourism development in Western Europe 50 years ago (Hall, 1998a). It is not going to be an easy task to placate the different demands for these two kinds of market. Additionally the notion of pristine countryside is the major attraction for most Western tourists. As tourism increases, however, this will have a profound impact on the environment and on the local population, who will have expectations of improved standards of living from tourism earnings as part of the development of an agrarian pluriactivity. The trend in many of the countries in the region has also been towards increasing urbanization (Hall, 1998b). Training and behaviour of the local communities will be extremely important, and there are already signs that without effective educational programmes, rural communities are not automatically receptive to tourists (Verbole, 1995).

The extent of the environmental challenges facing countries in the region should not be underestimated (it is predicted that it will cost Poland in the region of 30 billion euros to come up to EU environmental requirements, and at current spending that will take 20 years (Serafin et al., 2000).

Nevertheless, sustainability is now very much on the political agendas of all countries in the region, as may be seen from the examples cited in this chapter. In adopting the concept of biosphere reserves to explore the awareness and to identify the challenges faced at a local level, three case studies have been presented. It is evident from these cases that there is a high level of awareness of sustainability as well as strategies that provide a backbone for the daily management of these reserves.

Although all the reserves are at their initial planning stages regarding to tourism, a trend shared by most of the reserves is that they do not approach sustainability from a solid tourism perspective. For instance, in the Tatra Biosphere Reserve, sustainable and eco-tourism practices are not applied to enhance tourism industry needs and wants, and do not create the so-called 'tourism-centric' situation.

A key element in the successful management of all three reserves is the development of key concepts such as tourism carrying capacity and saturation levels. Due to a combination of inadequate resources, skills shortages and political indifference and in-fighting (Hall, 1999), declared intentions do not always translate into actual applications. A further serious hindrance is the lack of formalized tourism strategies in certain countries (for example the Czech Republic) (TTI, 2000).

The profound inequalities that exist in Central and Eastern Europe require the adoption of very carefully planned crises management strategies and overall policies for sustainable development (Hall, 1998a; Slee, 1998). Undoubtedly there is a base for creating eco-tourism products in the region, and for the tarnished image of environmental degradation to be replaced by that of unspoilt nature. There is, however, an urgent need for a harmonization of environmental regulations and agendas, all of which need to be supported by training and management development programmes (Iunius and Johnson, 2000). Finally this should be accompanied by mechanisms that guarantee transparency, adequate information and public participation, crisis scenario planning, and community support as well as providing administrative and control procedures for their enforcement and application.

References

Batisse, M. (1993) The silver jubilee of MAB and its revival. *Environmental Conservation* 20, 107–112.

Bioret, F., Cibien, C., Genot, J. and Lecomte, J. (1998) *A Guide to Biosphere Reserve Management: a Methodology Applied to French Biosphere Reserves*, MAB Digest 19. UNESCO, Paris.

Bridgewater, P.B. and Cresswell, I.D. (1998) The reality of the World Network of biosphere reserves: its relevance for the implementation of the convention on biological diversity. In: Workshop Proceedings of *Biosphere Reserves: Myth or Reality?* IUCN, Cambridge: pp. 1–6.

Diamantis, D. (2000) The concept of ecotourism: evolution and trends. *Current Issues in Tourism* 2(2/3).

Diamantis, D. and Ladkin, A. (1999) The links between sustainable tourism and ecotourism: a definitional and operational perspective. *Journal of Tourism Studies* 10(2), 35–46.

EC (1995) *Green Paper on Tourism.* DGXXIII, European Commission, Brussels.

Fink, S. (1986) *Crisis Management: Planning for the Inevitable.* American Management Association, New York.

Fletcher, J. and Cooper, C. (1996) Tourism strategy planning: Szolnok County, Hungary. *Annals of Tourism Research* 23(1), 181–200.

Hall, D.R. (1992) The challenge of international tourism in Eastern Europe.*Tourism Management* 13, 41–44.

Hall, D.R. (1998a) Central and Eastern Europe: tourism, development and transformation. In: Williams, A.M. and Shaw, G. (eds) *Tourism and Economic Development*, 3rd edn. John Wiley & Sons, Chichester, UK, pp. 345–373.

Hall, D.R. (1998b) Tourism development and sustainability issues in Central and South-Eastern Europe. *Tourism Management* 19(5), 423–431.

Hall, D.R. (1999) Destination branding, niche marketing and national image projection in Central and Eastern Europe. *Journal of Vacation Marketing* 5(3), 227–237.

Hall, D.R. and Kinnaird, V. (1994) Ecotourism in Eastern Europe. In: Cater, E. and Lowman, G. (eds) *Ecotourism: a Sustainable Option?* John Wiley & Sons, Chichester, UK, pp. 111–136.

Hunter, C. (1995a) Key concepts for tourism and the environment. In: Hunter, C. and Green, H. (eds) *Tourism and the Environment: a Sustainable Relationship?* Routledge, London, pp. 52–92.

Hunter, C. (1995b) On the need to re-conceptualize sustainable tourism development. *Journal of Sustainable Tourism* 3 (3), 155–165.

Hunter, C. (1997) Sustainable tourism as an adaptive paradigm. *Annals of Tourism Research* 24 (4), 850–867.

Ilieva, L. (1998) Development of sustainable rural tourism in Bulgaria. In: Hall, D. and O'Hanlon, L. (eds) Conference Proceedings *Rural Tourism Management: Sustainable Options.* SAC Auchincruive, UK, pp. 263–276.

Iunius, R. and Johnson, C. (2000) Human Resource Management in Romania. In: Lefevere, M., Hoffman, S. and Johnson, C. (eds) *International Human Resource Management in the Global Hospitality Industry.* The Educational Institute, East Lansing, Michigan.

Jelinkova, E. (1998) Science supporting local development from Czech point of view. Paper presented in the *III Biosphere Reserves Coordinators Meeting, Joensuu, Finland, September.* UNESCO.

Kokovkin, T. (1999) *Management Plan of Tarta Biosphere Reserve.* Tarta Council, Poland.

Kombol, T.P. (1998) Rural tourism and Croatia's islands. In: Hall, D. and O'Hanlon, L. (eds) Conference Proceedings *Rural Tourism Management: Sustainable Options* SAC Auchincruive, UK, pp. 391–407.

Mowforth, M. and Munt, I. (1998) *Tourism and Sustainability: New Tourism in the Third World.* Routledge, London.

Murphy, P. (1994) Tourism and sustainable development. In: Theobald, W. (ed.) *Global Tourism, the Next Decade.* Butterworth-Heinemann, Oxford, pp. 274–298.

Orams, M.B. (1995) Towards a more desirable form of ecotourism. *Tourism Management* 16(1), 3–8.

Paci, E. (1995) Importance du tourisme dans les économies des pays de l'Europe de l'est. *The Tourist Revue* 1/1995.

Phillips, A. (1998) Biosphere reserves and protected areas: what is the difference? In: Workshop Proceedings of *Biosphere Reserves: Myth or Reality?* IUCN, Cambridge, pp. 7–10.

Ratz, T. and Puczko, L. (1998) Rural tourism and sustainable development in Hungary. In: Hall, D. and O'Hanlon, L. (eds) Conference Proceedings *Rural Tourism Management: Sustainable Options*. SAC Auchincruive, UK, pp. 449–464.

Serafin, R., Tatum, G. and Heydel, W. (2000) ISO 14001 as an opportunity for engaging SMEs in Poland's environmental reforms. In: Hilary, R. (ed.) *ISO 14001, Case Studies and Practical Experiences*. Greenleaf Publishing, Sheffield, UK.

Sirse, J. (1995) The New Strategy for the Development of Tourism in Slovenia. *The Tourist Review* 1/1995.

Sirse, J. and Mihalič, T. (1999) Slovenian tourism and tourism policy – a case study. *The Tourist Review* 3/1999.

Slee, B. (1998) Soft tourism: can the Western European model be transferred into Central and Eastern Europe? In: Hall, D. and O'Hanlon, L. (eds) Conference Proceedings *Rural Tourism Management: Sustainable Options*. SAC Auchincruive, UK, 481–496.

Smeral, E. (1993) Emerging Eastern European tourism markets. *Tourism Management* 14(6), 411–418.

Staddon, C. (1999) Localities, natural resources and transition in Eastern Europe. *The Geographical Journal* 165(2) 200–208.

Suli-Zakar, I. (1993) The positive impact of rural tourism on the regional development of East-Hungary. *Communications from the Geographical Institute of Debrecen Lajos Kossuth University*, Hungary.

Toth, E. (1998) Local community involvement in the management of cultural landscape in the transitional zone in Aggtelek national park and biosphere reserve. Paper presented in the *III Biosphere Reserves Coordinators Meeting, Joensuu, Finland, September*.

TTI (Travel and Tourism Intelligence) (2000) *Tourism in Central and Eastern Europe*. Bedford Square, London.

Turncock, D.T. (1999) Sustainable tourism in the Romanian Carpathians. *The Geographical Journal*, London, pp. 192–199.

UNESCO (1984) Action plan for biosphere reserves. *Nature and Resources* 20(4), 1–12.

UNESCO (1996) *Biosphere Reserves: the Seville Strategy and the Statutory Framework of the World Network*. UNESCO, Paris.

Verbole, A. (1995) Pros and cons of rural tourism development: sustainability – a possible way out. Paper presented to the *Joint ECA-ECE Symposium on Rural Tourism*, Galilee, Israel.

Warner, J. (1999) Poland: the environment in transition. *The Geographical Journal* 165(2), 209–221.

WTO (World Tourism Organisation) (1995) *What Tourism Managers Need to Know: a Practical Guide to the Development and Use of Indicators of Sustainable Tourism*. WTO, Madrid.

WTO (World Tourism Organisation) (1996) *Tourism and Environmental Protection. WTO-ETAG, Joint Seminar, Heidelberg, Germany.* WTO, Madrid.

WTO (World Tourism Organisation) (1997) *Tourism 2000: Building a Sustainable Future for Asia-Pacific.* WTO, Madrid.

WTTC (World Travel and Tourism Council) (1999) *Millennium Vision Competitiveness Report.* WTTC, London.

WTTC/WTO/EC (World Travel and Tourism Council/World Tourism Organisation/Earth Council) (1995) *Agenda 21 for the Travel and Tourism Industry – Towards Environmentally Sustainable Development.* WTTC/WTO/EC, Oxford.

Yarnal, B. (1995) Bulgaria at a crossroads. *Environment* 37(10), 6–15, 29–33.

A Regional Look at Ecotourism Policy in the Americas

15

Stephen N. Edwards,[1] William J. McLaughlin[2] and Sam H. Ham[2]

[1]*Ecotourism Department, Conservation International, 1919 M Street, NW, Suite 600, Washington, DC 20036, USA;* [2]*Department of Resource Recreation and Tourism, University of Idaho, Moscow, ID 83844-1139, USA*

Introduction

One of the often touted approaches to accomplish conservation and sustainable development is ecotourism. Today's global conservation advocates continue to sell this social economic intervention as a viable way to protect biodiversity and parks. If this is true, after some 25 years of experimenting with ecotourism throughout the Americas (Honey, 1999; Fennell, 1999) one would expect to find: (i) well defined ecotourism policies; (ii) identifiable policy implementation mechanisms (i.e. strategic plans, guidelines, programmes); and (iii) established organizational linkages between nature conservation and tourism agencies within government, as well as other multi-sectoral organizational frameworks which are actively involved in balancing community and economic development with biodiversity conservation. If not, then perhaps ecotourism has fared no better than protected area management, where often the result of policy is at best a 'paper park or unenforceable environmental ideal'. To explore these questions we will use the findings of the 1998 *Comparative Study of Ecotourism Policy in the Americas* (Edwards *et al.*, 1999). The primary purpose of this study was to determine how governmental tourism agencies in the Americas define ecotourism and to identify the range of associated ecotourism policies they had created as of 1998. Our secondary purposes were to explore the relationship between government-adopted definitions of ecotourism and implementing strategy or policy, and to identify relationships with other governmental and non-governmental organizations.

Key Definitions that Guided the Study

For the study, ecotourism policy was defined to include:

- actions governments are engaged in as they carry out ecotourism policy (e.g. hiring consultants or using staff to develop reports on ecotourism, bringing together organizations and enterprises in their country to try to organize ecotourism, carrying out studies of ecotourism markets, drafting regulations, developing promotional materials, proposing legislation);
- policy outputs that governments or their partnerships have developed (e.g. strategic plans, ecotourism marketing plans, ecotourism guidelines, regulations);
- identifiable organizational mechanisms that address ecotourism (e.g. commissions, new divisions within organizations, new positions, partnerships).

This kind of definition was selected because it focuses attention on describing the types of ecotourism policy in existence and attempts to capture the evolutionary or developmental aspects of the policy-making process. Therefore this definition includes much of what has been called 'symbolic policy' (Steinberger, 1995). By this we mean it is only policy that respondents *claimed* to be their government's policy, regardless of whether or not it was effective, being used, or had legal backing. Nevertheless, these 'symbolic' actions are indeed inducing change in the present structure and delivery of ecotourism in the Americas.

Methods in Brief

Governmental tourism agencies in the Americas were the cases studied. Primary data were collected from individuals identified as being responsible for ecotourism in these agencies. These contacts were asked to provide their agency's definition of ecotourism and any ecotourism policy materials in existence or in preparation. The requested materials were normally provided by e-mail, fax or mail. A follow-up telephone interview also asked about other aspects of the agency, such as the agency's name, its position within government, its role in tourism policy development, and with whom it collaborates inside and outside government. In addition, supplemental data were collected whenever possible from the Internet, library sources and government document collections.

Defining which Countries to Include

Building on the World Tourism Organization's classification of official tourism organizations in the Americas (WTO, 1997), the governmental tourism agencies studied in this project were geographically organized into the following groupings:

1. USA and Canada
When addressed in this study, the USA and Canada represent:

- USA: 50 states, the US Virgin Islands and Puerto Rico;
- Canada: 10 provinces and 2 territories;
- National governments of the United States and Canada.

2. Latin America and Caribbean (LAC)
When addressed in this study, LAC represents:

- Mexico, Greenland, the Falkland Islands and St Pierre and Miquelon;
- Central America: 7 countries;
- South America: 13 countries, French Guiana;
- Caribbean: 28 countries and dependencies in the Caribbean.

Given the limited number of agencies we attempted a census. Unfortunately, we did not receive a response before our publication deadline from the following: Colorado (USA), South Carolina (USA), Texas (USA), Vermont (USA), Panama and St Maarten (Caribbean).

Creating a Database and Data Analysis

Data were scanned to create a digital format amenable to both qualitative and quantitative analyses. The textual information was examined using a contextual analysis program (NUDIST VIVO) to identify text with similar meaning. The program also served as a tool for storing and tracing the word groupings and keeping track of the meanings we assigned to the categories of information that we report in this study. For a detailed description of the stepwise process used see Edwards *et al.* (1999: Vol. 1, pp. 37–41).

Keeping the Findings in Context

In interpreting the results, it is important to consider four study limitations:

1. Information was collected from the perspective of an individual in a governmental tourism agency.
2. A complete census of all agencies was not achieved.
3. Limited incorrect translations of language may have occurred.
4. Only the policy information that was provided by key informants or was easily accessible via secondary sources was used in the analysis.

For example, when it came to characterizing ecotourism policy only 29 countries and dependencies in LAC and 25 states and provinces in USA and Canada provided documents that demonstrate their policies and activities in ecotourism. These documents served as the database we analysed. The types of material not included in ecotourism policy analyses were text units that described background information, history, market demographics, tourism data or discussions of existing tourism products. In spite of these limitations and based on the methodological procedures followed we believe that the results reported are trustworthy and fit the Americas (Erlandson *et al.*, 1993).

Key Findings

What is ecotourism?

One of the primary objectives of this study was to determine which governmental tourism agencies in the Americas have a written definition of ecotourism, and then compare and contrast the compiled definitions in order to classify underlying concepts and themes. From this analysis emerged the following findings:

- No single definition of ecotourism dominates the Americas. Instead, a range of definitions was identified, the majority of which are 'homegrown'. A 'homegrown' definition is one that an agency has developed to meet its needs or understanding of ecotourism as opposed to a 'standard' definition taken from another source, such as the tourism research literature or a professional tourism organization.
- Less than half of the LAC agencies (25 of 53 contacted) could provide a written definition of ecotourism, and 21 of these were 'homegrown' definitions.
- About a quarter of the agencies (17 of 66 contacted) in the United States and Canada provided a written definition, and 11 of the 17 provided were 'homegrown'.
- A qualitative analysis of the 42 ecotourism definitions identified by the 119 governmental tourism agencies contacted (supplemented by triangulation with ecotourism definitions identified in the literature and the cross-checking of the findings with practised professionals) resulted in the inductively developed, definitions-based, conceptual model of ecotourism shown in Fig. 15.1.

The elements included in this conceptual model resulted from an analysis of definitions provided by the tourism agency contacts. Because it illustrates the key 'categories of meaning' derived from the contextual analysis of ecotourism definitions it might be thought of as depicting the compilation of governmental ecotourism definitions in

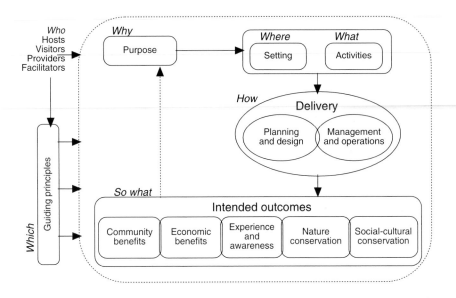

Fig. 15.1. The definitions-based model of ecotourism for the Americas.

the Americas in 1998. This model is explained in more detail in Edwards *et al.* (1999).

In essence, the elements of this model represent the kinds of questions policy makers need to ask as they set out to develop ecotourism policy:

Who needs to be involved in ecotourism policy development?

Which are the principles we want to guide our development of ecotourism?

Why will individuals and organizations want to be involved in ecotourism?

Where do we want ecotourism to take place?

What kind of activities should make up ecotourism?

How should we deliver ecotourism, if at all?

So what are the intended outcomes we want from ecotourism and to whom or what should they accrue?

Who is responsible for ecotourism?

In order to understand the titles and positions of individuals responsible for ecotourism within governmental tourism agencies, we compiled a list of the titles of the contact persons within each agency.

Individuals perceived to be responsible for ecotourism in the agencies contacted have: 'general administrative titles' (director, manager and project coordinator); 'environment/ecotourism titles' (director of the ecotourism section, outdoor recreationist specialist and environmental officer); 'planning and development titles' (tourism planner, director of community development, director of planning); 'research/technical titles' (research director, tourism specialist and technical adviser); and 'marketing titles' (advertising specialist, communications specialist, manager of travel industry marketing and manager of visitor information services). This range of position titles describes the level of variation in how ecotourism fits into the vertical and horizontal personnel structures of governmental tourism agencies in the Americas.

Variation in who is responsible for ecotourism in governmental tourism agencies occurred between the USA and Canada and LAC. For example, 25% of the agency contact persons responsible for ecotourism in the United States and Canada had marketing titles whereas in LAC less than half of that number (11%) had marketing titles. This is contrasted with the finding that 19% of the contact persons in LAC had environment/ecotourism titles compared with only 6% in the United States and Canada.

This broad range of titles and positions demonstrates that there is no uniform approach or universally established title or position for individuals responsible for ecotourism within governmental tourism agencies in the Americas.

Where does tourism and ecotourism fit in governmental organization?

In order to understand the context of governmental tourism agencies and the role they play in tourism policy, it is important to understand where they are located within their governments, as this often indicates their level of responsibility in the tourism policy making process.

- Forty-one per cent of the agencies that deal with tourism and ecotourism in LAC and 28% that deal with tourism and ecotourism in the USA and Canada are independent units of government. This difference may indicate that in LAC, tourism is perceived as having more importance than in the USA and Canada. This hypothesis is reinforced by the fact that 'tourism' is not even part of the name of more than half of the Canadian and US agencies contacted whereas it is a part of the name of over three-quarters of the LAC agencies contacted.
- LAC tourism agency names included words like environment, natural areas and parks, whereas USA and Canada agency names included words like sport and recreation. In both geographic cases

words like commerce and economic development were often
included in the name of the agency responsible for ecotourism.
- A larger percentage (34%) of the governmental tourism agencies
  contacted in the USA and Canada than those contacted in LAC
  (20%) reported collaborating with environment, fish and wildlife,
  natural resource and conservation agencies.

These findings indicate that the significance attached to tourism by
governments in the Americas varies. Equally important is that governmental structures evolve in numerous ways and for numerous reasons, and Canada in particular has adopted innovative approaches to
tourism policy – a reality that may not be accurately described in this
analysis. Therefore, interpreting positioning and naming of tourism
units within governmental structures as an indicator of power and significance has limitations.

Where does tourism and ecotourism fit when it comes to NGOs?

The types of organizations that governmental tourism agencies collaborate with demonstrate the range of sectors to which tourism is net-
worked.
- Governmental tourism agencies in LAC consistently listed international multilateral and bilateral organizations and banks as collab-
  orators, whereas these entities were rarely mentioned in the
  United States and Canada. One reason for the expanding role of
  international multilateral and bilateral organizations in the ecotourism policy arena in LAC is their involvement in funding,
  information sharing and technical support.
- In LAC there was a relatively high level (23%) of collaboration
  with environmental NGOs and associations, indicating that government tourism agencies are working actively with this sector. In
  the USA and Canada, however, only 6% of the contacts listed
  environmental and ecotourism oriented NGOs/associations as col-
  laborators. These findings suggest that the collaborative roles that
  environmental NGOs/associations play are different in the USA
  and Canada from those of their counterparts in LAC.
- Governmental tourism agencies, whether they are in the USA and
  Canada or LAC, collaborate with a broad range of entities in the private sector. These findings suggest that there are established channels
  for sharing of information and other mechanisms of collaboration
  between private ecotourism service providers and government.

Existing collaborative networks are established channels on which partnerships to develop ecotourism policy and deliver sustainable ecotourism can be based. They also could be a viable way for governmental

tourism agencies to provide information about desired outcomes to the private sector.

What options in implementing ecotourism policy are being used?

There are many possible roles and activities for governmental tourism agencies when it comes to ecotourism policy making (see Table 15.1 for examples of roles and Table 15.2 for activities). The variety of

Table 15.1. Range of tourism policy roles in LAC, USA and Canada.

Actively involved	Advise	Approve projects
Adapted/changed agency structure	Advocate policies	Assist in writing policy
	Answer to a commission	
Administer		
Central point for tourism policy formulation and administration	Conferences, workshops/courses	Create and write policy
	Consult	Created tourism management legislation for national parks
Collaborate with other government agencies	Consulted by government on matters related to tourism	
		Created tourism police; legal protection
Collaborate with others outside of government	Contribute ideas to government	
Commission studies		
Communicate with other departments	Coordinate with others	
Deeply involved in policy	Depends on issue	Develop projects
Depend on other government agencies to implement	Determine study areas	Draft legislation
	Develop legislation	
Environmental impact studies	Establish/review/monitor policy	Evaluate policy proposals from other agencies
Facilitate	Follow policies set by other agencies	Formulate and implement national tourism law/development plan
Facilitate collaborators and stakeholders	Formed/attend councils/taskforce	Fund policy writing
Find solutions to tourism issues	Formed/member of commissions, committees	Fund projects and studies
Follow governor's policy		

Table 15.1. *Continued*

Give input in product development issues	Guide training	Get people involved
Handle the negotiating process		
Implement policy	Initiate tourism policy	Involved in land use planning
Information/technical assistance provider	Inspection and review of projects	Involved in licensing
Initiate policy		
Lead/key role in policy formulation	Licences and standards for operators	Lobby government/pressure legislature
Mission is to help private sector growth	Monitor policy	
No role/not involved	Not involved in tourism policy	
Obtained funding for master plan	Official agency for tourism policy	Open policies to public comment
Participate in policy process	Produce development plans	Provide information/advice/ technical assistance
Partner with others	Promote and market	
Play lead/key role	Promote tourism in harmony with traditional culture	Provide input on policy
Primarily involved in promotions and marketing	Propose policy	Provide logistical help to all sectors
React to policy	Represent tourism in policy issues	Research policy
Recommend/suggest policy		Review environmental impact studies for new tourism projects
Regulate, govern and evaluate tourism policies	Represent tourism interests in policy issues	
Serve as mediator	Set tourism policy for entire country	Strategic/tourism planning
Serve on committees		Study environmental issues
	Set visitor entrance fees	
Tourism zoning		
Use consultants		
Work with stakeholders	Write/formulate policy	

Table 15.2. Range of ecotourism activities of government tourism agencies in USA, Canada and LAC.

Carry out/participate in clean up campaigns	Educate tourists, schools and local people
Fund/carry out research and studies	Provide information
Have formed taskforces/councils/ associations	Have identified need for standards and certification
Have prepared ecotourism manual	Have prepared guidelines and codes of ethics
Offer grant programmes	
Give talks	Involved in strategic planning for ecotourism
Lobby government for tourism/natural concerns	License ecotourism businesses
Prepare brochures, videos, maps and posters	Offer courses and training
	Promote and market ecotourism
Promote ecotourism activities	Promote partnerships and associations
Promote/support parks and natural areas	Provide technical assistance
Work to make tourism industry more environmentally friendly	Involved with financing and investing in ecotourism projects
Involved in establishing/maintaining protected areas and ecotourism sites	Participate in/host ecotourism related conferences and workshops
Provide economic assistance to communities and entrepreneurs	Work with other agencies, NGOs, projects, industry, private sector, local communities
Survey travellers on ecotourism	

items shown in these tables verifies the continuum of roles that interviewees mentioned, ranging from a more reactive position in which agencies provide input and react to policies established by other agencies, to a proactive position in which governmental tourism agencies are researching, writing, lobbying for and implementing ecotourism policy. More specifically we found the following:

● Most of the governmental tourism agencies contacted are aware of, and often engaged in, some form of ecotourism policy. However, relatively few have developed and officially adopted specific ecotourism implementation objectives, plans or programmes.

● Nearly three-quarters (72%) of governmental tourism agencies in LAC indicated that they have a legal mandate to initiate tourism policy. This clearly indicates that these are the agencies that are responsible for tourism policy in LAC. It is interesting to note that although nearly a third (30%) of governmental tourism agencies in USA and Canada claim a legal mandate to initiate tourism policy, slightly more than a third (35%) do not claim to have a legal mandate. A clear and explicit mandate to initiate tourism policy may be a key to getting these established tourism agencies more involved in the development of ecotourism policy.

● Agencies in all regions mentioned that they are involved in numerous types of ecotourism policy roles. These include being involved in land use planning, monitoring policy implementation, providing input on complementary policy such as conservation and economic development policy, partnering with other government agencies, providing technical assistance, and commissioning plans and studies. These findings show that governmental tourism agencies throughout the Americas play many different roles in tourism policy making, and that no standard or uniform role for them to develop and deliver ecotourism policy has emerged.

- Throughout the Americas many types of ecotourism policy documents exist or are under development. Examples of these include government reports, conference proceedings, discussion documents, legislation, tourism development plans, marketing plans, strategies, studies, surveys, fiscal reports, promotional literature and strategic planning reports. These findings demonstrate that there are usually multiple sources that describe and contain the ideas that underlie and the regulations that define a governmental agency's tourism policy. While this range of documents contains rich information, it is equally clear that creating a coherent, integrated ecotourism policy programme is a challenge.
- The diversity of ecotourism activities in which governmental tourism agencies are involved ranges from offering grant programmes and promoting ecotourism to preparing and implementing standards for certification of ecotourism entrepreneurs.

This array of activities documents the fact that tourism agencies may not have a written policy concerning ecotourism, yet may be very involved in other aspects of ecotourism policy.

Types of policy documents

The documents that were provided (see Table 15.3) included a broad diversity of types of documents. However, only those that met our criterion of having written ecotourism policy directives were analysed. These generally fell into the following categories: legislation, tourism plans, reports/discussion documents and speeches.

Legislation

Although legislation (e.g. draft legislation, laws, policy directives) is usually thought of as the most comprehensive and insightful example of ecotourism policy, in fact the depth of examples and range of approaches varied tremendously. For example, the laws provided by Yukon, El Salvador and Peru (Government of Yukon, 1997; El Salvador Diario Official, 1996; Resolución Suprema No. 0065–92-MITINCI, 1992) were quite limited, and if they did cover the themes from the model cat-

Table 15.3. Types of ecotourism policy documents provided in LAC, USA and Canada.

CD-ROMs	Conference and workshop
Discussion documents	proceedings
Ecotourism policy directives	Draft legislation
E-mail communication	Ecotourism studies
Fiscal reports	Fax communication
Government reports	Government policy documents
Issue papers	Investment literature
Letter communication	Legislation
National ecotourism plan	Magazine articles
Project financing proposal	Nature tourism survey
Proposals	Promotional literature
Tourism development plans	Strategic planning reports
Tourism policy statement	Tourism marketing plans
Tourism studies	Tourism strategies
Video	Travel guides
Working documents	White papers

egories, often it was with just one or two short and general phrases. Other legal documents, such as Bolivia's General Tourism Law and Brazil's Directives for a National Ecotourism Policy (Bolivia Reglamento General de Turismo, 1997; Grupo de Trabalho Interministerial MICT/MMA, 1994), provided excellent examples of in-depth and comprehensive ecotourism policy, yet covered most of the elements of the definitions-based model of ecotourism. However, when tourism legislation focuses only on large-scale tourism with a minor emphasis on ecotourism, or focuses on a limited segment of ecotourism (for example, Yukon's Wilderness Tourism Licensing Act 1997), then the policies tend to be far less involved and comprehensive. We consider legislated tourism laws to be one of the most long-term commitments to ecotourism policy. Legislation also indicates a relatively high level of government commitment, at least so far as is necessary to pass laws.

Tourism plans

Overall, tourism plans (e.g. ecotourism plans, tourism development plans, marketing plans, tourism strategies) proved to be the most comprehensive and detailed ecotourism policy statements. For example, they often include specific action plans and focus areas, both thematic and geographical. The development of tourism plans, like legislation, indicates a relatively high level of government commitment, particularly due to the investment made by the tourism agency and cross-sectoral involvement necessary for their development.

Reports and discussion documents

We analysed a broad range of reports and discussion documents (e.g. conference and workshop proceedings, government reports, white papers, discussion papers, and informal e-mail and fax communications) and found a tremendous range in detail, comprehensiveness and quality. Some discussion documents provided extensive historical and legislative background as well as action plans for the future and often included an in-depth exploration of themes included in the definitions-based ecotourism model. Others were lacking in specificity, and were general and unclear. While discussion documents are useful as a foundation for an agency to prepare ecotourism policy in that they provide essential background and often insightful suggestions, they are generally prepared by an individual or a small group within an agency, and may not yet have attained 'organizational buy-in' and broad governmental acceptance. However, our findings suggest that such documents are an important step and foundation in a policy process that could result in ecotourism legislation or plans.

Speeches

Several government tourism agencies provided written copies of speeches that had been given by their respective ministers or directors. Unlike other written policy materials, speeches usually addressed concepts in a general and limited manner. They tended to be short, and provide only an overview of policy direction. This is no surprise because speeches are usually constrained by time and are political in nature. The primary use of a speech in regard to ecotourism policy is that it sets the tone and general approach, as well as demonstrates the political support by the highest political levels for the establishment or continuation of ecotourism policy.

Range of tourism policy roles

Government tourism agencies' involvement (Table 15.1) in the policy process ranged from a more reactive, passive role in which agencies provide input and react to policies established by other agencies, to a proactive role in which government tourism agencies are researching, writing, lobbying for and implementing tourism policy. The agencies that strive for a more active role in the policy creation process might find benefit in contacting their counterparts that are already playing important roles in tourism policy in other state, provincial or national tourism agencies.

Conclusion

Policy makers in the Americas are actively considering and tailoring a vision and definition for ecotourism that relates to their perceived needs, and are engaging in a broad range of ecotourism development activities. However, with a few notable exceptions, there is still a lack of clearly defined ecotourism policy throughout the region.

In the Americas, ecotourism is clearly receiving attention and action at the governmental level. Ecotourism policy is being represented in a number of ways, from speeches and discussion documents to tourism plans and legislation. However, the number of states and countries that have reached the point of developing ecotourism policy is still limited, and the quality and breadth of these policy documents varies tremendously. There are a number of good examples that we came across, and we encourage policy makers to contact their neighbours to learn from their experience.

One area in which we find a clear link in establishing ecotourism policy was if government had clearly defined ecotourism for their agency, state or country. Defining ecotourism is a necessary first step in the ecotourism policy development process, as this definition provides the vision and framework for a common language and goals that need to be attained.

Although definitions of ecotourism in the Americas and in the literature varied widely, three key elements were common enough to warrant mention. Specifically, ecotourism, regardless of its scale, ought to produce at least three objectively verifiable outcomes: (i) it is a positive force for conservation (emphasizing protection and perpetuation of the very landscapes and features that attract the tourists), (ii) it benefits host communities economically and ensures that the people who must endure the social and environmental impacts of tourism development also share in its rewards), and (iii) it promulgates environmental awareness both among tourists and local communities. Thus, ecotourism may be distinguishable from other so-called 'forms' of tourism in its ideological orientation, more so than in any tangible qualities it might have.

Another key step is the establishment of a mandate for government tourism agencies to engage in policy creation. If these agencies are unable to take on the responsibility for influencing or establishing policy for ecotourism, then it is unclear who will.

In order for ecotourism to achieve its full potential for conserving natural and cultural systems, for providing economic and community development, and for providing positive experiences and education for both visitors and hosts, government tourism agencies in the Americas need to take on the issue of ecotourism and develop a vision, definitions, legal mandates, legislation and tourism plans. This process must include all relevant stakeholders in a meaningful way. Only when we have the benefit of a clear and shared vision, and a plan for how to get there, will we begin to obtain the many possible benefits of ecotourism.

Acknowledgements

The Organization of American States (OAS), Washington, DC, is gratefully acknowledged for providing partial funding for this research.

References

Bolivia Reglamento General de Turismo. (1997) Decreto Supremo No. 24583 de Fecha 25 de abril de 1997. Ministerio de Desarrollo Económico; Secretaria Nacional de Turismo. Bolivia Paso a Paso.

Edwards, S., McLaughlin, W. and Ham, S. (1999) *Comparative Study of Ecotourism Policy in the Americas: 1998*, Vols I–III. Contribution #872 of the Idaho Forest, Wildlife and Range Experiment Station, College of Natural Resources, University of Idaho. Organization of American States and the University of Idaho. Available online at: www.oas.org/tourism (also available in Spanish).

El Salvador Diario Official (1996) Decreto No. 779. Tomo No. 332 Ley de la Corporación Salvadoreña de Turismo.

Erlandson, D.A., Harris, E.L., Skipper, B.L. and Allen, S.D. (1993) *Doing Naturalistic Inquiry – a Guide to Methods*. Sage Publications, Newbury Park, California.

Fennell, D.A. (1999) *Ecotourism – an Introduction*. Routledge, New York.

Government of Yukon (1997) DRAFT Wilderness Tourism Licensing Act. Draft Government Policy Document. Whitehorse, Yukon.

Grupo de Trabalho Interministerial MICT/MMA (1994) Diretrizes para uma Política Nacional de Ecoturismo/Coordenaçao de Sílvio Magalhaes Barros II e Denise Hamú M. de La Penha. Embratur, Brasília.

Honey, M. (1999) *Ecotourism and Sustainable Development: Who Owns Paradise?* Island Press, Washington, DC.

Resolución Suprema No. 0065–92-MITINCI (1992) Reglamento de Organización y Funciones del Ministerio de Industria, Turismo, Integración y Negociaciones Comerciales Internacionales. Resolución Suprema No. 0065–92-MITINCI. 12 December 1992, Lima, Peru.

Steinberger, P. (1995) Typologies of public policy: Meaning construction and the policy process. In: McCool, D. (ed.) *Public Policy Theories, Models, and Concepts, an Anthology*. Prentice-Hall, Englewood Cliffs, New Jersey, pp. 220–233.

WTO (World Tourism Organization) (1997) Info from official tourism organizations – The Americas. Available at: www.world-tourism.org/americas.htm

Ecotourism Policies and Issues in Antarctica

16

Thomas Bauer[1] and Ross Dowling[2]

[1]*School of Hotel and Tourism Management, The Hong Kong Polytechnic University, Hung Hom, Kowloon, Hong Kong, SAR, China;* [2]*School of Marketing, Tourism and Leisure, Edith Cowan University, Joondalup, WA 6027, Australia*

Introduction

International tourist arrivals reached new heights during 2000 with an estimated 700 million arrivals around the globe. While many destinations are increasingly struggling to cope with the volume of visitors there is still one continent that is hardly visited at all – Antarctica. The continent was first sighted in 1820 but until 1894, no human had set foot on it. Today the continent is dotted with some 40 year-round scientific research stations, and tourist ships regularly visit the South Shetland Islands and the west coast of the Antarctic Peninsula as well as the Ross Sea. Between late November and early March, when ice conditions are less severe, the region becomes the setting for the world's most remote tourism operations. Tourists can also view the Antarctic grandeur through the windows of a Qantas aircraft on sightseeing flights from Australia and the very intrepid have the opportunity to trek to the geographic South Pole.

Mirroring the increase in the interest in nature-based tourism activities among the travellers of the world, the number of voyages to Antarctica increased significantly during the last decade of the 20th century. Antarctica is an interesting case study in the context of tourism policy making because its unique political situation prevents the usual procedures for tourism policy making from coming into play. The region is governed by the Antarctic Treaty System which, in Recommendation VIII-9/1975, acknowledges that 'tourism is a natural development in this Area and that it requires regulation' (Heap, 1990: 2602). Some environmental organizations have called for the establishment of Antarctica as a

'World Park' which would allow controlled tourism activities. In contrast, in Australia, the Wilderness Society and the Australian Conservation Foundation (ACF) have policies that would prohibit any tourism from taking place in Antarctica. This chapter outlines the setting in which tourism takes place, analyses the current management regime and investigates possible new directions for a sustainable tourism policy in the south.

Antarctica

Antarctica is a continent surrounded by the cold waters of the Southern Ocean (Fig. 16.1). It is centred on the geographic South Pole and covers 14 million km², nearly twice the size of Australia. It has the distinction of being the highest, coldest, windiest, driest and remotest of all the continents and its ice cover makes up 90–95% of the world's fresh water reserves. Temperatures can drop as low as −89.6°C as measured in July 1983 at Vostok in the Australian Antarctic Territory. As Sir Douglas Mawson discovered at Commonwealth Bay, wind gusts of over 200 miles an hour (320 km h⁻¹) are not uncommon. The interior of the continent is a lifeless, high altitude polar desert that receives only half the annual precipitation of Australia's driest locations. During the brief summer months the few ice-free coastal areas (less than 2% of Antarctica is not covered by ice) are home to a profusion of wildlife including penguins, flying seabirds and seals. It is here that tourists go ashore to marvel at the beauty of the place.

While Antarctica's natural history is many millions of years old (until 50 million years ago it was part of the great southern continent of Gondwana) its human history is much shorter. In the late 18th century, Captain James Cook circumnavigated the continent without sighting land. In the 19th century explorers such as James Clark Ross, Bellingshausen and Dumont d'Urville explored and mapped the region further. The first temporary human settlement did not take place until 1899 when Carsten Borchgrevink and his party spent a winter in two prefabricated wooden huts at Cape Adare. Many of the early Antarctic achievements have been overshadowed by the later endeavours of explorers like Roald Amundsen, Ernest Shackleton and Captain Robert Falcon Scott, and Sir Douglas Mawson. The race to the geographic South Pole in 1911/1912 as well as Shackleton's tribulations during his *Endurance* expedition have become the stuff of legends. Special events such as an Antarctic exhibition at the National Geographic Society in Washington, DC, keep the interest of the public in Antarctic affairs alive and contribute to an increase in interest in visits to the region.

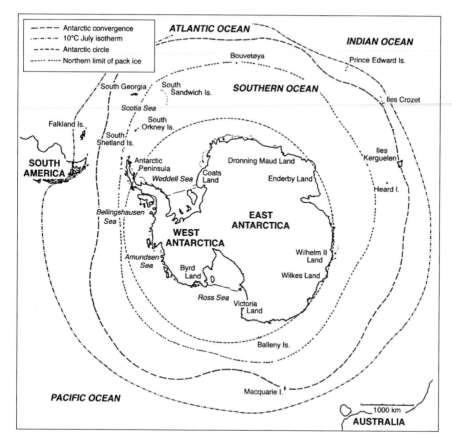

Fig. 16.1. Map of the Antarctic.

In contrast to all other continents, Antarctica lacks an indigenous human population, which adds to the uniqueness of the destination in two ways. Tourists cannot visit local markets, shopping centres, museums, cathedrals, hotels or pubs and there is no local population that could benefit economically from the visits of tourists.

Tourism in Antarctica

Antarctica is the only continent that is only accessible to the general public by joining an organized tour. Commercial tourism began in 1958 when two tourist voyages aboard an Argentinian vessel took place. The era of regular ship visits did not begin until 1966 when Lars Eric Lindblad of Lindblad Travel, New York, took passengers to the Antarctic Peninsula. During the 1970s an estimated 14,328 visited the continent by ship. This figure increased only slightly during the 1980s when 15,209 paying tourists went south. During the 1990s the situation changed substantially and by the end of the decade 84,173 ship-based passengers had visited the Antarctic. The end of the 20th century made the 1999/2000 season the most popular ever with 14,623 tourists making the trip south (Table 16.1).

As a consequence of the number of tourists visiting the Antarctic there has been an attendant increase in the number of sites visited (Fig. 16.2). This has risen from 36 in 1990 to nearly 200 just 7 years later (National Science Foundation, 1997). The main reason for the increase in ship-based tours during the 1990s was the ready supply of Russian research vessels. These had become available at reasonable charter rates after the Government of Russia could no longer afford to fully fund the oceanographic research activities of its research institutes. This in turn forced institutes such as the P.P. Shirov Insititute of Oceanology in Moscow to look for outside funding and when tour operators approached them with the idea to use these ice-strengthened vessels in the Antarctic tourist trade they were willing to do business. Today Russian flagged vessels make the majority of voyages to Antarctica (Table 16.1).

Impacts of Antarctic Tourism

Much has been written and said about the impacts of tourists on the Antarctic environment and tourism has frequently been portrayed in the popular media as an activity that is threatening the Antarctic environment. Antarctica is still a relatively pristine area but when discussing tourism impacts it should be kept in mind that the continent has not always had the level of protection it enjoys today. The exploitative activities of the early sealers and whalers as well as the building and operation of some of the scientific stations have previously despoiled the Antarctic environment. Today there is no argument that the beauty of Antarctica must be preserved but some caution is needed before tourists are blamed for a perceived deterioration of the environment of the region.

Antarctica is the least disturbed continent and the Australian Government in particular is working towards keeping it that way. They argue that growth in the tourism industry brings new challenges for wildlife, increasing the potential for disturbance and disease introduction. To overcome some of these problems they suggest that remediation of contaminated sites, removal of wastes and disused buildings, prevention of exotic species and disease introduction, and use of alternative energy systems should be considered by all national Antarctic programmes. The Australian research programme provides a

Table 16.1. IAATO preliminary estimate of Antarctic tourism, 2000–2001 (based on information provided by Antarctic tour operators to the IAATO Secretariat as of 28 August 2000).

Vessel	Charterer	Passenger capacity	Probable no. voyages	Estimated ave. load	Probable no. passengers
Peninsula voyages					
Explorer	Abercrombie and Kent/Explorer Shipping	96	6	80	480
Explorer	Victor Emanuel Nature Tours	96	1	80	75
Professor Molchanov	Aurora Expeditions	52	10	45	450
Kapitan Dranitysn	Quark Expeditions	100	9	90	900
Professor Multanovskiy	Quark Expeditions	52	9	40	360
Professor Multanovskiy	Oceanwide Expeditions	52	1	25	25
Clipper Adventurer	Clipper Cruise Line/New World Ship Management Co LLC	122	7	110	770
Clipper Adventurer	Zegrahm Expeditions Inc.	122	1	115	115
Hanseatic	Hapag-Lloyd	180	8	150	1200
Bremen (note the Continental trip mentioned below starts in the Ross Sea and ends in the Peninsula, plus there is one Peninsula voyage)	Hapag Lloyd	164	1	140	140
Caledonian Star	Lindblad Expeditions	110	6	100	660
Akademik Ioffe	Peregrine Adventures	117	11	70	770
Mariya Yermolova	Marine Expeditions	120	7	100	860

Continued

Table 16.1. Continued

Vessel	Charterer	Passenger capacity	Probable no. voyages	Estimated ave. load	Probable no. passengers
Mariya Yermolova	Cheeseman Ecology Safaris	120	1	87	87
Lyubov Orlova	Marine Expeditions	120	11	100	1320
Grigory Mikheev	Oceanwide Expeditions	36	9	29	261
Vista Mar	Plantours and Partner	280	3	260	780
S/Y Pelagic	Pelagic Expeditions	6	2	6	12
S/Y Golden Fleece	Golden Fleece Expeditions	10	3	10	30
Continental voyages					
Bremen	Hapag-Lloyd	164	1	130	130
Kapitan Khlebnikov	Quark Expeditions	108	3	100	300
Akademik Shokalskiy	Heritage Expeditions	46	2	46	92
Non IAATO members					
Marco Polo (Peninsula)	Orient Lines	800	6	490	2940
S/Y Sir Hubert Wilkins (Continental)	Ocean Frontiers Pty	20	2	18	36
Land-based programmes	Adventure Network Int'l				200
Yachts (~16)	Various		22	9	200
Totals			142		13,193

Table 16.2. Indicators of tourist impacts (after Davis, 1999).

Factors	Indicators
Crowding	Site ship numbers
	Site visitor numbers
Human impacts	Rubbish from old bases
	Litter on beaches
	Trampling of vegetation cover
Wildlife disturbance	Animals being displaced
	Birds leaving nest site
	Seals leaving haul-out site
	Egg snatching by predators

scientific basis for developing protocols and guidelines for use by environmental managers. Before any proposed activity commences in the Australian Antarctic Territory the proponent must prepare an environmental impact assessment (EIA) and submit it to the Australian Antarctic Division. This requirement also applies to any Australian anywhere in Antarctica. Such activities are assessed at one or more of three levels from a preliminary assessment to a comprehensive environmental evaluation.

Potential negative environmental impacts of tourism are raised in the literature but, frequently, little substantiation of arguments is being provided. When discussing the potential impacts of tourism the scale of these impacts has to be kept in perspective. For example, the

Fig. 16.2. Mountains and icebergs of the Antarctic Peninsula. This is the scenery that ecotourists have come to view (Ross Dowling).

fact that Antarctica is nearly twice the size of Australia is rarely mentioned in such discussions. Headland (1994: 279) puts Antarctic tourism activities into perspective when he points out that the effects of the tourist industry on the Antarctic may be estimated as 0.52% of the total human impact. The other 99.48% can be attributed to scientists and their support staff. It is also sometimes overlooked that Antarctic tourism is highly concentrated at several high-profile sites in the Antarctic Peninsula region and that the rest of Antarctica is practically never visited by tourists. This concentration of tourism activities raises questions over the potential for over-visitation of certain sites but one should not infer from this that all of Antarctica is under threat from tourist visits.

As Bauer (2001) notes, the abundance of many species of Antarctic wildlife is also often ignored. Chester (1993) quotes the Scientific Committee on Antarctic Research as estimating the populations of Antarctic wildlife as: 1.07 million pairs of breeding king penguins; 2.47 million pairs of Adelie; 7.49 million pairs of chinstraps; 314,000 pairs of gentoo; 3.68 million pairs of rockhoppers; 11.8 million pairs of macaronis and 195,000 pairs of emperor penguins. The same source estimates the number of seals as: 250,000–800,000 Weddell; 200,000 Ross; 30–70 million crabeater seals (half the world's pinniped population); 200,000–440,000 leopard seals; 600,000 southern elephant seals; and over 2 million Antarctic fur seals. These figures indicate that, unlike other prime wildlife destinations such as the Galapagos Islands, Antarctic wildlife populations are substantial.

Environmental Impacts

In a *Time* magazine article written in the late 1990s it is suggested that 'Antarctica-philes are at odds over the best way to protect the white continent from camera wielding sightseers' (Feizkhah, 1998: 40). Feizkhah argues that some environmentalists and government officials are uneasy about whether the Environmental Protocol is the most appropriate way to control tourism. On the other hand, she quotes Stanislaw Rakusa-Suszczewski, a marine biologist who heads Poland's Academy of Science's Department of Antarctic Biology and Henryk Arctowski station, who suggests that the notion that tourists threaten the environment is nonsense compared with illegal fishing in Antarctic waters.

Enzenbacher (1991: 90) sees the potential impacts of tourism as disruption of scientific programmes, trampling of mosses and lichens and disturbance of wildlife. Beck (1994: 380) points out that: 'All human activities in Antarctica, whether conducted by scientists, tourists, or others exert environmental impacts'. Erize (1987: 134) is of the opinion that Antarctic cruising and small boat operations have only a negligible impact on the environment. But after this positive note he warns that: 'It is in visiting natural areas that tourism may produce a considerable impact on wildlife. Careless tourists may easily disturb breeding colonies, of seabirds in particular, with the result that scared parent birds may temporarily desert chicks or eggs causing them to die of exposure'. He also notes that tourists may trample on scarce and fragile vegetation.

Stonehouse (1993: 331), writing about the results of the visitor/wildlife monitoring programme carried out under Project Antarctic Conservation, highlights the fragility of the Antarctic flora. He states that:

> Experiments in restricted areas confirmed quantitatively the extreme sensitivity of moss and lichen communities to even low incidence of trampling, indicating the need for strict visitor management in places where vegetation is at risk, and the need for further studies on the nature of trampling damage and possibilities for rehabilitation.

He also notes that passengers ashore are well behaved and points out that he has yet to see one 'drop litter, knowingly trample vegetation or interfere seriously with wildlife' (Stonehouse, 1994: 202).

Supporting this view, the Australian Antarctic Division's field equipment and training officer, Rod Ledingham, states 'The environmental impact [of tourism] is minuscule compared with that of long-term expeditions in national operations' (Whelan, 1996: 86) and he adds that 'the population of the region's government bases average ... about 200 times that of the tourists, none of whom drives a vehicle, eats, sleeps or excretes on the continent' (p. 87). Whelan (1996: 87) concludes that 'I believe that both scientists and tourists have the right to visit Antarctica, but everyone should adhere to the rigorous environmental standards required to protect our planet's last great wilderness'. These comments are in line with the observations of the authors who have participated in nine voyages to Antarctica as either guides and/or lecturers (Fig. 16.3).

In the *Greenpeace Book of Antarctica*, May (1988: 138–139) points out that tourism may put additional pressure on the natural environment and that 'fragile vegetation could easily be destroyed, and nesting and breeding grounds disrupted. Tourists could unwittingly spread bird or plant diseases and introduce new kinds of organisms to the Antarctic'. Hall (1992: 5) is of the opinion that 'undoubtedly, the most serious concerns surrounding tourism in Antarctica are focused on the potential impacts of tourism on the fragile Antarctic environment'. The Strategy for Antarctic Conservation produced by the IUCN (World Conservation Union) (1991: 55–56) takes a positive approach to the impacts of tourism on Antarctica. The strategy points out that:

Fig. 16.3. The Argentinian Research Station Almirante Brown in Paradise Bay (Ross Dowling).

'Tourism offers both benefits and threats to Antarctic conservation'. It lists the benefits as follows:

1. Visitors gain a greatly enhanced appreciation of Antarctica's global importance and of the requirements for its conservation.
2. Visits bring fulfilment to those seeking personal challenge and wilderness adventure.
3. Scientific activities may also benefit, since tourist visits can provide a useful link with the outside world and strengthen political support for Antarctic science.

The IUCN list of potentially undesirable impacts includes the following:

1. Disturbance at wildlife breeding sites.
2. Trampling of vegetation.
3. Disruption of routines at stations and of scientific programmes.
4. Environmental hazards of accidents and, resulting from them, time-consuming and costly search and rescue as well as environmental clean-up operations.

The strategy concludes that, in general, tourist operations have been conducted in a responsible manner and undesirable impacts have not been severe, especially when compared with the environmental impacts of scientific and associated logistical activity.

Research is needed to determine the full impact of tourism on the Antarctic environment, and on the activities of the scientific bases in order to establish the maximum carrying capacity of the area in terms of tourist numbers (Thomas, 1994).

Antarctic Tourism Policies

Unlike any other major 'landmass', Antarctica is not owned by anyone and hence many of the complexities of Antarctic tourism have their origins in this unique legal and political situation. Argentina, Australia, Chile, France, New Zealand, Norway and the United Kingdom all lay claim to parts of the continent but their claims to territorial sovereignty are not universally recognized. In particular the United States and Russia, two of the major countries in the discovery and exploration of Antarctica, do not lay or acknowledge any claims to Antarctic territory. To prevent international conflicts from erupting over the issue of ownership, the 12 countries that had established scientific bases in Antarctica during the highly successful and cooperative International Geophysical year 1957/1958 negotiated the Antarctic Treaty that came into force on 23 June 1961. By 2001, 44 countries had become parties to the Treaty.

The parties meet on an annual basis to discuss Antarctic issues and to make recommendations to their governments pertaining to the management of Antarctic affairs. The 44 Antarctic Treaty Parties represent an estimated 80% of the world's population. The linchpin of the Treaty is Article IV, which recognizes that the question of territorial sovereignty cannot be solved. It notes that the Treaty does not recognize, dispute or establish territorial claims and that no new claims shall be asserted while the Treaty is in force. The provisions of the Treaty apply to the area south of 60° latitude, including all ice shelves. The initial Treaty did not include any specific reference to tourism. However, over the years, several conventions and protocols were developed that affect the way tourism is carried out in the Treaty area.

Conventions, protocols and recommendations under the Antarctic Treaty System

Since coming into force in 1961 the Antarctic Treaty has developed from a single instrument into a system of conventions, annexes and recommendations. Collectively these are known as the Antarctic Treaty System (ATS). The *Handbook of the Antarctic Treaty System* (Heap, 1990) provides a comprehensive overview of the ATS. The following recommendations and conventions are seen as relevant in the context of tourism.

1. The Agreed Measures for the Conservation of Antarctic Fauna and Flora

These measures, annexed to Recommendation III–VIII, were adopted in 1964 by Antarctic nations in recognition of the importance of Antarctic conservation. 'The measures provide for overall protection of native animals and plants while establishing a system of managed protected areas' (Australian Antarctic Division, 1995: 23). The *Agreed Measures* are of importance for tourism because they influence where tourists can go and how they are required to behave while in Antarctica.

2. Convention on the Conservation of Antarctic Marine Living Resources (CCAMLR)

Through CCAMLR, the Treaty Parties afford protection to Antarctic marine living resources, including krill (*Euphausia superba*), one of the most important animals in the ecology of Antarctic seas. By protecting krill the Treaty Parties provided for the sustained supply of food to Antarctic marine mammals such as whales and seals, as well as for the various species of seabirds, including penguins. In the tourism context CCAMLR is significant because it ensures the survival of the bountiful Antarctic wildlife, one of the major attractions for tourists.

3. Convention for the Conservation of Antarctic Seals

This convention was accepted in 1972 to provide a means for the regulation of commercial sealing should it ever resume. The protection of the seals is of relevance to tourism because, like seabirds, seals form an important attraction for visitors.

4. Protocol on Environmental Protection to the Antarctic Treaty

The protocol was negotiated during 2 years of special meetings among the Antarctic Treaty Consultative Parties following the failure of the negotiations to establish the Convention on the Regulation of Antarctic Mineral Resources Activities. It was signed on 4 October 1991 in Madrid and entered into force on 14 January 1998. The Australian Antarctic Division (1995: 23) summarizes the key provisions of the Madrid Protocol as follows.

The Protocol places an indefinite ban on mining or mineral resource activity in Antarctica, designating the Antarctic as a natural reserve devoted to peace and science. It provides a multinational, codified set of environmental standards (Antarctica is the only continent for which this applies), and creates a new system of protected areas. The Protocol establishes environmental principles for the conduct of all activities, which must be assessed for their potential environmental impact before they are undertaken, and provides guidelines for conservation of Antarctic flora and fauna, managing and disposing of waste, and preventing marine pollution.

The Protocol has significant ramifications for the conduct of commercial tourism in Antarctica. Article 3 establishes the environmental principles that are at the core of the Protocol. These are elaborated and operationalized in Articles 1–8 in Annex I and describe the requirements for environmental impact assessments (National Science Foundation, 1995: 190, 202–205). Any activity that has more than a 'minor or transitory impact' is subject to the completion of an Initial Environmental Evaluation (IEE) and, where appropriate, a Comprehensive Environmental Evaluation (CEE) (Articles 2 and 3, Annex I of the Protocol). This means that prior to the start of the season, Antarctic tour operators have to submit their tour itineraries for the forthcoming season to the relevant government body (in Australia this is the Australian Antarctic Division) for approval. If they can show that their activities have no more than a transitory impact on the environment the approval will be granted. The Protocol is the single most important instrument regulating present and future Antarctic tourism activities, and potential future tourism developments such as air-supported, land-based tourism operations will have to meet environmental criteria established by the Protocol.

Citizens of all countries that are parties to the Antarctic Treaty are also subject to their country's national legislation governing their conduct in Antarctica. For example, in the United States the Marine Mammal Protection Act of 1972 prohibits US citizens from taking or importing marine mammals, or parts of marine mammals, into the US. Both accidental and deliberate disturbance of seals or whales may also constitute harassment under the Act. Further, the Antarctic Conservation Act of 1978 (US Public Law 95–541) was adopted by the US Congress to protect and preserve the ecosystem, flora and fauna of the continent, and to implement the Agreed Measures for the Conservation of Antarctic Fauna and Flora. The Act is legally binding for US citizens and residents visiting Antarctica and makes it unlawful to take native animals or birds, to collect any special native plant, to introduce species, to enter certain special areas (SPAs), or to discharge or dispose of any pollutants. Of special significance to tourists is the interpretation of 'take' which includes to remove, harass, molest, harm, pursue, hunt, shoot, wound, kill, trap, capture, restrain, or tag any native mammal or native bird, or to attempt to engage in such conduct. Under the Act, violations are subject to civil penalties, including a fine of up to US$10,000 and 1 year imprisonment for each violation.

Governments that are signatories to the Treaty can control the activities of their own citizens in Antarctica under their own laws devised within the Treaty. For example, New Zealand does not permit tourists to enter specified historic huts except in the presence of a government appointed warden (Stonehouse, 2001).

Industry self-regulation: IAATO

Antarctic tour operators have responded to the need to preserve the Antarctic environment by establishing their own industry association, the International Association of Antarctica Tour Operators (IAATO), and by developing their own code of conduct, which attempts to minimize the impact visitors have on the environment. The IAATO web site (www.iaato.org) gives the purpose of the organization as: 'A member organization founded in 1991 to advocate, promote and practice safe and environmentally responsible private-sector travel to the Antarctic'. Because of the remoteness of the continent, the enforcement of any government-imposed regulatory restriction is difficult and hence effective self-monitoring by operators is the best way to manage Antarctic tourism.

Travellers aboard IAATO member expeditions are reminded of the following regulations developed by the tour operators (see Fig. 16.4):

1. Do not disturb, harass or interfere with the wildlife.
2. Do not walk on or otherwise damage the fragile plants.
3. Leave nothing behind, and take only memories and photographs.
4. Do not interfere with protected areas or scientific research.
5. Historic huts may only be entered when accompanied by a properly authorized escort.
6. Do not smoke during shore excursions.
7. Stay with your group or with one of the ship's leaders when ashore.

In addition, IAATO guidelines for the Conduct of Antarctica Tour Operators require that tour operators:

1. Are familiar with the Antarctic Conservation Act of 1978 (US Public Law 95–541) and that they abide by it.
2. Are aware that entry to Specially Protected Areas (SPAs) and Sites of Special Scientific Interest (SSSIs) is prohibited unless permits have been obtained in advance.

Fig. 16.4. Ecotourists at Brown Bluff, Antarctic Peninsula. While there are strict regulations on the distance to which visitors can approach the wildlife, no such law applies to the often inquisitive penguins, which, in this case, ended up pecking at the pockets of the tourist (Ross Dowling).

3. Enforce the above mentioned Guidelines for Antarctic visitors in a consistent manner.

4. Hire a professional team of expedition leaders, cruise directors, officers and crew 75% of whom should have prior Antarctic experience.

5. Hire Zodiac drivers who are familiar with driving Zodiacs in polar regions.

6. Educate and brief the crew on the IAATO Guidelines of Conduct for Antarctic Visitors, the Agreed Measures for the Conservation of Antarctic Fauna and Flora, the Marine Mammal Protection Act of 1972 and the Antarctic Conservation Act of 1978, and make sure they are consistently enforced.

7. Ensure that for every 20–25 passengers there is one qualified naturalist/lecturer guide to conduct and supervise small groups ashore.

8. Limit the number of passengers ashore to 100 at any one place at any one time.

The implementation and effectiveness of these guidelines or codes of conduct depend on the level of understanding that paying visitors have of these guidelines, their level of agreement with them and their willingness to comply. The other crucial factor is the willingness of the operator to enforce the guidelines. Given that passengers have paid a lot of money for the privilege of visiting the Antarctic, this is

not always easy. Accounts to date suggest that tourists and tour operators alike have so far complied with the set industry guidelines and
with the Antarctic Treaty recommendations established by the Treaty
Parties to protect the environment. Beginning with Lars-Eric Lindblad
and continuing with IAATO members, tour operators have been
proactive in their measures to protect the resource on which their
businesses depend. As a result of the cooperation between international tourists, tour operators and Treaty Parties, Antarctic tourism is
today the best managed tourism in the world and other destinations
can learn much from the way it is conducted.

Future Planning and Management

Key general issues identified for Antarctic tourism include tourist
numbers and carrying capacities, concentration versus dispersal,
policing and enforcement and land-based tourism (Plimmer, 1994).
Specific site-based issues could include tourist perception of crowding, adverse environmental impacts and disturbance of wildlife populations and these could be monitored through a range of relevant
indicators (Davis, 1999; Table 16.2).

The current methods of management of Antarctic tourism are
reactive and general and lack a comprehensive approach to tourism
management within a wilderness. Because of Antarctica's large distance from any formal law enforcement agencies, the successful management of nature-based tourism depends on the commitment of
operators and the goodwill of tourists to voluntarily comply with specific visitor guidelines (Bauer, 1999).

In the past, the effective management of Antarctic tourism has
been achieved through treaties and guidelines. These have been con-
ceptually useful but offer no practical advice on how to avoid disturbing wildlife (Davis, 1999). Davis suggests that environmental planning
and management models such as the Limits of Acceptable Change
(LAC) (Stankey *et al.*, 1985) and the Recreational Opportunity
Spectrum (ROS) (Clarke and Stankey, 1979) should be introduced to
manage the future of Antarctic tourism. Such an approach is predicated on the concepts of identifying an area's existing resource and
social conditions, recognizing the acceptability of change, establishing
indicators of change, introducing the zoning of areas for different
tourist uses, and ensuring the practice of ongoing monitoring. The
LAC model is issue driven and therefore it is guided by indicators that
show how the particular issue is affecting the site under review. This
model lacks direct utility in the Antarctic as there are no social conditions and as far as the impacts on wildlife are concerned the only
acceptable option is for there to be no negative impact.

A number of other planning approaches could be utilized in the Antarctic including VIM, the visitor impact management planning framework (Graefe *et al.*, 1990); TOMM, the tourism optimization management model (McArthur, 1996); VAMP, the visitor activity management process (Nilsen and Tayler, 1997); and VERP, the visitor experience resource protection planning framework (Hof and Lime, 1997).

Another planning approach which has been implemented in a number of natural areas of the world and which could be applied to ecotourism planning and management in the Antarctic is the Environmentally Based Tourism (EBT) planning model (Dowling, 1993). It determines tourism opportunities through the identification of significant features, critical areas and compatible activities. Significant features are either environmental attributes which are valued according to their level of diversity, uniqueness or representativeness, or tourism features appreciated for their resource value. Critical areas are those in which environmental and tourism features are in competition and possible conflict although in the Antarctic it should be noted that this would apply to all sites. Compatible activities are tourism recreational activities that are considered to be compatible both with the bio-physical environment as well as with the members of the tour groups.

The essential elements of the model include its grounding in the sustainable development approach; that is, it is based on environmental protection in order to achieve environment–tourism compatibility. Other essential elements are that it is strategic and iterative, regionally based, incorporates land use zoning and is environmentally educative; that is, it fosters the environmental ethic. In essence it provides policy makers and planners with a number of zones designed to protect conservation values while fostering tourism activities and environmentally sensitive developments. Such an approach has utility for Antarctic tourism as it would foster tourist zones, small areas of concentrated attractions; recreation zones, natural areas that can accommodate compatible outdoor recreation activities; conservation zones, areas sustaining a combination of protection and use but with emphasis on the former; and finally sanctuary zones, areas requiring special preservation and therefore not suited for general human visitation. In one respect this approach already exists with SPAs and SSSIs etc. Sites of Special Tourist Interest have also been proposed but as yet none have been gazetted.

The future management of tourism in the Antarctic should serve two main objectives: the avoidance of adverse environmental impacts and the enhancement of tourist experiences. To be acceptable to tourists, management intervention should be low-key and persuasive. Explanation and education through interpretation is the key to affecting tourist behaviour in ways considered to be environmentally and socially acceptable. Tourist management measures have been described as involving a spectrum of approaches from soft, to intermediate and hard (Jim, 1989). Soft techniques are aimed at influencing user behaviour, intermediate techniques focus on redistributing use and hard techniques are those which are regimented and aim at rationing use. Such a spectrum of tourist management approaches is offered for consideration and possible application at individual sites within the Antarctic (Fig. 16.5).

Conclusions

Setting aside the question of whether tourists should be allowed to visit the Antarctic at all, the future management of Antarctic tourism will be attained by stakeholders recognizing the relationships that exist between environmental quality, the ecotourism experience and the overall viability of the tourism industry (Bauer, 1999). In all of these relationships it is the spirit of cooperation between governments and tour operators that is central to achieving the goal of sustainable tourism operation in the Antarctic. This approach is an exemplar for other environmentally sensitive regions. While responsible self-regulation in relation to Antarctic tourism is desired, it cannot be relied upon entirely to protect polar regions (Stonehouse, 2001).

Fig. 16.5. An interesting juxtaposition of a tourist, penguins, the cruise ship *Kapitan Dranitsyn* and a cross to mark the disappearance of three UK scientists lost in 1982 from a nearby station formerly run by the UK but now operated by the Ukraine (Ross Dowling).

Education is the most powerful tool currently available to manage Antarctic tourism, and tourist guides on Antarctic cruises play a pivotal role in the interpretation of the natural world. Thus, they need to be permanently updating their own information (Thomas, 1994).

> In the absence of more formal control measures, the 'Lindblad pattern' has for long provided the environment's most effective shield against tourist-induced damage. Each voyage becomes an 'expedition' with lectures, briefings and shore landings. Lecturers are often scientists or administrators...with long experience of Antarctic affairs. Passengers are briefed on the Antarctic Treaty and issued with a set of guidelines...covering behaviour ashore, possible hazards, the need to avoid interference with wildlife, and other points of conduct.
>
> (Stonehouse and Crosbie, 1995: 222)

Thus it will be through a combination of education and regulation that Antarctic tourism will continue to grow in future. As the number of visitors increases, it is clear that so will the impacts. But so will the number of people who, given a taste of the Antarctic, will return home as informed, caring and active individuals committed to fostering the reality that the Antarctic is a truly special place which requires the consideration of all people.

References

Australian Antarctic Division (1995) *Looking South: the Australian Antarctic Program in a Changing World.* Australian Antarctic Division, Kingston, Tasmania.

Australian Antarctic Division (2001) *Human Impacts.* www.antdiv.gov.au/information/aboutus/division.asp

Bauer, Th.G. (1999) Towards a sustainable tourism future: lessons from Antarctica. In: Weir, B., McArthur, S. and Crabtree, A. (eds) *Developing Ecotourism into the Millennium, Proceedings of the Ecotourism Association of Australia 6th National Conference, Margaret River, Western Australia, 29–31 October.* Bureau of Tourism Research, Canberra, pp. 75–78. <www.btr.gov.au/conf_proc/ecotourism98>

Bauer, Th.G. (2001) *Tourism in the Antarctic: Opportunities, Constraints and Future Prospects.* The Haworth Hospitality Press, Binghamton, New York.

Beck, P.J. (1994) Managing Antarctic tourism – a front burner issue. *Annals of Tourism Research* 21(2), 375–386.

Chester, S.R. (1993) *Antarctic Birds and Seals.* Wandering Albatross, San Mateo, California.

Clarke, R.N. and Stankey, G.H. (1979) *The Recreational Opportunity Spectrum: A Framework for Planning, Management, and Research.* US Department of Agriculture, Forest Service, Seattle, Washington.

Davis, P.B. (1999) Beyond guidelines: a model for Antarctic tourism. *Annals of Tourism Research* 26(3), 516–533.

Dowling, R.K. (1993) An environmentally based planning model for regional tourism development. *Journal of Sustainable Tourism* 1(1), 17–37.

Enzenbacher, D.J. (1991) A policy for Antarctic tourism: Conflict or cooperation? Unpublished Master of Philosophy thesis, Scott Polar Research Institute, University of Cambridge.

Erize, F.J. (1987) The impact of tourism on the Antarctic environment. *Environment International* 13(1), 133–136.

Feizkhah, E. (1998) Tourism on thin ice. *Time* Magazine, 4 May, pp. 40–42.

Graefe, A.R., Kuss, F.R. and Vaske, J.J. (1990) *Visitor Impact Management: the Planning Framework*, Vol. 2. National Parks and Conservation Association, Washington, DC.

Hall, C.M. (1992) Tourism in Antarctica: activities, impacts, and management. *Journal of Travel Research* 30(9), 2–9.

Headland, R.K. (1994) Historical development of Antarctic tourism. *Annals of Tourism Research* 21(2), 269–280.

Heap, J. (ed.) (1990) *Handbook of the Antarctic Treaty System*, 7th edn. Polar Publications, Scott Polar Research Institute, University of Cambridge.

Hof, M. and Lime, D.W. (1997) Visitor experience and resource protection framework in the National Parks system: rationale, current status, and future direction. In: McCool, S.F. and Cole, D.N. (eds) *Proceedings – Limits of Acceptable Change and Related Planning Processes: Progress and Future Directions,* Gen. Tech. Rep. INT-GTR-371. US Department of Agriculture, Forest Service, Rocky Mountain Research Station, Missoula, Montana; University of Montana's Lubrecht Experimental Forest, Rocky Mountain Research Station, Ogden, Utah, pp. 29–36.

IUCN (1991) *A Strategy for Antarctic Conservation.* International Union for the Conservation of Nature, Gland and Cambridge, UK.

Jim, C.Y. (1989) Visitor management in recreation areas. *Environmental Conservation* 16, 19–32.

May, J. (1988) *The Greenpeace Book of Antarctica*. Dorling Kindersley, London.

McArthur, S. (1996) Beyond the limits of acceptable change – Developing a model to monitor and manage tourism in remote areas. In: *Proceedings of Towards a More Sustainable Tourism Down Under 2 Conference.* Hosted by the Centre for Tourism, University of Otago, Dunedin, New Zealand, pp. 223–229.

National Science Foundation (1995) *Antarctic Conservation Act of 1978.* National Science Foundation, Arlington, Virginia.

National Science Foundation (1997) *Notes from the Seventh Antarctic Tour Operators Meeting.* National Science Foundation, Arlington, Virginia.

Nilsen, P. and Tayler, G. (1997) A comparative analysis of protected area planning and management frameworks. In: McCool, S.F. and Cole, D.N. (eds) *Proceedings – Limits of Acceptable Change and Related Planning Processes: Progress and Future Directions,* Gen. Tech. Rep. INT-GTR-371. US Department of Agriculture, Forest Service, Rocky Mountain Research Station, Missoula, Montana; University of Montana's Lubrecht Experimental Forest, Rocky Mountain Research Station, Ogden, Utah, pp. 49–57.

Plimmer, N. (1994) *Antarctic Tourism: Issues and Outlook*, PATA Occasional Paper No. 10. Pacific Asia Travel Association, San Francisco.

Stankey, G.H., Cole, D.N., Lucas, R.C., Peterson, M.E. and Frissell, S.S. (1985) *The Limits of Acceptable Change (LAC) System for Wilderness Planning*, General Technical Report INT-176. US Department of Agriculture, Forest Service, Intermountain Forest and Range Experiment Station, Ogden, Utah.

Stonehouse, B. (1993) Shipborne tourism in Antarctica. *Polar Record* 28(167), 330–332.

Stonehouse, B. (1994) Ecotourism in Antarctica. In: Cater, E. and Lowman, G. (eds) *Ecotourism: a Sustainable Option?* John Wiley & Sons, London, pp. 195–212.

Stonehouse, B. (2001) Polar environments. In: Weaver, D. (ed.) *The Encyclopedia of Ecotourism*. CAB International, Wallingford, UK, pp. 219–234.

Stonehouse, B. and Crosbie, K. (1995) Tourist impacts and management in the Antarctic Peninsula. In: Hall, C.M. and Johnston, M.E. (eds) *Polar Tourism: Tourism in the Arctic and Antarctic Regions*. John Wiley & Sons, Chichester, pp. 217–233.

Thomas, T. (1994) Ecotourism in Antarctica. The role of the naturalist-guide in presenting places of natural interest. *Journal of Sustainable Tourism* 2(4), 204–209.

Whelan, H. (1996) Antarctica's new explorers. *Australian Geographic* 42, 80–97.

Ecotourism Policy and Planning: Stakeholders, Management and Governance

17

David A. Fennell[1] and Ross K. Dowling[2]

[1]*Department of Recreation and Leisure Studies, Brock University, St Catherines, Ontario, Canada;* [2]*School of Marketing, Tourism and Leisure, Edith Cowan University, Joondalup, WA 6027, Australia*

The need for a book of this nature has come as a result of the fact that ecotourism continues to be looked upon as an agent of positive economic growth in communities and regions around the world. The interest in ecotourism does not appear to be waning, especially given the global move away from a dependency on primary industries, to a reliance, as suggested by Simpson, on the services sectors. The same was suggested in the Hungary case study by Diamantis and Johnson (with reference to the Aggtelek Biosphere Reserve), where there is a decline in traditional agriculture as well as population, along with limited economic opportunities, cultivated areas which have been abandoned, and declining services (health and education).

One of the book's emergent themes is the realization that policy development for ecotourism, indeed as it is for many other human initiatives, is contingent on solid understanding of the concept or phenomenon in question. The recognition that all forms of tourism are not the same, should not be developed the same, or marketed the same, has come through clearly in these chapters. In more than one case it was mentioned that policy formulation needs to stem from sound definitions of ecotourism, despite the fact that there is no one definitive and universally accepted statement on ecotourism. Consequently, the definitional issue is one which continues to hamper consistency in ecotourism development. (The same was true for tourism in general until the World Tourism Organization took it upon itself to organize meetings for the purpose of better articulating terms such as tourist, tourism, excursionist and visitor.)

© CAB *International* 2003. *Ecotourism Policy and Planning*
(eds D.A. Fennell and R.K. Dowling)

The findings of Edwards, MacLaughlin and Ham suggest that, in the case of the Americas, there are tremendous differences that exist between jurisdictions regarding ecotourism definitions. While many jurisdictions felt the need to develop their own 'homegrown' definitions (i.e. developed by the jurisdiction itself), the study also found that: (i) less than half of Latin America and Caribbean (LAC) agencies could provide a written definition of ecotourism; (ii) 25% of Canada and USA agencies' staff in ecotourism have a marketing title, 11% in LACs; and (iii) tourism is not even part of the name of more than half of the Canadian and USA agencies (75% in LACs). The authors summarize by suggesting that there is a link between effective ecotourism policy development and clearly articulated ecotourism definitions. In some jurisdictions, such as New Zealand, definition is hampered by the fact that there are no clear dividing lines between ecotourism and other forms of nature-based tourism (a prodromal crisis stage, as identified by Diamantis and Johnson, where a crisis exists in the definitional perspective of ecotourism). Definitions are too broad and too limited to capture the scope and scale of the market. The authors state that this is counterproductive to effective management of the industry, making it next to impossible to identify niche markets and foster product development.

Stakeholders

Many of the chapters in the book discussed stakeholders from a general standpoint, but also more specifically, with reference to key stakeholder groups. The main stakeholders involved in ecotourism were said to be governments at all levels, the private sector, non-governmental organizations, multilateral and bilateral donors, tourists and local communities. The key cog in the interactions which exist between these groups appears to be government, which is charged with the responsibility of balancing, as much as possible, the demands of all. In fact, an equitable balance is not always struck because these groups are often not seen to be equal, nor are their demands viewed as being equal. This may be as a result of access to government, shared mandates between stakeholder groups, availability of resources, lobbying techniques and so on. Because tourism is viewed essentially as an economic enterprise (instead of, for example, from the experiential side as is the case of recreation services), there is little question over the relationship that exists between private enterprise and government. Who is given a permit to undertake a development project is quite often a political decision. Money drives the agenda, and people want to be involved in situations and circumstances where money is made, and where power results from such relationships. The emphasis placed on money in ecotourism will thus place certain stakeholder groups, for example public and NGOs, in direct conflict over values and priorities regarding the development of the industry. An interesting notion, put forth by Holtz and Edwards, is that biodiversity is a phenomenon that deals with the interface between nature, commerce and social process: resource control and costs and benefits. It strikes us that whether we are talking about plants and animals, or economic development, we are in fact talking the same language. If something is valued, that thing will be subject to the forces of competition and control.

Holtz and Edwards also write that the public sector is often the central driving force behind tourism development and biodiversity conservation as a result of its dual mandate of use and conservation. While this is very much a traditional relationship, it has not always been an effective one in practice. Even in cases where branches of government have a conservation mandate, such as New Zealand's Department of Conservation, they must also be responsible for encouraging visitation. On top of this, Jenkins and Wearing write that there are a myriad of government departments, authorities and agencies who continue to operate in apparent isolation from one another, making decisions that might be productive, or might just as easily be counterproductive (see also the chapter by Hall).

Despite the position in which governments find themselves in relation to the management or facilitation of the goals of all of the various stakeholders, and the industry in general, it is still in a position of control. In addition, although some would say government must be in this position as a facilitator and organizational framework within society, there appear to be many emerging models which place decision making and control for tourism in a collaborative arrangement. This means the development of integrated systems which include donors, NGOs and local communities who are most likely to be the ones marginalized in tourism's power structure, and would then need to be included as equal partners in the development process (more on this to follow).

Beyond the relationships that are outlined in the Holtz and Edwards chapter (see the chapters by Hall, Bricker, and Dredge and Humphreys), there appears to be the need to better understand both the relative levels of influence and levels of interest on the part of various stakeholder groups regarding economic development and biodiversity conservation. For example, the simplistic diagrams (Fig. 17.1) illustrate that the private sector has a firm influence on economic development activities within the region, but only a peripheral interest in biodiversity conservation. On the other hand, the position of an NGO may be quite the opposite. The multitude of examples found throughout the book seem to indicate that these disparate positions act as one of the main constraints to effective policy development in

The power to influence development   Level of interest in biodiversity conservation

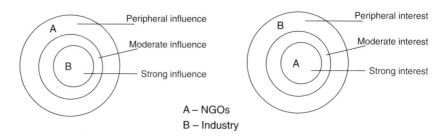

A – NGOs
B – Industry

Fig. 17.1. Levels of influence and interest.

ecotourism. Indeed the need for a fundamental shift in the power relationships which exist in ecotourism was identified by Jenkins and
Wearing, as well as Thompson and Foster, who suggest that the regulation of tourism in Kyrgyzstan is apparent through government
licensing. In reality, the authors illustrate that corruption is rampant,
as a result of the pressure from multinational organizations which
appear to have the most influence on tourism development in the
region. Recognizing the need to relinquish control is one thing; doing
it is quite another.

At least in some cases (Fiji and Australia), after rather long and
arduous journeys, there is more of a willingness to depart from older,
more traditional perspectives on policy. In the former case, Bricker
illustrates that ecotourism was instituted as an add-on to the more
standard sun, sea and sand attractions; something to be exploited if
tourists were interested in other things to do. More recently quite an
extensive organizational framework has been established linking the
Fiji Ecotourism Association with various government ministries, an
ecotourism advisory committee and a number of other stakeholder
groups. In the case of Australia, Dredge and Humphreys note that
local government in the Daintree area of the country is shedding its
traditional role as a provider and administrator of local services, and
becoming an active player in social and economic reforms. They suggest that the direction in which policy flows is very much a function
of inter-agency relations and the complexity of institutional environments. Perhaps more importantly, this means that content and direction of policy is as much a result of the local political discourse as it is
a product of the various stakeholder groups involved in its production. As Sofield and Li acknowledge, plans fail to be implemented as a
result of a lack of fit with the existing policies of a government.

Some authors stressed the importance of the tourism planning and
development board, which typically comprises many different stakeholder groups which represent a number of different positions. In

addition, importance was placed on the creation of a tourism strategy, which outlines the mission, vision and goals and objectives for tourism over specified periods of time. Of particular concern is the belief that such strategies need to be able to withstand the test of time, especially when time brings changes in political leadership within the region. Consequently, the board must have the authority to act, as much as possible, as an independent authority in order to be resilient to the massive swings in political ideology. This may also mean that representation and influence with such a board is not contingent upon financial backing or contributions (i.e. those who contribute most, get more decision making capacity). This is obviously a power situation which places stakeholders in majority and minority positions, and not one which should be advocated for ecotourism especially in communities that exhibit great capital and resource disparities. The issue is that governments are in a state of flux from one regime to another. The issues which are most pressing in tourism, at the broadest ecological and social levels, are generally unchanging across these regimes. Furthermore, we see in the work by Crouch and McCabe that policy makers are unaware of how different types of tourists consume resources and facilities of the destination; this is also not reflected in development of policy. They identify a vexing, yet important issue when suggesting that policy needs to be flexible enough to understand the diversity inherent in ecotourism and other forms of tourism, but also rigid enough to have meaning for all stakeholders involved. The tourist, they assert, is an integral partner who needs to be considered in attempts to develop policy.

Management Actions

Earlier in this chapter we mentioned that one of the roles of government was to act as a facilitator for economic development on one hand and to aid in conservation on the other (among other roles). From the perspective of the former, the state must intervene in its attempts to achieve an optimal allocation of resources and to control unfettered growth. The importance of an emerging 'guiding hand' in economic thought/decision making has been reinforced through the field of ecological economics, which puts in perspective the importance of development in the context of a broader social, economic and ecological agenda, in contrast to more traditional theories of growth and consumption. Since the 1970s, researchers from a number of domains have sought to advance the thinking on poverty, health, security, justice and environmental quality in the face of over-population, urbanization and economic expansion (e.g. at the most recent meeting of the International Society for Ecological Economics in Sousse,

Tunisia (April 2002), this agenda was intensified through an examination of the many global issues of environment and development, especially the challenges for local and international governance).

Such a focus on environment and economics arms decision makers with the ability to interpret the social and ecological implications of growth in economic terms. In her chapter, Mihalič notes that today's environmental economic theories may be broadly grouped as systems theories, growth theories or behavioural theories. The theory of externalities, a systems theory, posits that the main reason for ecological disturbances is a result of the belief that the environment is cost free. The classic example of a negative externality is when a firm dumps sewage into a body of water, thus reducing the water's potential to support other firms who rely on the water system for commercial and recreational fishing, swimming and other forms of outdoor recreation. Mihalič says that polluters should bear the social costs of their effects (i.e. that they should internalize all external effects). More and more, there appears to be a move in tourism to consider many of the different perspectives advocated in the chapter by Mihalič. These include environmental taxes, subsidies, fees and contributions, tourist certificates, eco-labels and ethics.

The basis of the problem lies in the fact that there appears to be an absence of an acceptable and rational environmental ethic in the way business is conducted, and in consumer behaviour. This quite likely stems from the notion that ignorance is a function of a lack of sufficient research and education, as suggested by Mihalič. The notion that 'Only good men will follow good laws' is an important one and underscores the importance of good, right and moral leadership among those whose responsibility it is to develop and manage ecotourism. Yet there is a cultural and ethical relativism that pervades the global scene, exacerbating attempts to conceptualize not only ecotourism, but also the broader issues related to the fundamental relationships between humans and nature. This dilemma is most effectively illustrated in the chapter by Sofield and Li who remind us that not all cultures of the world share the same environmental attitudes and values. A dominant philosophy in China holds that because nature is imperfect, humans have the responsibility to improve it. The ecocentric philosophers of the Western world would no doubt think the direct opposite.

In their chapter on China, Sofield and Li elected to operationalize the Recreational Opportunity Spectrum (ROS), which stands as one of the most often used of a range of preformed planning and management frameworks, designed to consider how recreational experiences might best be maximized through the classification of different settings. Through the ROS, different types of experiences and recreational uses are matched with appropriate settings in providing a spectrum of alternatives for individuals in and between natural areas. This was thought most appropriate for the Chinese case study, which would allow for the integration of ecotourism to occur in altered settings alongside a more conventional Western approach to the concept. Furthermore, many of the authors in this compendium focused on the biosphere reserve concept as a means by which to place ecotourism in environments that continue to serve as living, working landscapes. In Antarctica, Bauer and Dowling proposed the use of the Environmentally Based Tourism (EBT) planning model, which is strategic, iterative, regionally based, incorporates land use zoning and is environmentally educative.

The EBT planning model is a strategic planning approach to environment – tourism planning in five stages. These are a statement of objectives, survey and assessment, evaluation, synthesis and proposals. The five stages can be expanded into ten processes. The first stage consists of one process, the statement of objectives. It begins with a background analysis of the environment–tourism relationship in order to produce the basic direction for the succeeding stages. The direction is determined by the objectives or planning goals which have emerged from the environment–tourism relationship review.

It is important to note that the objectives are imported into the framework from the survey of the study area and its environment–tourism issues and they are not arbitrarily fixed for all applications. However, as a general guide, a number of planning zones are defined which are designed to protect conservation values while fostering tourism developments and activities.

These zones are identified and described based on an approach in which the land and water areas of a region are classified according to their need for protection and compatibility with tourism. They include a range of zones from 'sanctuary', areas requiring special preservation; 'conservation', areas sustaining a combination of protection and use but with emphasis on the former; 'recreation', natural areas that can accommodate compatible outdoor recreation activities; and 'development', small areas of concentrated touristic attractions. All other areas in the study region are designated as areas with other uses.

This provides a guide for future environmental planning, tourism planning and regional development planning. Zoning also assists in managing the tension between preservation and use and more importantly seeks ways of fostering tourism in natural areas. The main argument against zoning is possibly rigid and inflexible prescriptions for use; however, the zones are used as general guides rather than rigid prescriptions.

The second stage consists of survey and assessment. This includes two processes: the description and assessment of environmental attributes and the description and assessment of tourism resources. The third stage is one of evaluation of significant features, critical areas and compatible activities together with suggestions for appropriate strategies and controls. Following the evaluation the resultant information is amalgamated in the fourth stage of synthesis in which the planning zones are allocated. The final stage outlines the proposals and includes the preparation and presentation of the final zoning plan as well as its implementation.

Because of its independent political status, Antarctica emerged as a unique case study in the book. This region stands as a 'last frontier', which has enabled (more likely forced) stakeholder groups from many countries to agree on some common guidelines for the appropriate use of the continent for tourism purposes. Tour operators, for example, must submit their itineraries the year before to a relevant government body. These itineraries must show that activities will at most have a transitory impact in order for these to be accepted. The heavy reliance on industry self-regulation through adherence to a proactive mentality and code of ethics has meant that the region is perhaps the best managed site in the world. Bauer and Dowling say it is the spirit of cooperation between governments and tour operators that is central to sustainable tourism in Antarctica. This model has application in other world regions, especially those biological and geological hotspots which exist around the world.

Perhaps one of the most important realizations, which has intensified over the last 20 years, is the notion that governments and other stakeholder groups (most notably industry) do not necessarily have all the facts in their decisions to develop, manage or exploit resources. The precautionary principle is an anticipatory principle that suggests to politicians that 'scientific progress does not justify the delay of measures preventing environmental degradation' (Gollier *et al.*, 2000: 231). O'Riordan and Cameron (1994: 12) define the precautionary principle as 'a culturally framed concept that takes its cue from changing social conceptions about the appropriate roles of science, economics, ethics, politics and the law in pro-active environmental protection and management'. In this regard, precaution has been extended to include six basic concepts: (i) preventive anticipation; (ii) safeguarding ecological space; (iii) that the restraint adopted is not unduly costly; (iv) duty of care, or onus of proof on those who propose change; (v) promotion of the cause of intrinsic natural rights; and (vi) paying for past ecological debt.

Due to the fact that there appears to be a great deal of uncertainty in decisions that are made about the environment, management must be adaptive enough to take these uncertainties into consideration. We simply cannot accurately predict yields and consequences of actions enough to proceed without solid empirical data. This has prompted various authors to suggest that the precautionary principle should prompt decision makers and various other groups to be open to the following three priorities. First, to include a wider participation in groups producing scientific advice; second, to allow these groups to have equal decision making power; and third, to encourage the debates about scientific research and various procedures to be open and publicly accessible. As regards ecotourism policy, the precautionary principle may be of use as an established guideline for policy makers who must make the most appropriate decisions on tourism development, understanding that any decisions which are made on behalf of ecotourism will have ripple effects throughout communities and other industries.

Policy Development, Complexity and Governance

Citing Hall and Jenkins (1995: x), Hall suggested that ecotourism policy may be defined as 'whatever governments choose to do or not to do with respect to ecotourism'. This, he contends, involves action, inaction, decisions, non-decisions, choice and process. As such, there are no simple, straightforward solutions emerging from the application of different principles and policies, especially across space. Added to this is the notion that policy is, more often than not, nested within a broader set of institutional arrangements, between a number of different political entities. The fact that there are no departments of ecotourism challenges policy makers, as we have seen in cases such as New Zealand where functions and philosophies of agencies are quite different. This has led many authors to infer that policy in ecotourism is not a rational, linear process. Indeed, the movement through various planning stages, implementation and evaluation has been anything but seamless.

Complexity is a word that is creeping into the vernacular more and more in reference to tourism and ecotourism. Complexity is viewed as the antithesis of reductionism and of the linear, equilibrium theory of nature which suggests that there is a predictable pattern of events that determines cause and effect relationships (Scoones, 1999). The unpredictability of the world has catalysed governments and other organizations to be more open-minded and flexible, allowing for interactive and cooperative management between various public, private and not-for-profit sectors, through multiple methods and at a variety of different scales.

Complexity allows stakeholders to better understand the intricacies of systems – human and ecological – their interrelationships and their inherent feedbacks that allow systems to 'learn' and thus affect their future behaviour (e.g. through an understanding of positive and negative impacts, in social, economic and ecological realms). Consequently, there is a recognition that social systems cannot be analysed independently of natural systems (Berkes and Folke, 1998) and, depending on setting and other characteristics, systems are unique, with no optimal solution. An understanding of complexity and the nature of systems provides a new impetus to implement governance – which may be viewed as the amalgam of institutions, policy and management – according to the conditions of unique settings. This means less of a focus on traditional policy and more of a willingness to place value on new, innovative and integrated policy schemes.

Complexity further suggests that there is at least the means (not necessarily the motivation though) to place into perspective the necessary processes and frameworks to better understand the heterogeneity of ecotourism as a complex phenomenon. At the grandest level, there appears to be a good deal of consensus on the fact that sustainability is the most realistic and rational paradigm by which to clearly define the possibilities and limits of ecotourism. Sustainability has also strongly influenced a number of different governance regimes, which continue to provide the structural basis for decision making in society as a whole. Flexibility is one of the fundamental dimensions of the governance model, which allows for, as stated above, collaborative management structures, the support for regional diversity and the encouragement of citizen engagement, at many scales (Fennell, 2002).

At the other end of the spectrum, however, is the need for operators or other service providers to be not only active players in the operationalization of policy but also shapers of policy. Unfortunately, this is a group that has been viewed as a stakeholder that must only adhere to policy and guidelines. Fennell (2002) writes that the literature on recreation programme planning, and the practice which goes along with this, is well suited to the field of ecotourism. Until recently no such mechanism existed for the field, contributing to a great deal of disjointedness and misapplication of programme-related concepts. Following a traditional systematic programme planning framework (which has been applied from missile testing in the US army to recreational programmes), programming for ecotourism entails adherence to many simple but well planned principles. These are geared not only towards the satisfaction of participants, but also to better understanding of how the organization may be accountable to itself through increased knowledge of, for example, capital, equipment, effective use of time, behaviour of employees, or clients. Simply stated, programming is the process of organizing resources and opportunities for other people, for the purpose of meeting their needs. In general, this includes:

- *Programme planning*: including mission and vision statements for the organization, goals and objectives, and programme strategies and approaches.

- *Needs and assets*: an overview of nature-based tourist motives and needs, along with an inventory of attraction and assets for nature-based tourism.
- *Programme design*: includes the programme's structure (areas, settings, lodging, mobility), interpretation, leadership, guides, professional development and risk management.
- *Programme implementation*: includes, for example, product life cycle, marketing, staff training, budgeting and public relations, and implementation strategies.
- *Evaluation*: formative and summative evaluations, accreditation and certification, as well as consideration of a series of different models of evaluation.

The systems approach, therefore, provides the means by which to link a programme realm with a broader environmental realm that includes governance based on a sustainable ecotourism ethic (Fig. 17.2). Feedback from exceptional ecotourism programmes will thus help to push the bounds of appropriate policy further, in the same way that policy, articulated through tourism boards or other related bodies, affects the service provider in the field. The resulting dynamic provides perhaps a different and more integrative structure to capture the true heterogeneity of the industry and at the same time contribute to a tension between the highest levels of authority and the workings of the industry on the ground. However, the systems perspective also allows for the incorporation of other system-dependent entities (e.g. environmental management and marketing) to be woven into the fabric of a programme plan.

As suggested above, this more holistic approach to ecotourism allows decision makers to be as integrative and adaptive as possible in resource management, in recognizing that many social and natural phenomena need to be examined as a linked whole, using processes and techniques that cut across different sectors and scales. In drawing this link there is the belief that governments, tourism associations, community groups, tourism operators and other interested parties might be better able to understand the social, ecological and economic conditions of a setting in the development of ethically and sustainably based ecotourism programmes. Furthermore, the evaluative element has often been ignored in the policy process. For example, Parker (1999) cites Anderson (1994) in illustrating that policy is a function of three stages: policy formation, policy adoption and policy implementation. It would seem logical to take this process one step further in suggesting a policy evaluation stage, which would effectively build an iterative component into the policy development process.

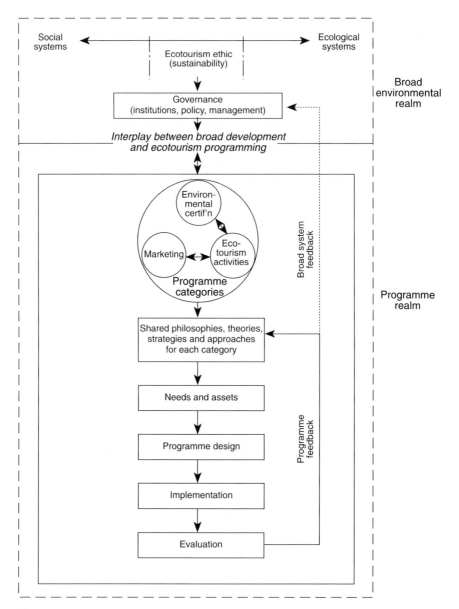

Fig. 17.2. Integrated ecotourism programme planning.

Conclusion

Perhaps like never before in the history of humanity, there appears to be an urgency towards travel for the purpose of experiencing the natural history of a region. We are bombarded daily with messages regarding biodiversity and habitat loss, endangered species, pollution and so on. Whether these messages catalyse us to visit some of the world's most precious places is unknown. If true, such a purpose has contributed to an ever-increasing demand for ecotourism throughout the world. In the race to develop – in the race to have more – the industry has often compromised the integrity of people and resources for financial ends. It strikes us, however, that ecotourism policy should not just be for development, but also for the purpose of controlling development or halting it altogether in places where human intrusion is not recommended or valued. In this context, not every corner of every region in the world should be open for business; a premise that policy makers should perhaps consider in their efforts to balance use and preservation. Accordingly, there needs to be the realization that tourists are not all the same in intent, motivations and expected outcomes.

On the other side of the coin, decision makers across the board need to be open-minded, fair, responsible and rational in their deliberations. This means in our best attempts to balance land use and conservation we must recognize that development in certain forms is needed. Refusing to clear land for the construction of a state-of-the-art ecolodge, for example, is not only potentially harmful to the local economy but also to ecotourism in general. This means that stakeholder groups, who may advocate a strong ecocentric philosophy, do not always have to be adversarial and dogmatic in their thinking. Relationships often change according to the demands of different situations and settings. For example, McMahon (2002) writes that, in the US, environmental groups have started to buy large parcels of land for conservation purposes in order to halt urban sprawl (where virgin lands may be lost to development indefinitely). The interesting aspect of this initiative is that such groups, who cannot afford to pay for this land, have been asking logging companies who can pay for the land for assistance. These long-term relationships allow loggers to work the land, jobs are saved, governments get cheap open space, and in return certain parts of the forests are protected altogether while other regions are turned over to the loggers. As those of us in resource management have come to understand, such a relationship would never have been considered in the past.

It is worth noting as a final thought that ecotourism, and those who act on its behalf, have a responsibility to ensure that this form of tourism stands apart as a barometer or benchmark from which to influence the development of other forms of tourism. This means testing new approaches and philosophies; the adoption of greener accommodation, transport and programmes; and the willingness to place the resource base foremost in decision making, which will ultimately stimulate a more ecologically and socially based form of tourism that will have lasting benefits for many years.

References

Berkes, F. and Folke, C. (1998) Linking social and ecological systems for resilience and sustainability. In: Berkes, F. and Folke, C. (eds) *Linking Social and Ecological Systems: Management Practices and Social Mechanisms for Building Resilience.* Cambridge University Press, Cambridge, pp. 1–25.

Fennell, D.A. (2002) *Ecotourism Programme Planning.* CAB International, Wallingford, UK.

Gollier, C., Jullien, B. and Treich, N. (2000) Scientific progress and irreversibility: an economic interpretation of the 'Precautionary Principle'. *Journal of Public Economics* 75, 229–253.

McMahon, P. (2002) Logging deal might save forest. *USA Today.* Thursday 28 March, x.

O'Riordan, T. and Cameron, J. (1994) The history and contemporary significance of the precautionary principle. In: O'Riordan, T. and Cameron, J. (eds) *Interpreting the Precautionary Principle.* Earthscan, London, pp. 12–30.

Parker, S. (1999) Ecotourism, environmental policy, and development. In: Soden, D.L. and Steel, B.S. (eds) *Handbook of Global Environmental Policy and Administration.* Marcel Dekker, New York.

Scoones, I. (1999) New ecology and the social sciences: what prospects for a fruitful engagement? *Annual Reviews Anthropology* 28, 479–507.

Index

Note: page numbers in *italics* refer to figures and tables

Browse Read and Buy

www.cabi.org/bookshop

ANIMAL & VETERINARY SCIENCES
BIODIVERSITY   CROP PROTECTION
HUMAN HEALTH   NATURAL RESOURCES
ENVIRONMENT   PLANT SCIENCES
SOCIAL SCIENCES

 CABI *Publishing*
A division of CAB International

Online BOOK SHOP

Subjects

Search

Reading Room

Bargains

New Titles

Forthcoming

Order & Pay Online!

MasterCard

VISA

AMERICAN EXPRESS

Crop Pollination by Bees
Keith S. Delaplane and Daniel F. Mayer

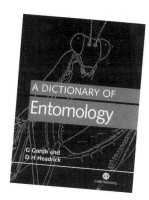
A DICTIONARY OF Entomology
G Gordh and D H Headrick

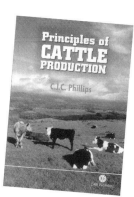
Principles of CATTLE PRODUCTION
C.J.C. Phillips

Seeds THE ECOLOGY OF REGENERATION IN PLANT COMMUNITIES 2nd EDITION
Edited by Michael Fenner

★ FULL DESCRIPTION   ★ BUY THIS BOOK   ★ BOOK OF THE MONTH

Tel: +44 (0)1491 832111   Fax: +44 (0)1491 829292